# MANUFACTURING, MODELLING, MANAGEMENT AND CONTROL 2004
## (MIM 2004)

*A Proceedings volume from the IFAC Conference,
Athens, Greece, 21 –22 October 2004*

Edited by

## G. CHRYSSOLOURIS and D. MOURTZIS
*Laboratory for Manufacturing Systems and Automation,
Department of Mechanical Engineering and Aeronautics,
University of Patras, Greece*

Published for the

INTERNATIONAL FEDERATION OF AUTOMATIC CONTROL

by

ELSEVIER LIMITED

ELSEVIER Ltd
The Boulevard, Langford Lane
Kidlington, Oxford OX5 1GB, UK

Elsevier Internet Homepage
http://www.elsevier.com

Consult the Elsevier Homepage for full catalogue information on all books, journals and electronic products and services.

IFAC Publications Internet Homepage
http://www.elsevier.com/locate/ifac

Consult the IFAC Publications Homepage for full details on the preparation of IFAC meeting papers, published/forthcoming IFAC books, and information about the IFAC Journals and affiliated journals.

First edition 2005

**Library of Congress Cataloging in Publication Data**

A catalogue record for this book is available from the Library of Congress

**British Library Cataloguing in Publication Data**

A catalogue record for this book is available from the British Library

ISBN 0-08-044562 4
ISSN 1474-6670

*These proceedings were reproduced from manuscripts supplied by the authors, therefore the reproduction is not completely uniform but neither the format nor the language have been changed in the interests of rapid publication. Whilst every effort is made by the publishers to see that no inaccurate or misleading data, opinion or statement appears in this publication, they wish to make it clear that the data and opinions appearing in the articles herein are the sole responsibility of the contributor concerned. Accordingly, the publisher, editors and their respective employers, officers and agents accept no responsibility or liability whatsoever for the consequences of any such inaccurate or misleading data, opinion or statement.*

Transferred to digital print, 2008
Printed and bound by CPI Antony Rowe, Eastbourne

**To Contact the Publisher**

Elsevier welcomes enquiries concerning publishing proposals: books, journal special issues, conference proceedings, etc. All formats and media can be considered. Should you have a publishing proposal you wish to discuss, please contact, without obligation, the publisher responsible for Elsevier's industrial and control engineering publishing programme:

Christopher Greenwell
Senior Publishing Editor
Elsevier Ltd
The Boulevard, Langford Lane
Kidlington, Oxford
OX5 1GB, UK

Phone: +44 1865 843230
Fax: +44 1865 843920
E.mail: c.greenwell@elsevier.com

General enquiries, including placing orders, should be directed to Elsevier's Regional Sales Offices – please access the Elsevier homepage for full contact details (homepage details at the top of this page).

# IFAC CONFERENCE ON MANUFACTURING, MODELLING, MANAGEMENT AND CONTROL 2004

*Sponsored by*
International Federation of Automatic Control (IFAC)
Technical Chamber of Greece

*Organized by*
Laboratory for Manufacturing Systems and Automation (LMS), Department of Mechanical Engineering and Aeronautics, University of Patras, Greece

*International Programme Committee* (IPC)

| | |
|---|---|
| Chryssolouris, G. | Chair |
| Monostori, L. | Co-Chair |
| Roll, K. | Vice-Chair from Industry |

| | | |
|---|---|---|
| Alting, L. (Denmark) | Kimura, F. (Japan) | Shpitalni, M. (Israel) |
| Boer, C. (Italy) | Kjellberg, T. (Sweden) | Suh, N. (USA) |
| Bouzakis, K. (Greece) | Klocke, F. (Germany) | Teti, R. (Italy) |
| Browne, J. (Ireland) | Kopacek, P. (Austria) | Tichkiewitch, S. (France) |
| Bruns, W.F. (Germany) | Kovacs, G. (Hungary) | Ueda, K. (Japan) |
| Bueno, R. (Spain) | Kumara, S. (USA) | Van Brussel, H. (Belgium) |
| Byrne, G. (Ireland) | Kusiak, A. (USA) | Van Houten, F. (Netherlands) |
| Carrie, S.A. (UK) | Krause, F.L. (Germany) | Vernadat, F. (France) |
| Chen, D. (France) | Lu, S. (USA) | Veron, M. (France) |
| Dornfeld, D. (USA) | Molina, A. (Mexico) | Wertheim, R. (Israel) |
| Erbe, H. (Germany) | Nagao, T. (Japan) | Westkaemper, E. (Germany) |
| Filip, F. (Romania) | Nemes, L. (Austria) | Whitman, L. (USA) |
| Galantucci, L. (Italy) | Nof, S. (Germany) | Wiendahl, H. (Germany) |
| Geiger, M. (Germany) | Panetto, H. (France) | Wieringa, P. (Netherlands) |
| Hatamura, Y. (Japan) | Peklenik, J. (Slovenia) | Williams, D. (UK) |
| Huang, Y.C. (Taiwan) | Santochi, M. (Italy) | Wortmann, C.J. (Netherlands) |
| Inasaki, I. (Japan) | Sasiadek, J. (Canada) | Zaremba, B.M. (Canada) |
| Jovane, F. (Italy) | Schuh, G. (Germany) | Zuehlke, D. (Germany) |
| Kals, H. (Netherlands) | Seliger, G. (Germany) | Zuest, R. (Germany) |

*National Organizing Committee* (NOC)

| | |
|---|---|
| Chryssolouris, G. | Chair |
| Dounis, J. | Vice-Chair from Industry |

| | | |
|---|---|---|
| Alavanos, I. | Houssos, E. | Papazoglou, V. |
| Bouzakis, K. | Kanarachos, A. | Protonotarios, E. |
| Fassois, S. | Mourtzis, D. | Sfantzikopoulos, M. |
| Groumpos, P. | Papadopoulos, G. | Tzafestas, S. |
| Haidemenopoulos, Gr. | Papaioannou, S. | Venieris, I. |

*Conference Coordinator:*
Mourtzis, D.

# PREFACE

The beginning of the 21$^{st}$ Century finds manufacturing research and technology development more relevant than ever before. The globalisation of the economy, the increasing shortage in natural resources, and the importance of the environmental issues are some of the challenges that manufacturing will be faced with, in the upcoming years. Recent developments notably, in the telecommunications and information technology sectors, present manufacturing with new opportunities. If properly utilised, they can lead to lower cost, faster time-to-market and higher product quality. Modelling, Management and Control of Manufacturing Systems take a new meaning in this new technological era. The wide spread of the Internet, the electronic commerce and the other Internet based business activities, allow manufacturing companies to become global players, irrespective of their size and location. The adaptation of the manufacturing environment to this new wave of technological development represents a major challenge to the manufacturing research community. Simultaneously, the manufacturing education paradigm has to change rapidly in order to provide industry with a new breed of engineers who understand, utilise and develop new technologies in the context of the production environment.

In this context the papers of this conference, included in these proceedings, address topics such as Manufacturing Systems Modelling and Simulation, Manufacturing Processes Modelling, Manufacturing Systems Planning and Control, Lean Production and Agile Manufacturing, Concurrent Engineering, Supply Chain Management, Logistics and Manufacturing, Data Management, Virtual Reality and Manufacturing, Life Cycle Design and Manufacturing and Rapid Manufacturing.

George Chryssolouris

# CONTENTS

ELSEVIER

IFAC
PUBLICATIONS
www.elsevier.com/locate/ifac

# MANUFACTURING SIGNATURE OF TURNED CIRCULAR PROFILES

**Colosimo B.M., Moroni G., Petrò S., and Semeraro Q.**

*Politecnico di Milano, Dipartimento di Meccanica,
Milano, Italy, giovanni.moroni@polimi.it.*

This paper focuses in defining an empirical model for the manufacturing signature left by
a turning process when roundness is the geometric tolerance of interest. In particular, the
proposed method is based on a statistical analysis of Discrete Fourier Transforms and is
able to overcome limitations related to the arbitrary definition of the boundary between
manufacturing signature and random error. Furthermore the method is able to deal with
autocorrelation that characterizes sequentially measured points. *Copyright © 2004 IFAC*

Keywords: auto correlation, Discrete Fourier Transform, machining, models, regression,
signature, tolerance.

## 1. INTRODUCTION

Quality of mechanical components is more and more often related to geometric tolerances, e.g., roundness, flatness, parallelism, position. Usually, geometric tolerances are checked using a Coordinate Measurement Machine (CMM), which has as major advantages accuracy and flexibility. A CMM essentially samples a set of points on a surface, and then, using appropriate algorithms, it estimates a tolerance. Traditional approaches for measuring geometric tolerances obtained on a manufactured component are independent from the specific process adopted for realizing that feature. In fact, a tolerance is most of the times computed by means of a generic measuring strategy, where the number and the position of the measuring points are empirically determined (usually they are uniformly or randomly distributed on the ideal feature that has to be inspected). Of course, as the number of measured points increases, inspection costs increase as well while uncertainty in estimated tolerance decreases: hence, sampling strategy must be designed in order to balance these two factors.

A closer analysis of the specific process involved in manufacturing the feature can add cleverness in designing the measuring strategy, through a proper selection of the number and the position of the measuring points. This added cleverness can result in a reduction of inspection costs for a fixed level of uncertainty. Cleverness coming from the knowledge of the manufacturing process is related to the definition of the empirical model representing the "signature" left by the process on the feature machined. This manufacturing signature is the systematic pattern that characterizes all the features machined with that process. In fact, measured surfaces (or profiles) often present a systematic pattern and a superimposed random noise: the first is mainly due to the process used in specific operation

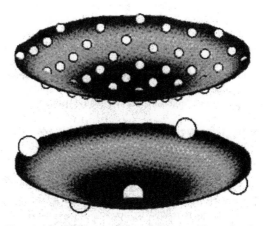

Fig. 1. Sampling a planar hollow surface: in the higher part, usual practice; below, distribution suggested by the signature.

conditions, while the second is due to unpredictable factors which are usually known as "natural variability". The systematic pattern constitutes what is called "the signature" of the process (Wilhelm *et al.*, 2001). To better understand why signature identification can improve the sampling strategy, consider as an example a planar surface that, due to the specific process used, is systematically hollow (Fig. 1 reports the systematic pattern once random noise is filtered out from data). A traditional, "blind" sampling strategy will distribute sampling points uniformly on the whole surface while cleverness coming from the signature identification will suggest a different distribution of points (mainly on the edges and in the middle of the nominal surface).

Besides advantages induced on sampling strategy, the identification of the manufacturing signature can also improve identification of interventions that are required on the machine. In fact, signature characteristics may be unavoidable (as the cylindrical helix left by turning), or related to machine-tool's defects (as anomalous wear of a single ball in a bearing sustaining the spindle). When these second type of problems are identified through signature analysis, machine problems can be easily identified and predictive maintenance can be scheduled.

This paper deepen problems related with the identification of manufacturing signatures. In particular it faces the problem of identifying signature left by a turning process when roundness is the geometric tolerance of interest. The approach presented identifies signature as the empirical model built through a statistical analysis of a set of turned specimens. Compared with previous studies presented on this topic, the method proposed in this paper is able to overcome limitations related to the arbitrary definition of the boundary between signature and random error. Furthermore the method is able to deal with autocorrelation that characterizes sequentially measured points. This work represents the first step toward the development of a general method to evaluate a geometric tolerance associated with a feature with the help of its "history", i.e. the machining process it went through.

*1.1 State of the art*

Advantages related with identification of "manufacturing signature" have already been explored in some papers available in the literature. We will briefly describe four recently proposed models: two of them are based on the use of Discrete Fourier Transforms (DFT) to define the signature while the last two models use general analytical functions and "eigenshapes", respectively.

With reference to approaches based on DFT representation, Capello and Semeraro (2001a) propose the use of signature in an economic model (Capello and Semeraro, 2001b) aimed at choosing the number of sampling points. The objective function is selected in order to balance inspection costs and costs related to possible errors in quality inspection, due to uncertainty in tolerance estimate. First of all, they develop analytical relations linking DFT coefficients and least squares substitute geometry for cone, cylinder, straight line, plan, and circle. For these features, DFT representation allows to calculate average radius, inclination of the cylinder axis, inclination of the normal to the plane, and so on. The authors, then derive the empirical statistical distribution of the estimate errors as a function of the number of sampled points, while considering the effect of aliasing on DFT components which define the substitute geometry; aliasing impact is evaluated basing on the signature. The optimal number of points that has to be measured is thus computed considering the trade off between error computed in estimating tolerance of interest vs. inspection costs.

Another approach based on DFT analysis, is proposed by Cho and Tu (2001). Authors' intention is to create a "database" which can allow to simulate sampling of roundness profiles. Since the adoption of DFT can be geometrically interpreted as a sum of sinusoids, the manufacturing signature is described in this paper by the statistical distribution of amplitudes (i.e. the absolute value of the coefficients of the DFT) of harmonics contained in a profile. The authors analyze the case of cylindrical parts made by turning, and, using experimental data, they identify a beta distribution for each harmonic amplitude (where parameters of the beta distribution differ harmonic from harmonic).

Finally, models which do not use DFT to describe profiles are presented by Henke *et al.* (1999) and by Summerhays *et al.* (2001). According to the first model, a profile (or a surface) subject to a geometric tolerance can be considered as a sum of analytical functions, chosen by the analyst depending on the specific case faced. For instance, the authors develop the case of cylindrical surfaces (holes), and they propose a model which defines radius variations using sinusoidal functions in circumferential direction, and Chebyshev polynomials in axial direction. An experimental stage is presented to verify the adequacy of the model. The second method defines the signature using "eigenshapes", i.e., shapes obtained from sampled points after applying the statistical technique known as "principal components". A real surface or profile is then constituted by a weighted sum of eigenshapes. Finally, the authors propose an heuristic algorithm which solves the problem of choosing the sampling points pattern (but not the number) basing on the manufacturing signature.

None of this works proposes an objective criterion to distinguish the signature from random error or noise component. Capello and Semeraro (2001a, 2001b) are mainly interested in the problem of choosing the optimal number of sampling points, and do not explicitly face the problem of separating the signature from the random noise. Cho and Tu (2001) suggest the use of the first fifty harmonics to identify signature in circular profiles, without providing an analytical explanation. Similarly, Henke *et al.* (1999) propose twelve components (analytical function or eigenshapes) to describe cylindrical surfaces.

*1.2 Paper objectives*

The aim of this paper is to design a robust procedure for signature identification in which the empirical model that represents the signature is determined through a statistical criterion aimed at separating systematic behavior (signature) from random noise. The adoption of the approach presented has as major advantages:
- objectiveness: discrimination between signature and random noise does not rely on subjective judgment of the analyst, as in the paper by Cho and Tu (2001) and Henke *et al.* (1999);
- generality: the approach presented is not related to a specific geometry but it can be easily extended to different processes and/or tolerances.

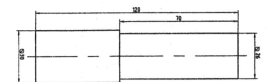

Fig. 2. Turned components used in the study.

To validate the method proposed, an experimental study is presented in the paper. It is based on identifying systematic behavior in roundness profiles obtained by turning. Applying the model proposed to the specific case studied, residual diagnostic reveals that autocorrelation is a further problem in the identification of the signature (i.e., the resulting noise is often not "white"). As a by-product of this paper, a method for identifying the signature when autocorrelation affects the random noise is presented.

## 2. EXPERIMENTAL DETAILS ON THE REAL CASE STUDIED AND SIGNATURE DEFINITION METHODOLOGY

A series of 36 turned components have been produced as in Figure 2. A component is a simple steel cylinder machined by turning. Cutting parameters used during machining are:

- Cutting speed: 163 m/min;
- Feed: 0.2 mm/rev;
- Cutting depth: 1 mm (machining requires two steps of cutting).

The larger diameter part (left side in Fig. 2) was left raw, because this part was used only for clamping on the lathe and on the CMM. The chosen material was a C20 carbon steel, which was supplied in ⌀30 mm rolled bars.

Each cylinder was sampled by a Zeiss Prismo Vast HTG, whose MPE is 2 μm. The inspection plan was developed according to ISO/TS 12180 (ISO/TC 273, 2003), a technical specification concerning cylindricity. This normative defines the "perfect operator", measurement conditions that allow us to say that any estimate error in geometric tolerancing does not depend on sampling strategy. This norm does not suggest a sample points number, but it is strictly related to ISO/TS 12181 (ISO/TC 273, 2003), which refers to roundness: in this norm, the number of points to sample depends on the wavelength of the gaussian form filter chosen to separate form from waviness and roughness. Having chosen a 0.8 mm wavelength filter, and considering that at least seven points are required for each wavelength cut-off, the perfect operator is as follow (for a ⌀26 mm, 50.8 mm length cylinder):

- Sampled points per circle: 748;
- Sampled points per generatrix: 515;
- Scanning speed: 10 mm/s;
- Probe ball tip radius: 0.5 mm;
- Total points number: 385220.

Sampling was performed by continuous scanning by generatrices; the measurement of a single part thus requires 2h 40min per cylinder.

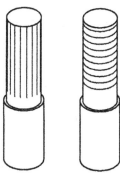

Fig. 3. Sampling strategy and extracted profiles. On the left, sampling by scanning along generatrices; on the right, roundness profiles are extracted at equi-spaced levels.

From each cylindrical surface sampled, ten roundness profiles were extracted: these profiles will be the subject of the following analysis (Fig. 3).

### 2.1 Choice of functions family constituting signature

As in previus papers on signature, we will assume that a profile (or a surface) can be modeled as a weighted sum of analytical components (which constitutes the signature), plus a random error. With reference to roundness profiles, this means that we may write the following model:

$$r = r(\theta) = r_0 + \beta_1 f_1(\theta) + \\ + \beta_2 f_2(\theta) + ... + \beta_k f_k(\theta) + \varepsilon(\theta) \quad (1)$$

so the radius is a function of the angle only as represented in Figure 4.

DFT comes in handle when searching for a good family of functions to describe circumferences. In fact, the well known geometrical interpretation of DFT is given by:

$$r(\theta) = r_0 + \sum_{\tau=1}^{(n-1)/2} \mathrm{Re}(F(\tau))\cos(\tau\theta) - \mathrm{Im}(F(\tau))\sin(\tau\theta) \quad (2)$$

where $F(\tau)$ is the $\tau$-th coefficient of the DFT and $n$ the number of sampling points.

As it can be observed in Figure 5, DFT representation is particularly adapt to describe turned profiles because it is able to capture periodic patterns that usually characterize these circular profiles. It has to be outlined that DFT representation does not distinguish signature from error $\varepsilon(\theta)$.

In particular, $\varepsilon(\theta)$ is usually related to high frequency components in the DFT, so the sum in equation (2) should be divided in two parts, the first of which contains $m$ harmonics constituting signature, while the remaining components constitute the noise:

$$r(\theta) = r_0 + \sum_{\tau=1}^{m} \left( \mathrm{Re}\big(F(\tau)\big)\cos(\tau\theta) - \mathrm{Im}\big(F(\tau)\big)\sin(\tau\theta)\right) +$$
$$+ \sum_{\tau=m+1}^{(n-1)/2} \left( \mathrm{Re}\big(F(\tau)\big)\cos(\tau\theta) - \mathrm{Im}\big(F(\tau)\big)\sin(\tau\theta)\right) \qquad (3)$$

Recall the objective is separating signature from error by using some statistical criterion and this can be done considering that harmonic components related to the signature should appear systematically on circular profiles, i.e., the associated amplitudes should be significantly different from zero.

To recognize which components satisfy this condition, relation existing between DFT and regression can be used.

### 2.2 Relation between DFT and regression

Consider typical regression equation:

$$\mathbf{y} = \mathbf{X}\boldsymbol{\beta} + \boldsymbol{\varepsilon} \qquad (4)$$

where $\mathbf{y}$ is a $[n \cdot 1]$ vector containing the "response", i.e. experimental data, $\mathbf{X}$ is a $[n \cdot m]$ matrix containing "predictors", $\boldsymbol{\beta}$ is a $[m \cdot 1]$ vector containing coefficients that have to be estimated (i.e. the weights of predictors used in the model), and $\boldsymbol{\varepsilon}$ is a $[n \cdot 1]$ vector containing "residuals" not explained by the predictors themselves, that is the "background noise" of the phenomenon observed.

A particular kind of regression is "harmonic regression" where the predictors matrix is given by:

$$\mathbf{X} = \begin{bmatrix} 1 & \cos(\theta_1) & \sin(\theta_1) & \dots & \cos(k\theta_1) & \sin(k\theta_1) \\ 1 & \cos(\theta_2) & \sin(\theta_2) & \dots & \cos(k\theta_2) & \sin(k\theta_2) \\ \dots & \dots & \dots & \dots & \dots & \dots \\ 1 & \cos(\theta_n) & \sin(\theta_n) & \dots & \cos(k\theta_n) & \sin(k\theta_n) \end{bmatrix} \quad (5)$$
$$k = (n-1)/2$$

Here, $\theta_i$ is given by:

$$\theta_i = 2\pi \frac{i-1}{n} \qquad (6)$$

where $n$ is the number of sampled points. In the case of roundness profiles, $\theta_i$ represents the angular coordinate of the $i$-th sampled point, and the response is the point radius (Fig. 4).

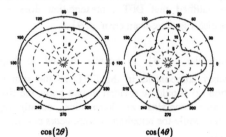

$$\cos(2\theta) \qquad\qquad \cos(4\theta)$$

Fig. 5. Lobed profiles in DFT.

The use of harmonic regression is strictly related to the use of DFT representation. In fact, given the DFT model reported in equation (2), for each $\tau$ the following relationships hold:

$$\begin{aligned} \beta_1 &= r_0 \\ \beta_{2\tau} &= \mathrm{Re}\big(F(\tau)\big) \qquad (7) \\ \beta_{2\tau+1} &= -\mathrm{Im}\big(F(\tau)\big) \end{aligned}$$

where, $\beta_i$ represents a coefficient in the harmonic regression model described in (4).

### 2.3 Statistical approach to separate noise from signature

Using relations presented in (7), we can separate signature from noise: signature is defined by the harmonics characterized by coefficients $\beta_i$ which are significantly different from zero.

This problem, known as variable selection in the statistical literature, should be theoretically faced by fitting all the possible regression models (where each regression model is characterized by a different subset of candidate predictors), and computing for each model a performance index. This performance index represents a trade off between the ability of the model in describing variability observed in the data (efficiency) and the parsimony of the model (i.e., to avoid overfitting, a model with a small number of predictors should be preferred). Among the performance indexes used to this aim, $R_{adj}^2$ and Mallows's $C_p$ (Montgomery et al. 2001) statistics are usually considered.

Unfortunately, when the number of candidate predictors is large, fitting all the possible regression models becomes computationally intractable, and the literature on regression analysis suggests the use of heuristic methods as the stepwise regression (Montgomery et al. 2001, p.310).

Considering our application of harmonic regression to identify signature in circular profiles, using $n=748$ points measured for each profile (as suggested by the "perfect operator" standard), we should in principle fit around $7.4030 \times 10^{224}$ regression models and chose

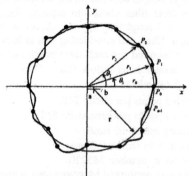

Fig. 4. Parametric model $r = f(\theta)$ for a roundness profile (Capello, Semeraro, 2001a)

the best among them. It follows that the heuristic method known as stepwise regression (Montgomery *et al.* 2001) is the only viable solution to identify signature in our case. This is the statistical criterion we suggest for separate the manufacturing signature from random noise.

### 2.4 Applying stepwise regression to our experimental results

Before applying the stepwise regression, we have to understand whether the signature left on the machined part is dependent or not from the specific level of the circular profile on the specimen (Fig. 3). To solve this issue, an ANOVA (ANalysis Of VAriance) test was conducted on coefficients of the first 20 harmonics, that are usually associated with signatures. As a matter of fact the first harmonics are associated to low frequency components, that usually characterize systematic pattern; while random error is usually associated with high-frequency components. As shown in Figure 6 for coefficient of the second harmonic, the distribution of this coefficient is dependent from the level of the circular profile considered. Therefore a different signature should be identified for circular profile at each level. The technological reason underneath this behaviour is due to a different inflexion of the workpiece while moving far from the spindle.

Therefore the signature identification algorithm was applied independently level by level. For each level, the variable selection algorithm (stepwise regression) was eventually concluded by diagnostic checking of residuals obtained after regression fitting. This step is required to assess that the regression model identified is adequate, i.e., it leaves unexplained just a random white noise. A typical behaviour of the sample autocorrelation function for residual obtained after the identification of the signature (obtained after applying stepwise regression) at one level (e.g., level 1) can be observed in Figure 7. The presence of autocorrelation in residuals obtained does not allow to consider results from regression analysis valid.

To overcome this problem, we adopt the method proposed by Montgomery *et al.* (2001), suggested when residuals follow a first order autoregressive model. The approach proposed suggests to estimate coefficient $\phi_1$ characterizing the first order autoregressive model obtained for residuals, and then use the following equations to transform data sampled and predictors, respectively:

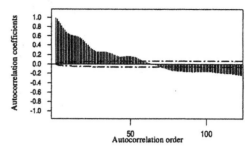

Fig. 7. Residuals autocorrelation for level 1.

$$
\begin{aligned}
y_i' &= y_i - \phi_1 y_{i-1} \\
x_{i,j}' &= x_{i,j} - \phi_1 x_{i-1,j} \\
&\forall i \in [1,m] \quad \forall j \in [1,n]
\end{aligned}
\tag{8}
$$

After this transformation phase, the algorithm requires a further computation of regression analysis. If residuals obtained after this final step are uncorrelated, the algorithm can stop, otherwise a further iteration is required. Unfortunately, autocorrelation in our residuals corresponds to an autoregressive model of the fourth order, and hence the algorithm suggested by Montgomery *et al.* (2001) can not be applied. However, considering the actual meaning of such a strong autocorrelation, gapping strategy should be adopted without affecting information included in data sampled. In fact, autocorrelation of the fourth order represents a situation in which information included in the $i^{th}$ observation is almost sufficient to predict information contained in the following $i+1^{th}$, $i+2^{th}$, $i+3^{th}$ and $i+4^{th}$ data measured on the circular profile. Therefore a clever sampling strategy should reduce the number of measured points by skipping some adjacent measurements on a given profile, since these measurements can be considered redundant (strategy known as gapping). As showed in Figure 8, the order of the autoregressive model decreases with the order of gapping considered. In particular, a good compromise between information contained in data and the order of the gapping strategy is attainable with gapping of order ten, i.e., sampling one data each ten observations available in the original set of measured points. In this case, residuals are hence represented by a first order autoregressive model and the algorithm by Montgomery *et al.* (2001) can be directly adopted. Applying the method proposed to circular profiles obtained level by level, harmonics which resulted significant, i.e. related to the

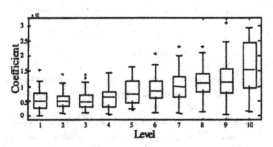

Fig. 6. Second coefficient dependence from level.

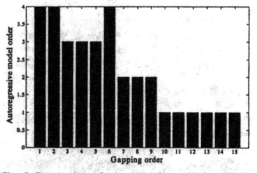

Fig. 8. Decreasing of autoregressive model order for residuals as gapping order increases.

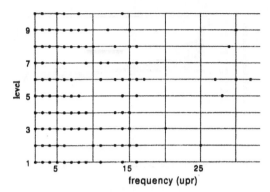

Fig. 9. Selected predictors

signature, are characterized by a dot in Figure 9. Therefore, we can conclude that the turning signature is identified by around fifteen components in the DFT representation assumed. This conclusion seems to correct the empirical rule proposed by Cho and Tu (2001), who suggested the use of fifty harmonic to identify roundness in circular profiles. Furthermore, the conclusion of our study allows to suggest a strong reduction in the number of points that should be measured to identify this signature. In fact by invoking the Shannon theorem for sampling, around thirty one sampling points should be enough to "read" this manufacturing signature.

## 3. CONCLUSIONS AND FUTURE AIMS

Advantages in identifying the signature of a process have been addressed in the literature. However the definition of a criterion to identify the boundary between signature (i.e. systematic pattern) and random noise is left to analyst experience in the model proposed in the literature. This paper goes beyond the state of the art in this direction, by suggesting a statistical approach to separate noise from signal. The method proposed is based on considering the relationship between Discrete Fourier Transform and harmonic regression, and then adopting statistical approaches traditionally used in the field of regression to identify the model which is able to capture and represent the systematic variability in data collected (i.e., the signature of a process). As shown in the experimental study presented, the identification of the signature can usefully translates into a reduction of the measuring points required to identify the signature. This reduction can hence induce savings in inspection costs.

Although applied to a specific process (turning) and to a specific tolerance (roundness), the approach proposed is quite general and can be easily extended to different signatures. Furthermore, the effect of cutting parameters and workpiece size on signature is a topic that is being investigated at this time.

Further advantages of signature analysis should be investigated in the near future. Firstly, scheduling of predictive maintenance for the machine-tool can rely on information coming from the signature. To go further in this direction, a deeper study of the relationship between components of the signature identified and the machine-tool conditions is required. A starting point in this direction is the study by Tu *et al.*, (1997), in which an analytical analysis of relationship between defects in spindle rotation and roundness profiles observed, is reported.

Eventually, the use of different representations of the signature should be explored. As an example, the use of wavelets functions instead of DFT represents a promising direction for further research.

## ACKNOWLEDGEMENTS

This work was carried out with the funding of the Italian M.I.U.R. (Ministry of Education, University, and Research).

## REFERENCES

Capello, E. and Q. Semeraro (2001a), The harmonic fitting method for assessment of the substitute geometry estimate error. Part I: 2D and 3D theory, *International Journal of Machine Tools & Manufacture*, **41**,, pp. 1071-1102.

Capello, E. and Q. Semeraro (2001b), The harmonic fitting method for assessment of he substitute geometry estimate error. Part II: statistical approach, machining process analysis and inspection plan optimization, *International Journal of Machine Tools & Manufacture*, **41**, pp. 1103-1199.

Cho, N. and J. Tu (2001), Roundness modelling of machined parts for tolerance analysis, *Precision Engineering*, **25**, pp. 35-47.

Henke, R.P., K. D. Summerhays, J. M. Baldwin, R. M. Cassou, C. W. Brown (1999), Methods for evaluation of systematic geometric deviations in machined parts and their relationships to process variables, *Precision Engineering*, **23**, pp. 273-292.

ISO/TC 213 (2003), *ISO/TS 12180 - Geometrical product specification (GPS) – Cylindricity*, part 2, International Organization of Standardization, Geneva, Switzerland.

ISO/TC 213 (2003), *ISO/TS 12181 - Geometrical product specification (GPS) – Roundness*, part 2, International Organization of Standardization, Geneva, Switzerland.

Montgomery, D.C., E.A. Peck and G.G. Vining, (2001), *Introduction to linear regression*, Wiley Interscience.

Summerhays, K.D., R.P. Henke, J.M. Baldwin, R.M. Cassou and C.W. Brown (2001), Optimizing discrete point sample patterns and measurement data analysis on internal cylindrical surfaces with systematic form deviations, *Precision Engineering*, **26**, pp. 105-121.

Tu, J.F., B. Bossmanns and S.C.C. Hung (1997), Modeling and error analysis for assessing spindle radial error motions, *Precision Engineering*, **21**, pp. 90-101.

Wilhelm, R.G., R. Hocken and H. Schwenke (2001), Task Specific Uncertainty in Coordinate Measurement; *CIRP Annals*, **50**, pp. 553-563.

# MODELING AND ERGONOMICS EVALUATION OF THE AUTOMOBILE ACCESSIBILITY MOVEMENT

**Lempereur[1] M., Pudlo[1] P., Gorce[2] P., Lepoutre[1] F-X.**

[1]*Laboratoire d'Automatique, de Mécanique et d'Informatique industrielles et Humaines UMR CNRS 8530*
*Université de Valenciennes et du Hainaut-Cambrésis*
*Le Mont Houy*
*59313 Valenciennes, France*

[2]*Laboratoire d'Ergonomie Sportive et Performance EA 31-62*
*Université de Toulon et du Var*
*Avenue de l'université*
*83957 La Garde, France*

Abstract: The generated felt discomfort during the automobile accessibility movement is a purchase criterion of the customers. The concurrent engineering implies the ergonomists and designers to use digital mock-up and computer-aided design. The objective of our research is to develop a computer human model able to simulate the automobile accessibility movement. The simulation model computes the joint angles of the lower body from the right foot trajectory of tall subjects with an optimization technique. The constraints taken into account are stemmed from a biomechanical analysis of the experimental joint angles. In results, there is a good correlation between the experimental and simulated flexion/extension angles. *Copyright © 2004 IFAC*

Keywords: Optimization problem, minimization, robotics, automobile, movement.

## 1. INTRODUCTION

In order to validate new automobile prototypes, the car manufacturers use full-scale mock-up. Several tests of entering and exiting movements are performed to evaluate the discomfort during the automobile accessibility movement by a significant number of subjects. However, for financial reasons and production time, these tests are only done at the end of the design phase. More and more, the will of the car manufacturers is to include the evaluation phase of discomfort during the ingress/egress movement as early as possible in the design phase (Tessier, 2000). So, the engineers and ergonomists have recourse to the digital mock-up and to the computer-aided design (Porter, *et al.*, 1993). Various

human simulation models exist such as RAMSIS, SAFEWORK or Jack. They incorporate a statistical model which considers the multivariate correlation of anthropometric dimensions that define human size and shape. They also have an inverse kinematics method for assisting designers in selecting postures and simulating samples motions (Chaffin, 2001). The inverse kinematics algorithms are now being developed and implemented in the Digital Human Models to assist the designers in manually positioning and moving the humanoid, but the resulting postures and motions are still very robotic and inhuman. This has limited the use of DHM technology to analyzing a sequence of single tasks that often are static rather than simulating dynamic movements.

The objective of our research is to develop a computer human model able to predict and evaluate a realistic entering/exiting movement. To predict the movement, 2 steps are required : the anthropometrical adaptation and the simulation. The adaptation aims at to personalize the trajectories of the feet and hands, the trunk mass centre trajectory and the spatial orientation of the trunk to a new subject and to a new vehicle. The simulation generates a human motion in order to fit the adapted trajectories and to generate physiological joint angles.

The aim of this paper is to present the simulation model. This inverse model based on optimization computes the joint angles from the trajectories of the feet and the hands. In this study, only the entering movement of the right leg of tall subject will be studied and simulated. Nevertheless, the simulation methodology will be the same for the others body segments and the others subjects.

This paper is organized as follows : the first section describes the humanoid model and the simulation model, the section 2 presents the obtained results.

## 2. METHOD

### 2.1. Humanoid model

The automobile accessibility movement is a three-dimensional movement (Andreoni and Rabuffetti, 1997). During the ingress movement, for example, most French drivers bend leftward the trunk while they flex and abduct the left leg. Their heads show a little leftward bending like the trunk, and a rightward rotation to watch at the vehicle geometry. During the egress movement, the subjects continuously turn rightward and extend the trunk while extending the left lower limb placed on the ground. Their heads were taken out before the rest of the body. So the planar models can not be applied. Therefore, the humanoid model has to be three-dimensional. These models allow us to study the movements in the sagittal, frontal and horizontal plane.

The suggested model is based on a control method of multi-chains system. The assumption is to control a body of reference to which are connected independent simple open kinematic chains. The method was applied to various movements such walking under perturbation (Gorce, *et al.*, 2001), stepping motion over an obstacle (El Hafi and Gorce, 1999), grasping (Gorce, *et al.*, 1994) and manipulation (Gorce and Rezzoug, 2000).

Our biomechanical model is composed of 4 simple open kinematic chains constituting the arms and the legs (Lempereur, *et al.*, 2003). These four chains are connected to a common rigid body : the trunk and the head, despite the fact that the trunk has some flexion/extension, twisting and lateral bending movements. Each kinematic chain is made up of 7 degrees of freedom (3 dof for hip and shoulder, 2 dof for knee and elbow, 2 dof for ankle and wrist). The figure 1 shows the biomechanical model of the humanoid.

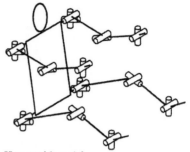

Fig. 1. Humanoid model.

The convention of Denavit and Hartenberg (1955) is adopted for modeling the humanoid. The figure 2 shows the kinematic models of the upper and lower body.

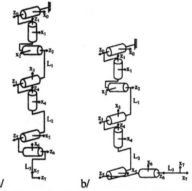

a/      b/

Fig. 2. a/ Upper body model. b/Lower body model.

The table 1 presents the parameters of Denavit and Hartenberg for the upper and lower body.

Table 1. Denavit and Hartenberg parameters for the upper and lower body.

| i | $\alpha_i$ | $a_i$ | $q_i$ | $d_i$ | value of $q_i$ |
|---|---|---|---|---|---|
| 1 | -90° | 0 | $q_1$ | 0 | 0 |
| 2 | -90° | 0 | $q_2$ | 0 | -90° |
| 3 | -90° | $-L_1$ | $q_3$ | 0 | -90° |
| 4 | -90° | 0 | $q_4$ | 0 | -90° |
| 5 | 90° | 0 | $q_5$ | $-L_2$ | 0 |
| 6 | 90° | 0 | $q_6$ | 0 | 90° |
| 7 | 0 | A | $q_7$ | B | 0 |

$$(A,B) = \begin{cases} (-L_3 ; 0) \text{ if upper body} \\ (0 ; L_3) \text{ if lower body} \end{cases}$$

For the lower (respectively upper) body, $q_1$, $q_2$ and $q_3$ are the flexion/extension, medial/lateral rotation and abduction/adduction angles of the hip (resp. shoulder), $q_4$ and $q_5$ represent the flexion/extension and medial/lateral rotation angles of the knee (resp. elbow), $q_6$ and $q_7$ are the flexion/extension and abduction/adduction angles of the ankle (resp. wrist). The lengths $L_1$, $L_2$ and $L_3$ are those of the thigh (resp. arm), the leg (resp. forearm) and the foot (resp. hand).

## 2.2. Construction of the simulation model

The automobile accessibility movement is a three-dimensional movement and a complex movement. The entering/exiting movement will be studied by phases. This multi-phases approach has already used in biomechanics such as in postural control (Gorce, 1999), in gait (Nilsson, *et al.*, 1985) or in manipulation (Gorce and Rezzoug, 2000). The simulation model is based on optimization. The optimization technique minimizes, at each time t and for each phase, the Euclidian distance between the adapted position $x_d$ and the current position of the right foot. It may be expressed as $\min \|x_d - f(q_i)\|$, where $\|\cdot\|$ is the Euclidian distance, f the forward geometric function and $q_i$ the joint angles. Some constraints are added to this objective function. For each phase, the constraints taken into account are the variation of the joint angles, the maximum variation between two consecutive instants of each experimental joint angles, the sign of the joint angles and the joint limits.

Different steps are required in order to find this constraints. The successive steps are the recording of the movement, the estimation of the joint centres, the filtering of the data, the computation of joint angles, the cutting of the movement, the biomechanics analysis of the joint angles and the determination of the constraints. The figure 3 presents the methodology to obtain the constraints by phase.

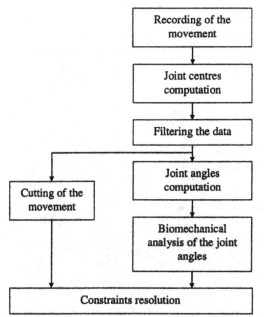

Fig. 3. Methodology to find the constraints.

After optimization, the entering movement of the right leg is simulated with the humanoid model with the trunk mass centre trajectory, the spatial orientation of the trunk and the joint angles computed with the optimization technique.

## 2.3. Recording of the ingress/egress movement

An experimental procedure and device were setup in order to record the automobile accessibility movement (Lempereur, *et al.*, 2004).

The experimental device is made of a simplified vehicle and an optoelectronic system (Vicon).

The experimental vehicle is composed of a seat and a steering wheel. The sill and the roof rail are also represented (figure 4). But this simplified vehicle has no door.

The system Vicon records the position of 42 reflecting markers (diameter 14 mm) placed on the skin of the subject. Eight infrared cameras are around the vehicle. The sampling rate is 120 Hz. The figure 4 presents a entering movement.

Fig. 4. Entering movement.

Twenty-one subjects (23.2 years ± 3.73, 176.43 cm ± 6.51, 68.31 kg ± 5.98) took part in the experiments. These subjects are share out between three categories of height : small (inferior to 1.75m), medium (between 1.75 and 1.80) and tall (superior to 1.80m). The table 2 indicates the number of subject by category.

Table 2. Number of subject by category

| Small | Medium | Tall |
|-------|--------|------|
| 8 | 8 | 5 |

## 2.4. Joint centres estimation

The joint centre of the elbow, wrist, knee and ankle are defined in the middle of the joints (Cappozzo, *et al.*, 1995). Whereas the hip and the shoulder (glenohumeral joint) joint centre are estimated with the functional method (Leardini, *et al.*, 1999). It consists in the following steps : during the movements of rotation, we suppose that the markers on the thigh or the arm describe spheres and they have a common centre of rotation : the joint centre of the hip (Piazza, *et al.*, 2001) or the shoulder (Veeger, 2000).

## 2.5. Filtering the data

The 42 reflecting markers trajectories and the new trajectories of the joint centres are filtered with a second order Butterworth without lag. This filter is usually used in biomechanics (Allard, *et al.*, 1990). The cut-off frequency is computed from the spectral analysis of trajectories. The cut-off frequency is given in order to preserve more than 98% of the energy.

### 2.6. Joint angles computation

The joint angles are computed for each joint (trunk with respect to the laboratory reference system, hip, knee, ankle, shoulder, elbow and wrist). For that, the Euler method is used from a local coordinate system defined for each body segment (Wu, *et al.*, 2002). The sequence of rotation is : flexion/extension, abduction/adduction, medial/lateral rotation.

### 2.7. Cutting of the automobile accessibility movement

The automobile accessibility movement is split into two phases : the entering and the exiting movement. As a general rule, the entering movement starts when the subject takes off his right heel, just before entering. It finishes when the subject is in the final sitting position, ready to drive, hands on the steering wheel, feet on the floor vehicle. The exiting movement starts, when the subject, in sitting position, takes off his left heel of the vehicle floor and finishes when he strikes his right heel on the ground.

For example, the figure 5 presents the phases of the right leg movement during the ingress movement for the tall subjects.

| Right heel off | | Right foot strike on floor | | Final sitting position |
|---|---|---|---|---|
| Phase P1 | Phase P2 | Phase P3 | Phase P4 | |
| | Right foot above the sill | | Subject seated | |

Fig 5. Right leg movements phases during the ingress movement for tall subjects.

## 3. RESULTS

### 3.1. Constraints of each phase

The biomechanical analysis of the joint allowed us to define the different constraints of the four phases of the entering movement of the right leg of tall subjects.

Tables 3-6 present these constraints.

#### Table 3. Constraints of the phase P1.
Constraints of P1

$q_1$ increase

$q_4$ decrease

$q_5$ decrease

$q_6$ increase

$|q_i(t) - q_i(t-1)| < V_i$

$q_3 > 0$

$q_4 < 0$

$q_{i,min} < q_i < q_{i,max}$

#### Table 4. Constraints of the phase P2.
Constraints of P2

$q_1$ decrease

$q_4$ increase

$q_5$ decrease

$q_6$ increase

$|q_i(t) - q_i(t-1)| < V_i$

$q_3 < 0$

$q_4 < 0$

$q_{i,min} < q_i < q_{i,max}$

#### Table 5. Constraints of the phase P3.
Constraints of P3

$q_1$ increase

$q_5$ decrease

$q_6$ decrease

$|q_i(t) - q_i(t-1)| < V_i$

$q_3 < 0$

$q_4 < 0$

$q_{i,min} < q_i < q_{i,max}$

#### Table 6. Constraints of the phase P4.
Constraints of P4

$|q_i(t) - q_i(t-1)| < V_i$

$q_{i,min} < q_i < q_{i,max}$

$V_i$ is the maximum variation between two consecutive instants of each experimental joint angles. $q_{i,min}$ and $q_{i,max}$ are the joint limits.

### 3.2. Joint angles

After optimization, the simulated trajectories of the right foot fit to the experimental ones. Indeed, the maximum gap between the simulated and experimental trajectories is 5 mm.

Figures 6-8 show the simulated and experimental joint angles of the lower body during the entering movement of a tall subject measuring 1.89m. His trunk mass centre trajectory and the spatial orientation of the trunk come from the experiments.

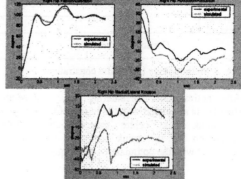

Fig. 6. Right hip angles.

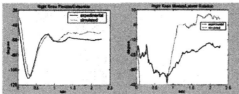

Fig. 7. Right knee angles.

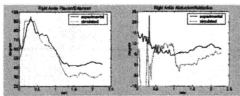

Fig. 8. Right Ankle angles.

The table 7 presents the correlation coefficients for the relationships of the experimental and simulated joint angles of the right leg.

Table 7. Correlation coefficients for the relationships of the experimental and simulated joint angles of the right leg.

| $q_1$ | $q_2$ | $q_3$ | $q_4$ | $q_5$ | $q_6$ | $q_7$ |
|------|------|------|------|------|------|------|
| 0.99 | 0.56 | 0.92 | 0.91 | 0.72 | 0.96 | -0.42 |

The results show a good correlation for the hip flexion/extension angle, the hip abduction/adduction angle, the knee flexion/extension angle and the ankle flexion/extension angle. A important difference exists between the experimental and simulated hip medial/lateral rotation and knee medial/lateral rotation angles. For the knee medial/lateral rotation, this maximum difference is near 30° whereas for the hip medial/lateral rotation, it is more than 40°. This result could be explained by a lack of constraints for $q_2$, $q_5$ and $q_7$. The biomechanical analysis shows a mismatch between the 5 subjects for these joint angles.

## 4. CONCLUSION

The entering movement of the right leg is simulated from the right foot trajectories by an optimization method. This optimization, applied to each phase of the movement, minimizes the Euclidian distance between the experimental and current position of the foot. The constraints are determined from a biomechanical analysis of the experimental joint angles. The constraints take into account the variation of the joint angles, the maximum variation between two consecutive instants of each experimental joint angles, the sign of the joint angles and the joint limits. The future step will to determine others constraints in order to have good correlation for the hip medial/lateral rotation, the knee medial/lateral rotation and the ankle abduction/adduction angles. Some discomfort indexes could be defined for the joint angles which have a correlation coefficient superior to 0.90.

## 5. REFERENCES

Allard P., Blanchi J. P., Gautier G. and Aïssaoui R., (1990). Technique de lissage et de filtrage de données biomécaniques. *Science & Sports*, 5, 27-38.

Andreoni G. and Rabuffetti M., (1997). New approaches to car ergonomics evaluation oriented to virtual prototyping. EURO-BME Course on Methods & Technologies for the Study of Human Activity & Behaviour, March 19-20, Milano, Italy.

Cappozzo A., Catani F., Croce U. and Leardini A., (1995). Position and orientation in space of bones during movements : anatomical frame definition and determination. *Clinical Biomechanics*, 10(4), 171-178.

Chaffin D. B., (2001). Digital Human modeling for vehicle and workplace design. Society of Automotive Engineers Inc., ISBN: 0-7680-0687-2.

Denavit J. and Hartenberg R. S., (1955). A kinematic notation for lower pair mechanism based on matrices. *Journal of Applied Mechanics*, 22, 215-221.

El Hafi F. and Gorce P., (1999). Behavioural approach for bipedal robot stepping motion gait. *Robotica*, 17(5), 491-501.

Gorce P., (1999). Dynamic postural control method for biped in unknown environment. *IEEE Transactions on Systems, Man, And Cybernetics*, 29(6), 616-626.

Gorce P., El Hafi F. and Lopez Coronado J., (2001). Dynamic control of walking cycle with initiation process for humanoïd robot. *Journal of Intellignet and Robotic Systems*, 31, 321-337.

Gorce P. and Rezzoug N., (2000). Numerical method applied to object tumbling with multi-body systems. *Computational Mechanics*, 24(6), 426-434.

Gorce P., Villard C. and Fontaine J. C., (1994). Grasping, coordination and optimal force distribution in multifingered mechanisms. *Robotica*, 12(2), 243-251.

Leardini A., Cappozo A., Catani F., Toksvig-Larsen S., Petitto A., Sforza V., Cassanelli G. and Giannini S., (1999). Validation of a functional method for the estimation of hip joint centre location. *Journal of biomechanics*, 32(1), 99-103.

Lempereur M., Pudlo P., Gorce P. and Lepoutre F.-X., (2003). A biomechanical computational model to simulate accessibility movement for car ergonomics evaluation. IEEE Computational Engineering in Systems Applications., July 9-11, Lille.

Lempereur M., Pudlo P., Gorce P. and Lepoutre F.-X., (2004). An inverse kinematics method based on optimization to simulate the automobile accessibility movement. SAE Digital Human Modeling For Design and Engineering Symposium, June 15-17, Oakland University, Rochester, Michigan.

Nilsson J., Thorstensson A. and Halbertsma J., (1985). Changes in leg movements and muscle activity with speed of locomotion and mode of progression in humans. *Acta Physiol Scand*, 123, 457-475.

Piazza S. J., Okita N. and Cavanagh P. R., (2001). Accuracy of the functional method of hip joint center location: effects of limited motion and varied implementation. *Journal of Biomechanics*, 34(7), 967-973.

Porter J. M., Case K., Freer M. T. and Bonney M. C., (1993). Computer-Aided Ergonomics Design of Automobiles. In Automotive Ergonomics, Eds. Peacock and Karwonski, Taylor & Francis, 43-78. ISBN: 0748400052.

Tessier Y., (2000). "Vers des mannequins numériques integrés dans la conception de produits", Les modèles numériques de l'homme pour la conception de produits. Lyon-Bron, 2000.

Veeger H. E. J., (2000). The position of the rotation center of the glenohumeral joint. *Journal of Biomechanics*, 33(12), 1711-1715.

Wu G., Siegler S., Allard P., Kirtley C., Leardini A., Rosenbaum D., Whittle M., D'Lima D. D., Cristofolini L. and Witte H., (2002). ISB recommendation on definitions of joint coordinate system of various joints for the reporting of human joint motion--part I: ankle, hip, and spine. *Journal of Biomechanics*, 35(4), 543-548.

ELSEVIER

IFAC

PUBLICATIONS
www.elsevier.com/locate/ifac

# THEORETICAL INVESTIGATION OF THE GRINDING WHEEL EFFECT ON GRIND HARDENING PROCESS

**Chryssolouris George and Salonitis Konstantinos**

*Laboratory for Manufacturing Systems and Automation (LMS)*
*Dept. of Mechanical Engineering and Aeronautics*
*University of Patras*
*Greece*

Abstract: Grind hardening process utilizes the heat dissipation in the cutting area for inducing metallurgical transformations on the surface of the ground workpiece. In the present paper, a preliminary theoretical model for simulating the power flux generated during the grind-hardening process is presented. The study focuses in investigating the effect of conventional corundum wheels and parameters considered are the grain size and the wheel porosity (structure). The model's predictions indicate that the most appropriate wheels for generating high heat rates in the contact area, and therefore be ideal for grind-hardening operations, should be relatively close-structured with fine grains. Preliminary experiments verified the theoretical predictions. *Copyright © 2004 IFAC*

Keywords: process models, simulation, processing techniques, grind-hardening, wheels

## 1. NOMENCLATURE

| | |
|---|---|
| $A$ | Geometric contact area |
| $A_a$ | Actual area of contact |
| $A_g$ | Wear flat area per grain |
| $a_e$ | Depth of cut |
| $b$ | Grinding wheel width |
| $d_e$ | Equivalent diameter |
| $d_g$ | Average grain diameter |
| $F_t$ | Tangential cutting force |
| $k_1, k_2$ | Linear coefficients |
| $l_g$ | Geometrical contact length |
| $M$ | Grain size from grinding wheel specification |
| $n$ | Number of active grains |
| $P_c''$ | Grinding power flux |
| $p_m$ | Average contact pressure |
| $Q_w$ | Heat generation rate |
| $S$ | Grinding wheel structure number |
| $u_c$ | Cutting speed |
| $u_w$ | Feed speed |
| $V$ | Finite grinding wheel volume |
| $V_g$ | Grain volume |
| $V_g(\%)$ | Volumetric concentration of abrasive grains |
| $\mu$ | Friction coefficient |

## 2. INTRODUCTION

Grind-hardening process is a special grinding process that can simultaneously harden and grind roughly a workpiece. The process is based on the utilization of the process generated heat for inducing a suitable temperature field on the workpiece, capable of producing high surface hardness. This is achieved as the dissipated heat and the subsequent quenching of the workpiece induce martensitic transformation to the workpiece surface.

Grind-hardening process is a relatively new process that was introduced by Brinksmeier and Brockhoff (1996). The main process parameters are the workpiece speed, the depth of cut, the cutting speed, the workpiece material and the grinding wheel type. The effect of the first three process parameters on Hardness Penetration Depth (*HPD*) has been investigated both experimentally (Brockhoff, 1999; Tsirbas *et al.*, 1999; Chryssolouris *et al.*, 2001) and theoretically (Tsirbas, 2002; Chryssolouris *et al.*, 2005). The suitability of a workpiece material to be ground-hardened depends on its composition and mainly on its carbon content and has been investigated thoroughly for a number of other surface hardening process. The results of these studies can be easily transferred to grind hardening process. However, the effect of the grinding wheel on the grind-hardening process has not been investigated.

The proper selection of process parameter values and grinding wheel type will enable the generation of sufficient heat at the contact zone and thus, achieve the heat treatment of the workpiece. The scope of the present paper is to preliminary investigate the effect of two major grinding wheel parameters (structure and grain size) on the grind hardening process. The power flux generated in the contact area is an important parameter of the surface layer hardening and was therefore chosen for checking the grinding behavior and the heat generation of the different wheel specifications. For the present study, the research has been focused on corundum wheels.

## 3. THEORETICAL INVESTIGATION

The theoretical approach presented hereafter can be utilized for the prediction of the heat generation rate due to sliding of the grinding wheel on the workpiece surface. In order to simplify the investigation and to reduce the large number of possible grind hardening operations, the investigation was limited on a two-dimensional surface grind hardening process with simple kinematics (fig. 1). The results however can be easily transferred to other grinding operations.

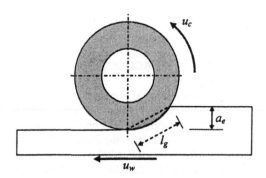

Fig. 1. Surface grinding conditions used for the present investigation

During grinding process a certain amount of energy is converted into heat which is concentrated in the grinding zone. This heat generation rate can be estimated from equation:

$$Q_w = F_t \left( u_c \pm u_w \right) \qquad (1)$$

where $Q_w$ is the heat generation rate, $u_c$ is the peripheral speed of the grinding wheel, $u_w$ is the feed speed and $F_t$ is the tangential component of the cutting force. Sign + is considered for up-grinding processes whereas − for down-grinding processes.

Since in most cases $u_c >> u_w$, it can be safely assumed that:

$$Q_w = F_t \cdot u_c \qquad (2)$$

This heat is consumed during material removal and sliding of the grinding wheel on the workpiece surface. The heat generation due to plastic deformation is negligible (Lavine, 2000) and therefore it can be safely assumed that heat is produced solely by the action of the sliding forces. The sliding forces are calculated from the following equation (Malkin, 1989):

$$F_t = \mu \cdot p_m \cdot A_a \qquad (3)$$

where $\mu$ is the friction coefficient between the workpiece material and the abrasive grains, $p_m$ is the average contact pressure of the abrasive grains on the workpiece and $A_a$ is the actual area of contact between the abrasive grains and the workpiece.

*Estimation of the average contact pressure.* The average contact pressure $p_m$ can be experimentally defined, and according to Malkin (1989), the average contact pressure is a linear function of the cutting curvature difference. For the case of peripheral grinding wheel speed $u_c$ being significantly larger than the feed speed $u_w$, the average contact pressure is given from the following equation (Chryssolouris et al., 2005):

$$p_m = k_1 \frac{4u_w}{d_e u_c} + k_2 \qquad (4)$$

where $d_e$ is the equivalent diameter, $k_1$ and $k_2$ are linear coefficients that are experimentally defined.

*Estimation of the actual contact area.* The *actual area of contact* between the grains and the workpiece depend on the grinding wheel composition. Grinding wheels are composed of grains, bonding material and a lot of air enclosures. The specification of a grinding wheel (figure 2) describes comprehensively its composition.

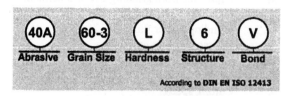

Fig. 2. Grinding wheel specifications

It is assumed that the heat is generated only between the grains and the workpiece material and therefore the bonding material is neglected. For the calculation of the actual area of contact, the number of grains adjacent to workpiece surface and the wear flat area per grain have to be determined.

The number of active grains can be determined considering a finite volume including all the grains in the contact area. This volume will have its three dimensions equal to contact length, grinding wheel width and grain height (Figure 3). The grains are considered spherical, thus the height of each grain will be equal to grain diameter. The volume percentage of the grains (referred also as volumetric concentration of abrasive grains $V_g(\%)$) included in volume $V$ is determined from equation:

$$V_g(\%) = \frac{n \cdot V_g}{V} \times 100 \qquad (5)$$

Where $n$ is the number of active grains, $V$ is the finite volume and $V_g$ is the volume of every grain.

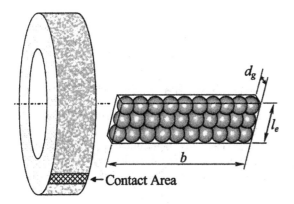

Fig. 3. Finite volume for the estimation of the number of the active grains

The volumetric concentration of abrasive grains in the wheel is related with the wheel marking structure number $S$. The structure number of the grinding wheel indicates the degree of porosity that the wheel presents. The equation relating the volume percentage of grain and the structure number is given by Malkin (1989):

$$V_g(\%) = 2 \cdot (32 - S) \qquad (6)$$

Combining equations (5) and (6) results in the active number of grains:

$$n = \frac{12 \cdot (32 - S)}{100} \cdot \frac{l_g \cdot b}{\pi d_g^2} \qquad (7)$$

where $l_g$ is the geometrical contact length, $b$ is the grinding wheel width and $d_g$ is the average grain diameter. The geometrical contact length is calculated from the process kinematics as:

$$l_g = \sqrt{d_e \cdot a_e} \qquad (8)$$

The average grain diameter is correlated with the grain size number $M$ found in the grinding wheel marking with the following equation (Malkin, 1989):

$$d_g = 15.2M^{-1} \qquad (9)$$

The above equation approximates the grit dimension $d_g$ as 60% of the average spacing between adjacent wires in a sieve whose mesh number equals the grit number $M$.

The actual contact area, as already mentioned, is determined as the product of the number of active grains times the wear flat area per grain $A_g$:

$$A_a = n \cdot A_g \qquad (10)$$

Each grain is considered to penetrate into the workpiece up to a depth equal to half its height. Therefore the wear flat area per grain is assumed to be half the grain surface.

*Estimation of heat generation rate.* Once the actual contact area is estimated, the sliding force is determined from eq. (3) and subsequently the heat generation rate is estimated from eq. (2).

*Estimation of power flux.* The power flux is finally determined by dividing the heat generated rate with the contact area as shown in equation (11).

$$P_c'' = \frac{Q_w}{A} \qquad (11)$$

where $P_c''$ is the grinding power flux, $A = (l_g*b)$ is the geometric contact area between the grinding wheel and the workpiece.

## 4. THEORETICAL RESULTS

The model was solved for estimating the dependence of the heat generation rate on the grinding wheel structure for various depths of cut, cutting speeds and feed speeds. Figures 4, 5 and 6 indicate that grind-hardening with grinding wheels having greater values of structure, thus having increased porosity, results always in the generation of less power flux in the contact area.

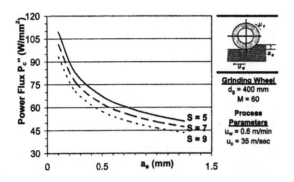

Fig. 4. Power flux dependence on depth of cut for various grinding wheel structures

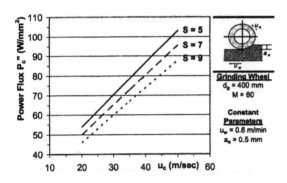

Fig. 5. Power flux dependence on cutting speed for various grinding wheel structures

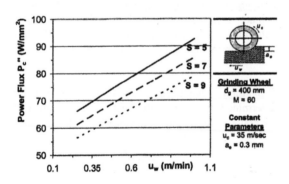

Fig. 6. Power flux dependence on feed speed for various grinding wheel structures

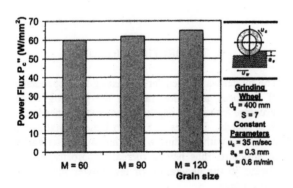

Fig. 7. Power flux dependence on grain size

The theoretical model was further solved for various grain sizes. The model predicted that when using grinding wheels with finer grits, the power flux is slightly increased (figure 7).

## 5. EXPERIMENTAL VERIFICATION

For the experimental verification of the model, a number of experiments were designed. Since the heat generation in the contact area can not be measured directly, the tangential cutting forces were measured with a piezo-electric force sensor. For the determination of the power flux, the following equation was used.

$$P''_{c,exp} = \frac{F_{t,exp} \cdot u_w}{A} \qquad (12)$$

Where $P''_{\varsigma exp}$ is the grinding power flux experimentally determined, $u_w$ is the workpiece feed speed, $F_{t,exp}$ is the experimentally measured tangential forces and $A$ is the geometric contact area.

Surface up grind-hardening tests were performed on St.37 specimens at the IWT premises with 5 different grinding wheels manufactured by TYROLIT. The experimental setup is presented in table 1 and the grinding wheels' specifications that were used are listed in table 2. Once the experiments were conducted, the power flux was determined based on the cutting force measurements. In figures 8 and 9 the experimentally determined power flux on the contact area when grind-harden with wheels having different structure number and grain size is shown.

Table 1 Experimental setup

| Process Parameter | Value |
|---|---|
| Cutting Speed | $U_c = 35$ m/sec |
| Feed Speed | $U_w = 0.6$ m/min |
| Depth of Cut | $a_e = 0.3$ mm |

Table 2 Grinding Wheel specifications

| Grinding wheel | Grain Size | Structure |
|---|---|---|
| A 60 L7 V | $M = 60$ | $S = 7$ |
| A 90 L7 V | $M = 90$ | $S = 7$ |
| A 120 L7 V | $M = 120$ | $S = 7$ |
| A 60 L5 V | $M = 60$ | $S = 5$ |
| A 60 L8 V | $M = 60$ | $S = 8$ |

Figure 8 reveals that the use of grinding wheels with more close structure, i.e. big structure number, results in lower power flux values. This is in tandem with theoretical model's predictions as shown in figures 4, 5 and 6.

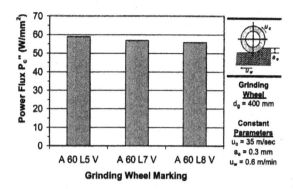

Fig. 8. Experimentally determined power flux for grind-hardening with grinding wheels having different structures

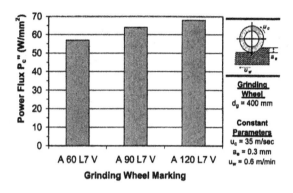

Fig. 9. Experimental determined power flux for grind-hardening with grinding wheels having different grain size

Furthermore, figure 9 shows that grinding wheels composed with coarser grits – low grain size number – results in smaller power flux values. The theoretical predictions, as pointed in figure 7, are in agreement with these experimental results.

## 6. CONCLUSIONS

The theoretical model developed for the investigation of the grinding wheel effect, proved that the composition of the grinding wheel affects greatly the amount of heat generated.

Close structured wheels, thus grinding wheel with small structure values, produce higher heat generation rate which is beneficial for grind hardening applications. This indicates that the most appropriate grinding wheels should have high concentration of abrasive grains (small structure marking). However, packing limitations due to grain size and grain distribution results in an upper limit of the grain concentration in the grinding wheel. The maximum packing density can be achieved when applying moderate pressures without causing grain

crushing (Malkin, 1989). Higher limiting packing densities are obtained with coarser grains than with fine and less symmetrical shapes.

The model predicted that a grinding wheel composed with fine grits will induce slightly more heat in the contact area when compared with a wheel composed of coarser grits.

A number of preliminary experiments were performed that proved the validity of the theoretical predictions.

## 7. FUTURE WORK

The present study was focused in the investigation of the grinding wheel effect on the grind-hardening process, taking however into consideration only the structure and the grain size of the wheel. The effect of bonding material, bonding hardness and grain material are currently investigated and an updated model will be published shortly. Furthermore, the model will be modified accordingly as to predict the behaviour of CBN wheels as well.

Finally, the effect of the process parameters on the *HPD* will be investigated. Since the above model predicts the heat generation rate, a heat partition model will be integrated in the theoretical approach presented, as to account for the heat dissipated to the grinding wheel, the chips, the coolant fluid and the environment.

## 8. ACKNOWLEDGEMENTS

The work reported in this paper was partially supported by CEC / Growth Programme (GRD1-2001-40535), "Development of low energy and eco-efficient grinding technologies".

## 9. REFERENCES

Brinksmeier, E. and T Brockhoff (1996). Utilisation of Grinding Heat as a New Heat Treatment Process. Annals of the CIRP, Vol. 45/1, pp. 283 – 286.

Brockhoff, T. (1999). Grind-hardening: A comprehensive View. Annals of the CIRP, Vol. 48/1, pp. 255 – 260.

Chryssolouris, G., K. Tsirbas, and S. Zannis (2001). An experimental investigation of Grind-Hardening. Proceedings of the 34th CIRP International Seminar on Manufacturing Systems, pp. 121 – 123.

Chryssolouris, G., K. Tsirbas and K. Salonitis (2005). An analytical, numerical and experimental approach to grind hardening. SME Journal of Manufacturing Processes, Accepted to be published.

Lavine, A.S. (2000). An exact solution for surface temperature in down grinding. <u>International Journal of Heat and Mass Transfer</u>, **Vol.43**, pp. 4447 – 4456.

Malkin, S. (1989). <u>Grinding Technology: Theory and Applications of Machining with Abrasives.</u> Ellis Horwood, Chichester.

Tsirbas, K., D. Mourtzis, S. Zannis and G. Chryssolouris (1999). Grind-hardening modelling with the use of Neural Networks. <u>Proceedings of the AMST'99 International Conference on Advanced Manufacturing Systems and Technology</u>, pp. 197 – 206.

Tsirbas, K. (2002). <u>An analytical and experimental investigation of the grind-hardening process</u>. PhD thesis, University of Patras Editions, Patras.

ELSEVIER
IFAC
PUBLICATIONS
www.elsevier.com/locate/ifac

# POSTURE BASED DISCOMFORT MODELLING USING NEURAL NET

**George Chryssolouris, Menelaos Pappas, Vassiliki Karabatsou**

*Laboratory for Manufacturing Systems & Automation (LMS),
Department of Mechanical Engineering & Aeronautics,
University of Patras, Rio-Patras 26110, Greece*

Abstract: This paper discusses the development of generalized models, for the estimation of the human feeling of discomfort during a task execution, using both measured biomechanical parameters and subjective answers on experimental questionnaires. The objective is to estimate the discomfort value of any human type on body joints for a wide range of joint-postures. Several modelling methods, such as regression and fuzzy logic have been followed in the past in this research field. In the present work a multi-layer Neural Network model has been developed for each joint, presenting a new approach for the estimation of human discomfort. The models could be used to support the ergonomic design of industrial products and processes, in which the human intervention is crucial, improving sufficiently their quality. *Copyright © 2004 IFAC*

Keywords: Artificial intelligence, ergonomics, modelling, neural networks.

## 1. INTRODUCTION

One of the major aspects in industry is to design products that would be more suitable and comfortable to human operators. For instance, safety and comfort are nowadays two of the basic targets of the automotive manufacturers (Schamale *et al.*, 2002). In order to evaluate a human's feeling of discomfort, during a task or process execution, methods that could define and even more predict this feeling, are needed. In this way, designers will be able to evaluate if the motion of a specific person, in a specific environment, provides him/her with a feeling of discomfort. The aim is to reduce this unwanted feeling for persons at any age and for any anthropometrics. It is worth pointing out that discomfort and repetitive movements involve bad postures adoption, and can additionally have some psychosomatic effects on health (Pheasant, 1999).

As a result, a number of different modelling approaches have been introduced to the research community, in the field of human discomfort prediction and evaluation. Experimental data for the joint angles of isocomfort (JAI), in sitting and standing males, based on perceived comfort ratings for static joint postures, had been collected. Based on the experimental results, regression equations were derived for each joint posture, in order to represent the relationships among different levels of joint deviation/joint posture and corresponding normalized comfort scores. The results of this study have put forward that static postures cause greater discomfort for the hip joint, and less discomfort for the elbow than that for the other joints under study (Kee and Karwowski, 2001). A regression model for the prediction of the perceived discomfort, in respect to the joint movement, has been developed also, using a central composite design, based on the response surface method. Joint angles of the upper body, with seven degrees of freedom and the weight of a load, were treated as independent variables, while the perceived discomfort measured, using a magnitude estimation technique, was used for the response variable (Jung and Choe, 1996). A Fuzzy Logic approach, to model relations among human perception, human characteristics and workspace structure, has been presented. A model was produced, where the drivers' perceived comfort, when handling interior controls, is in relation to anthropometrics and control positions (Hanson *et al.*, 2003). A quantitative evaluation methodology of postural stresses has been presented, using a postural coding system and neural network approach for typical automobile assembly tasks. Initially, psychophysical discomfort values were determined for varying postures, at five body joints: wrist, shoulder, neck, back and leg, using free modulus magnitude estimation (Lee *et al.*, 1999).

This work aims at showing the feasibility to define effective human motion discomfort related models both from the subjective evaluation of discomfort and from the objective physiological and biomechanical measurements, by using neural networks (NN). This will result in the accurate estimation of human stress when using several alternative designs of new products.

## 2. DATA ANALYSIS

### 2.1 Experimental data.

The modelling of the feeling of discomfort could not be based on a theoretical study, due to the fact that the answer in the question "How much discomfort do you feel?" could vary, even if both the anthropometrics and the body posture are the same, because it also depends on the nature and the mood of the subject. These parameters could not be represented by a natural law so as to define a theoretical model. Thus, experimental/empirical models were developed in order to describe this phenomenon.

For this purpose, a number of experiments have been performed, by the Institute of Ergonomics (LfE) at the Technical University of Munich, for measuring the feeling of discomfort, on specific human body joints, in several anatomically possible postures. Five subjects were used to measure their feeling of discomfort, for different joint-postures. The different anthropometric characteristics of each subject are shown in Table 1. Ten body joints were selected to be studied, namely the left and right wrists, elbows, shoulders, hips and knees.

Table 1. Subjects' anthropometric characteristics

| Subject ID | Gender | Age (years) | Height (cm) | Weight (kg) |
|---|---|---|---|---|
| 1 | M | 25 | 182 | 69 |
| 2 | M | 21 | 188 | 100 |
| 3 | M | 26 | 172 | 65 |
| 4 | F | 26 | 180 | 60 |
| 5 | M | 22 | 173 | 67 |

After positioning every subject in several joint-postures, into the range of motion of the corresponding joint, their feeling of discomfort documented in a questionnaire. The same procedure was repeated for all ten selected joints. At each joint posture, the rotation angles by x, y and z-axis were measured. The local coordinate systems of all the joints under study, together with their anatomically possible rotations are presented in Fig. 1.

The Category Partitioning Scale CP-50 method (Shen and Parsons, 1997) was used to quantify the feeling of discomfort. After each experimental set, all the subjects select the category of discomfort that matched their feeling and then refine their judgment by choosing a number within the selected category. The recorded discomfort values were grouped per joint. In order to set all the discomfort values under the same scale, they were normalized per joint into the range [0, 1], feeling the maximum discomfort when the value is equal to 1. After each

experimental execution a data set was arisen consisting of the subject's characteristics (age, height and weight), the measured rotation angles of the corresponding joint's positioning and the resultant discomfort value.

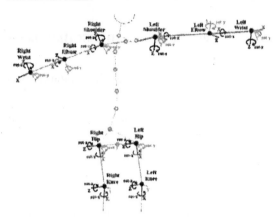

Fig. 1. Studied joints and their anatomically possible rotations

The valid ranges of the models' independent variables are presented in Table 2 and Table 3. These ranges vary from joint to joint and have been created by the minimum and the maximum values of the available experimental data sets. Into these ranges the prediction is more accurate and reliable. If the user's input data fall outside the valid ranges, it may not be a problem, however, it should be noted that when the inputs to a model are significantly different from the data, used for the model development, the accuracy of the results might be questionable.

Table 2. Ranges of rotation angles for each joint

| | Joint | rot-x | | rot-y | | rot-z | |
|---|---|---|---|---|---|---|---|
| | | min | max | min | max | min | max |
| Left | Wrist | - | - | -60 | 60 | -30 | 30 |
| | Elbow | -112,5 | 67,5 | -115 | 0 | - | - |
| | Shoulder | -120 | 45 | -80 | 70 | -30 | 120 |
| | Hip | -30 | 30 | -30 | 36 | 0 | 90 |
| | Knee | -30 | 30 | 0 | 120 | - | - |
| Right | Wrist | - | - | -60 | 60 | -30 | 30 |
| | Elbow | -67,5 | 112,5 | -115 | 0 | - | - |
| | Shoulder | -45 | 120 | -70 | 80 | -30 | 120 |
| | Hip | -30 | 30 | -36 | 30 | 0 | 90 |
| | Knee | -30 | 30 | 0 | 120 | - | - |

Table 3. Ranges of anthropometrics

| Age (years) | | Height (cm) | | Weight (kg) | |
|---|---|---|---|---|---|
| min | max | min | max | min | max |
| 21 | 26 | 172 | 188 | 60 | 100 |

### 2.2 Selection of influence parameters.

The determination of the factors that influence the feeling of discomfort during a movement is a quite complex procedure. The discomfort that someone feels only in one joint, isolating it from the rest of the body, is very difficult. Moreover, this feeling depends not only on the current posture, but also on the previous ones. Furthermore, the selection of the influence factors in a generalised modelling work should be made so as to be able for the models to evaluate a large number of tasks and definitely should be always quantitative in order to be easily identified through experiments.

Based on the previous thoughts the age, the height and the weight of each subject were measured before each set of experiments (physiological parameters). Moreover the rotation angles of each one of the selected joint-postures were recorded (biomechanical parameters). Before using these parameters as input to the modelling work, the diagrams that show the variation of the feeling of discomfort changing only one parameter. These diagrams were created in order to investigate either the linearity or the complexity of the discomfort's dependency of the measured quantities. An example of such diagrams in the case of left elbow joint is placed hereinafter.

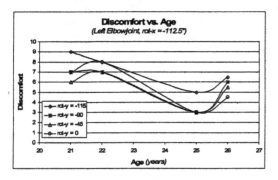

Fig. 2: Variation of the feeling of discomfort while the subject's age is changing

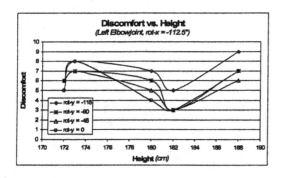

Fig. 3: Variation of the feeling of discomfort while the subject's height is changing

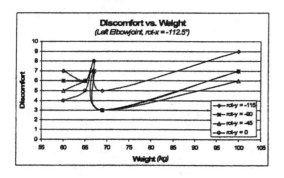

Fig. 4: Variation of the feeling of discomfort while the subject's weight is changing

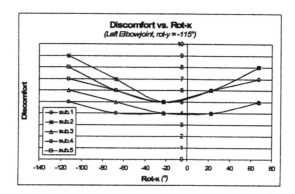

Fig. 5: Variation of the feeling of discomfort while the joint's rot-x is changing

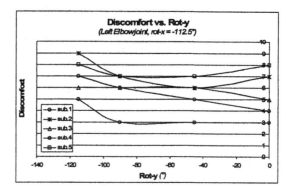

Fig. 6: Variation of the feeling of discomfort while the joint's rot-y is changing

As it is clearly shown in the above diagrams the relation between the discomfort feeling and the influence factors is highly non-linear. Thus, the neural network modelling approach was selected considering as independent variables (inputs) the subject's age, height and weight together with the rotation angles of the corresponding joint, while the feeling of discomfort, measured during the experiments, was used as the response variable (output). The number of rotation angles, and consequently the total number of input parameters, is related to the degrees of freedom of the corresponding joint.

## 3. METHOD DESCRIPTION

Discomfort is associated with psychological, physiological and biomechanical factors that produce feelings, such as pain, stiffness and fatigue. In the present work, discomfort is investigated in terms of biomechanical and physiological parameters. Appropriate posture-based models, were developed for the estimation of discomfort in each selected joint, using neural networks. The capability of neural nets to capture dynamic or non-linear relationships, makes the proposed modelling approach more appropriate compared with other methods (Dayhoff, 1990). Highly interconnected networks of nonlinear neurons (e.g. tangent) are shown to be extremely effective in computing. (Hopfield and Tank, 1985).

### 3.1 Neural Network architecture

The determination of the number of hidden layers and neurons into them was the first and most important step in order to create the most efficient network. After a number of trials and errors, changing the number of hidden layers and the number of containing neurons into them, a network of one hidden layer and three neurons into it, proved to be the choice with the most accurate results.

Next step was the selection of the connections type among neurons of two contiguous layers, as well as among the neurons within a layer. The nodes of the layers are connected by using the feed forward method. The sequential update of weights and bias was performed with the gradient descent with momentum function. After several trials, the hyperbolic tangent sigmoid transfer function was selected for all the neurons of the network. Moreover, the dot product weight function and the sum network input derivative function have been used into all the neurons.

As far as the training method concerns, the backpropagation was selected due to its high effectiveness on training of multi-layer neural nets (Radi and Poli, 1999). Finally, the selection of the learning method was critical for obtaining good generalization in neural networks with limited training data. The training of the neural networks was performed by using the Bayesian regularization function and the performance was measured according to the mean squared error function..

### 3.2 Models processing and results

Each body-posture is defined by a certain number of joint-postures. Every joint-posture is defined by rotation angles of which the number depends on the degrees of freedom of the corresponding joint. The rotation angles of the joint under study, together with the anthropometrics of the subject, are loaded as input parameters to the corresponding neural network and give as result the discomfort value of the specific frame. An example of the neural network structure concerning the left elbow joint (2-DoF) is presented in Fig. 7. Similar neural network models were developed for the rest joints that have been studied.

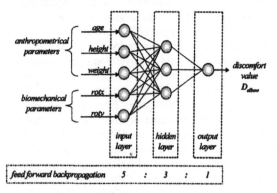

Fig. 7: Neural Network structure for the estimation of the left elbow joint discomfort values

Before proceeding to the neural networks' training, the experimental data sets were split in two groups, the 'training' and the 'simulation' data sets. As an example, in the neural network's training of the left elbow joint, the experimental data (anthropometrics, rotation angles and discomfort values) of subjects 1, 2, 3, and 4 were arbitrarily selected to feed the model. The diagram that is presented in Fig. 8 together with the correlation coefficient ($R^2$) value, which is equal to 0.8579, show a satisfactory quality of approximation of the experimental 'training' data sets during the neural network training phase.

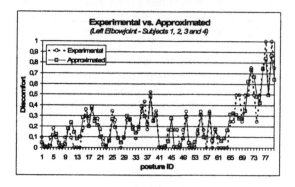

Fig. 8: Approximation of the discomfort values during the Neural Network training

After the network's training, the experimental data of subject 5 were used for the evaluation of the models' accuracy and produced good results. In order to demonstrate the accuracy of the models the discomfort values for the left elbow joint of subject 5 are presented in Fig. 9, in every measured posture, as they came up from the experimental phase (dashed line) and also as they were produced/estimated by the NN model after giving as input the appropriate input data (continuous line). As it is shown, no major deviations occurred between the experimental 'simulation' and estimated discomfort values for the same person. The correlation coefficient ($R^2$) between the experimental and estimated discomfort values is equal to 0.5753 indicating the ability for generalization and the accuracy of the model, considering that modelling/prediction of the human discomfort feeling is a quite difficult task due to the great number of influence parameters that could not be able to analyse (e.g. subject's mood or fatigue).

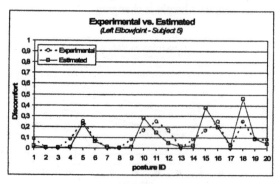

Fig. 9: Experimental vs. estimated discomfort values of the left elbow joint

Similar results have been also produced both for the training and simulation procedures, for all the joints (Table 4) marking out neural networks as a promising modelling approach for the estimation of the human feeling of discomfort. Moreover, the number of iterations (epochs) before stopping the training procedure, are presented in the last column of this table. This value represents the requested time for models development.

Table 4. Training and simulation results for all the joints

| | Joint | Training | Simulation | Epochs |
|---|---|---|---|---|
| Left | Wrist | 70 data sets $R^2 = 0.9466$ | 15 data sets $R^2 = 0.4308$ | 71 |
| | Elbow | 80 data sets $R^2 = 0.8579$ | 20 data sets $R^2 = 0.5753$ | 200 |
| | Shoulder | 267 data sets $R^2 = 0.6284$ | 60 data sets $R^2 = 0.4743$ | 69 |
| | Hip | 151 data sets $R^2 = 0.9182$ | 30 data sets $R^2 = 0.6702$ | 124 |
| | Knee | 52 data sets $R^2 = 0.8535$ | 8 data sets $R^2 = 0.4807$ | 63 |
| Right | Wrist | 70 data sets $R^2 = 0.9503$ | 15 data sets $R^2 = 0.4250$ | 76 |
| | Elbow | 80 data sets $R^2 = 0.8716$ | 20 data sets $R^2 = 0.5578$ | 161 |
| | Shoulder | 267 data sets $R^2 = 0.6233$ | 60 data sets $R^2 = 0.3923$ | 89 |
| | Hip | 151 data sets $R^2 = 0.9143$ | 30 data sets $R^2 = 0.6165$ | 158 |
| | Knee | 52 data sets $R^2 = 0.8861$ | 8 data sets $R^2 = 0.5880$ | 65 |

The determination of the number of hidden layers and neurons into them was the first and most important step in order to create the most efficient network. After a number of trials and errors, changing the number of hidden layers and the number of containing neurons into them, a network of one hidden layer and three neurons into it, proved to be the choice with the most accurate results.

## 4. CONCLUSIONS

The results of this modelling work could be used for the evaluation of a design with respect to the human motion. The joint-level neural network model could be used to estimate the feeling of discomfort for different humans in terms of some anthropometrics and body postures.

The fact that the development of these posture-based models were not based on specific recording motion data but on a wide range of static joint-posture data makes them generalized enough giving quite good estimation of the feeling of discomfort, during the execution of several human tasks. Moreover, this paper came up to the result that the same neural network structure could be used for several joints, independently of their location in the human body or their degrees of freedom.

Further topics in this field constitute the enhancement of the models quality by providing new experimental data for training purposes and also the development of additional neural network models for the rest joints of the human body. Finally, another interesting idea is the development of a more generalized model that would estimate the feeling of discomfort for the whole body. This model will synthesize the results of the joint-level models that developed in the frame of this work in order to give an estimated value of the discomfort feeling in the whole body of the subject. Input of this generalized model will be a body-posture instead of joint-posture that is the input for the models of the current study.

## 5. AKNOWLEDGEMENTS

This study was partially supported by the IST RTD project REALMAN/IST-2000-29357, funded by the European Commission.

## REFERENCES

Dayhoff, J. (1990). Neural Network Architectures: An Introduction. Van Nostrand Reinhold Co. New York.

Hanson, L., W. Wienholt and L. Sperling (2003). A control handling comfort model based on fuzzy logics. *International Journal of Industrial Ergonomics*, **31**, pp. 87-100.

Hopfield, J.J. and D.W. Tank (1985). Neural' computation of decisions in optimization problems. *Biological Cybernetics*, **52(3)**, pp. 141-152.

Jung, E.S., J. Choe (1996). Human reach posture prediction based on psychophysical discomfort. *International Journal of Industrial Ergonomics*, **18**, pp. 173-9.

Kee, D., W. Karwowski (2001). The boundaries for joint angles of isocomfort for sitting and standing males based on perceived comfort of static joint postures. *Ergonomics*, **44(6)**, pp. 614-648.

Lee, I., M. Chung and S. Kim (1999). Quantitative assessment of postural stresses during automobile assembly tasks using a neural network method. *Proceedings of the 14th Occupational Ergonomics and Safety Conference*, pp. 43-47.

Pheasant, S. (1999). *Bodyspace. Anthropometry, Ergonomics and the Design of Work*, (Taylor & Francis. (2)). London.

Radi, A. and R. Poli (1999). Genetic Programming Discovers Efficient Learning Rules for the Hidden and Output Layers of Feedforward Neural Networks. Genetic Programming, *Proceedings of EuroGP*, pp. 120-134.

Schamale, G., W. Stelzle, T. Kreienfeld, C.D. Wolf, T. Hartel and R. Jodicke (2002). COSYMAN: A simulation tool for optimization of seating comfort in cars. *Proceedings of the VDI Digital Human Modeling Conference*, pp. 301-311.

Shen, W. and K.C. Parsons (1997). Validity and reliability of rating scales for seated. *International Journal of Industrial Ergonomics*, **20**, pp. 441-461.

ELSEVIER
IFAC
PUBLICATIONS
www.elsevier.com/locate/ifac

# SIMULATION SUPPORTED ANALYSIS OF A DYNAMIC RESCHEDULING SYSTEM

**András Pfeiffer[1], Botond Kádár[1], Balázs Csanád Csáji[1], László Monostori[1,2]**

[1]*Computer and Automation Research Institute Hungarian Academy of Sciences Kende u. 13-17, Budapest, H-1111, Hungary*
[2]*Department of Production Informatics, Management and Control, Faculty of Mechanical Engineering, Budapest University of Technology and Economics, Budapest, Hungary E-mail: pfeiffer@sztaki.hu*

Abstract: The paper discusses the job shop scheduling problem and schedule measurement techniques, especially outlining the methods that can be applied in a dynamic environment. The authors propose a periodic rescheduling method by taking the rescheduling interval and schedule stability factor as input parameters into consideration. The proposed approach is tested on a simulated environment in order to determine the effect of stability parameters on the selected performance measures. *Copyright © 2004 IFAC*

Keywords: Dynamic Scheduling, Rescheduling, Simulation, Stability

## 1. INTRODUCTION

The broad goal of manufacturing operation management, such as a resource constrained scheduling problem, is to achieve a co-ordinated efficient behaviour of manufacturing in servicing production demands while responding to changes in shop-floors rapidly and in a cost effective manner.

In theory the aim is to minimize or maximize a performance measure. Regarding complexity, the job-shop scheduling problem (and therefore also its extensions), except for some strongly restricted special cases, is an NP-hard optimization problem (Baker, 1998; Williamson, et al., 1997).

### 1.1 Dynamic scheduling

The above mentioned job-shop scheduling is a static case, where all the information is available initially and it does not change over time. Most of the solutions in the literature concerning scheduling concentrate on this static problem. However, in many real systems, this scheduling problem is even more

difficult because jobs arrive on a continuous basis, henceforth called dynamic job shop scheduling (DJSS). According to Rangsaritratsamee, et al. (2004), previous research on DJSS using classic performance measures like makespan or tardiness concludes that it is highly desirable to construct a new schedule frequently so recently arrived jobs can be integrated into the schedule soon after they arrive.

Scheduling techniques addressing the dynamic – in the current case job shop – scheduling problem are called dynamic scheduling algorithms. These algorithms can be further classified as reactive and proactive scheduling techniques. Depending on the environment, there may be deviations from the predictive schedule during the schedule execution due to unforeseen disruptions such as machine breakdowns, insufficient raw material, or difference in operator efficiency overriding the predictive schedule.

The process of modifying the predictive schedule in the face of execution disruptions is referred to as reactive scheduling or rescheduling (Szelke and Monostori, 1999).

The practical importance of the decision whether to reschedule or repair has been noted in (Szelke and Kerr, 1994), while an additional categorization of scheduling techniques relating to the stochastic or deterministic characteristics of the problem can be found in (Kádár, 2002).

## 1.2 Rescheduling strategies

From the practical point of view in scheduling it is not possible to create schedules in every minute, however, the (theoretically) best performance of the whole system could be realized if schedule could be able to adapt to any changes, disruptions occurring in real-time. Most industrial planning and scheduling systems create schedules at idle time of the production e.g. at nights, while creating schedules for larger job-shop mostly requires a lot of computational time.

Schedule modification can be executed in given time periods (periodic rescheduling strategy), or related to specified events occurring during schedule execution (event-driven rescheduling strategy). Combining the two methods hybrid rescheduling strategy can be defined under which rescheduling occurs not only periodically but also whenever a disturbance is realized in the system (e.g. machine failures, urgent orders).

Define the time at which a new schedule is constructed as the rescheduling point and the time between two consecutive rescheduling points as the rescheduling interval (RI). At each rescheduling point, all jobs from the previous schedule that remained unprocessed are combined with the jobs that arrived since the previous rescheduling point and a new schedule is built.

Vieira, et al. (2000) presents new analytical models that can predict the performance of rescheduling strategies and quantify the trade-offs between different performance measures. Three rescheduling strategies are studied in a parallel machine system: periodic, event-driven and hybrid, similarly to the work of Church and Uzsoy (1992). They realized that there is a conflict between avoiding setups and reducing flow time, and the rescheduling period affects both objectives significantly, which statement is coincident concluded by Rangsaritratsamee, et al. (2004).

## 1.3 Schedule stability measurements

It is important to outline, that while rescheduling will optimize efficiency using classic performance measures (makespan or tardiness) the impact of disruptions induced by moving jobs during a rescheduling event is mostly neglected. This impact is frequently called stability (Rangsaritratsamee, et al., 2004; Cowling and Johansson, 2002). In related previous works, the number of times rescheduling

takes place was used by Church and Uzsoy (1992) as the measure of stability and it was suggested that a more frequent rescheduling means a less stable schedule. Other approaches defined stability in terms of the deviation of job starting times between the original and revised schedule and the difference of job sequences between the original and revised schedules. One of the shortcomings of these approaches is that they ignore the fact that the impact of changes increases as they are made closer to the current time. Rangsaritratsamee, et al. (2004) propose a method which addresses DJSS based on a bicriteria objective function that simultaneously considers efficiency and stability, and so let the decision maker to strike a compromise between improved efficiency and stability. In the approach, two dimensions of stability are modelled. The first captures the deviation of job starting times between two successive schedules and the second reflects how close to the current time changes are made.

The reaction to the realised disruption generally takes the form of either modifying the existing predictive schedule, or generating a completely new schedule, which is followed until the next disruption occurs (Kempf, et al., 2000). The first technique is described in (Vieira, et al., 2000; Rangsaritratsamee, et al., 2004), while the second is presented in (Bidot, et al., 2003; Cowling and Johansson, 2002). The importance of stability is outlined in the selected studies (see "monotonic and non-monotonic approach" in (Bidot, et al., 2003), or "2D stability" in (Rangsaritratsamee, et al., 2004). The most important point is that while scheduling will optimize the efficiency measure, the strategy generates schedules that are often radically different from the previous ones. From the practical point of view the scheduling technique mentioned first seems to be better, while in industrial applications constructing completely new schedules during schedule execution process must be avoided.

## 2. EVALUATION OF PRODUCTION SCHEDULES

The quality of factory scheduling, generally, has a profound effect on the overall factory performance. As stated in (Kempf, et al., 2000), an important aspect of the schedule measurement problem is whether an individual schedule or a group of schedules is evaluated. Individual schedules are evaluated to measure its individual performance. For a predictive schedule, the result may determine whether it will be implemented or not. There might be different reasons for evaluating a group of schedules. One of them is to compare the performance of the algorithms with which the different schedules were calculated. The comparison of different schedule instances against different performance measures is an other option in the evaluation of a set of schedules for the same problem.

According to Kempf, et al. (2000), relative comparison assumes that for the same initial factory state two or more schedules are available, and the task is to decide which is better. An absolute measurement of schedule quality consists in taking a particular schedule on its own and deciding how „good" it is. This requires some set of criteria or benchmarks against which to measure.

Regarding the predictive schedules, a set of decisions is made on the base of estimates on future events, without knowing the actual realizations of the events in question until they actually occur. Taking this fact into consideration, Kempf, et al. (2000) differentiates between the static and dynamic measurements of predictive schedules. A static measurement means the evaluation of the schedule independently of the execution environment.

Contrary to static measurement, the dynamic measurement of a predictive schedule is more difficult. In this case, beyond the static quality of the schedule, the robustness of the schedule against uncertainties in the system should also be taken into consideration.

Regarding the evaluation classes listed above, a dynamic measurement of individual predictive schedules will be presented in the following sections.

## 3. SIMULATION IN DYNAMIC SCHEDULING

Simulation captures those relevant aspects of the production planning and scheduling (PPS) problem, which cannot be represented in a deterministic, constraint-based optimization model. The most important issues in this respect are uncertain availability of resource, uncertain processing times, uncertain quality of raw material, and insertion of conditional operations into the technological routings.

The features provided by the new generation of simulation software facilitate the integration of these tools with the production planning and scheduling systems. Additionally, if the simulation system is combined with the production database of the enterprise it is possible to instantly update the parameters in the model and use the simulation parallel to the real manufacturing system supporting and/or reinforcing the decisions on the shop-floor.

The reason of the intention to connect the scheduler to a discrete event simulator was twofold. On the one hand, it serves as a benchmarking system to evaluate the schedules on a richer model; on the other hand, it covers the non-deterministic character of the real-life production environment. Additionally, in the planning phase it is expected that the statistical analysis of schedules should help to improve the execution and support the scheduler during the calculation of further schedules

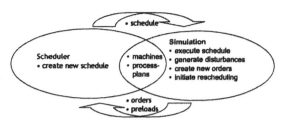

Figure 1. The rescheduling process initiated form the simulation side

In the proposed architecture the simulation model replaces a real production environment, including both the manufacturing execution system and the model of the real factory.

Simulation also generates continuously new orders into the system, while these new orders are scheduled and released by the scheduler.

The outline of the developed architecture is presented in Figure 1. Rescheduling action can be initiated when an unexpected event occurs or if a main performance measure bypasses a permissible threshold.

The dynamics of the prototype problem have been constructed to preserve realism as closely as possible and make the problem manageable for analysis.

This way simulation is capable for interaction with a specified scheduler, because all the required parameters are available any time for both systems, and so formulating an environment for further analysis on e.g. order pattern or sensitivity on significant parameters.

## 4. PROPOSED METHOD

The study analysis the impact of the rescheduling interval and the rate of schedule modification on classical performance measures as system load, efficiency as well as stability in a single machine prototype system.

### 4.1 Efficiency

The system to be scheduled is a single machine system with continuous job arrivals, but without any due date limitations. According to Baker (1974), the current scheduling problem can be classified as a single machine sequencing case with independent jobs and without due dates. In these situations the time spent by a job in the system can be defined as its flow time and the "rapid turnaround" as the main scheduling objective can be interpreted as minimizing mean flow time. The objective function is calculated as follows:

$$\overline{F} = \frac{1}{n}\sum_{j=1}^{n}\left(c_j - r_j\right) \qquad (1)$$

where

$\overline{F}$ is the mean flow time

$n$ is the number of total arrivals

$r_j$ is the point in time at job $j$ entered the system

$c_j$ is the completion time of job j, calculated when job j leaves the system

### 4.2 Stability

In our study stability is calculated for each available job in the system during schedule calculation by giving penalty (PN) values, using the relation *penalty = starting time deviation + actuality penalty*. Starting time deviation is the difference between the start time of the job at the new and previous rescheduling points. Actuality penalty is related to a penalty function associated with deviation of the start time of the job from the current time. Penalty values are only calculated in case starting time deviation is greater than 0. A schedule with less penalty value can be considered as a more stable schedule. The mean value of stability $\overline{PN}$ is calculated for all schedules as follows:

$$\overline{PN} = \frac{1}{n_{pn}} \sum_{j \in B} \left[ |t'_j - t_j| + \frac{100}{\sqrt{t_j - T}} \right] \quad (2)$$

where

$B$ is the set of available jobs $j$ that have not begun processing yet and $|t_j' - t_j| > 0$

$n_{pn}$ is the number of the elements in $B$

$t_j$ is the estimated start time of job j in the current schedule

$t_j'$ is the estimated start time of job j in the successive schedule

$T$ is the current time

### 4.3 Schedule Stability Factor

When minimizing the objective function Equation (1), in a single machine case the optimal dispatching rule to be selected is SPT (shortest processing time) detailed in (Baker, 1974). In the current case we use a truncated shortest processing time (TSPT) rule, in which the schedule stability factor (SF) can be introduced as the measure of the importance of schedule continuity or monotony. SF is the continuity rate of the schedule creation. In case SF equals zero, the new schedule may completely differ from the previous one, in case SF equals 1 the "old" jobs in the successive schedule must have the same position as in the previous one.

### 4.4 Schedule Creation

SPT based scheduling means, that the priorities of the available activities are calculated by taking only the length of the processing time into consideration. On the other hand, the TSPT rule we introduce – see Equation (3) – generates schedules using SF in order to override the priorities of the activities given by the

SPT rule, this way ensuring a more stabile schedule. Each priority must have an integer value and it is calculated as follows:

$$prio'_j = \left( prio_j \times SF + prio_{j,SPT} \times (1 - SF) \right)_{INT} \quad (3)$$

where

$A$ is the set of available jobs $j$ that remained unprocessed in the previous schedule

$prio_j'$ is the modified priority of job j ($j \in A$) in the successive schedule

$prio_j$ is the priority of job j in the previous schedule

$prio_{j,SPT}$ is the temporary priority of job j calculated using SPT rule

At each rescheduling point the following scheduling procedure is executed:

1. new jobs are added to set $A$
2. create a priority list of jobs in set $A$ by using SPT rule
3. compare current and previous priorities for "old" jobs and calculate new priorities using Equation (3)
4. add remaining priorities to new jobs and sort the list by priority, calculate penalties using Equation (2)
5. apply successive schedule and continue the schedule execution until the next rescheduling point defined by RI, then return to 1.

## 5. ANALYSIS AND EXPERIMENTAL RESULTS

The above mentioned method was tested on a simulated single machine prototype system in order to measure the characteristics of stability measures in a simple environment.

The simulation system was developed using eM-Plant object oriented, discrete event driven simulation tool, which will be helpful during the extension of the current problem to larger, job shop problems.

In single machine case, minimizing mean flow time we applied SF and RI as input at given shop utilization levels. As output we considered $\overline{F}$, $n_{pn}$ and total penalty which is the sum of all $\overline{PN}$ values multiplied by $n_{pn}$ calculated at the end of each simulation run.

It was experimentally determined that the results from the first 2000 arrivals should be eliminated from computations to remove transient effects. Hence, each simulation run in this study consisted of 12000 arrivals of which the final 10000 were used to compute the performance and stability measurements reported. Each experiment was replicated 10 times to facilitate statistical analysis.

The interarrival time ($b$), i.e. the average time between arrivals for jobs and are generated from exponential distribution with mean calculated using Equation (4):

$$b = \frac{\bar{p} \times n_o}{U \times m} \qquad (4)$$

where

$\bar{p}$ is the mean processing time per operation

$n_o$ is the number of operations in a job, in the current case equals 1

$U$ is shop utilization level

$m$ is the number of machines in the system, in the current case equals 1

## 5.1 Experiment 1

The main goal of Experiment 1 was to analyse the impact of system utilization level on $\bar{F}$, where SF was set to 0. Figure 2 shows, that both the system utilization and RI have a significant effect on $\bar{F}$. In the following experiment, where stability is examined we would like to use a relatively high utilization level in order to provide as much work-in-process as possible.

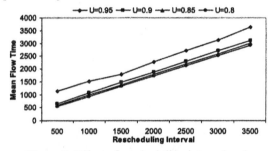

Figure 2. Effect of rescheduling interval and utilization level on mean flow time

As it is expected, extremely high utilization level lead to undesirable system instability, namely increasing the standard deviation of the resulted values and worsening the quality of the experimental results. The maximum acceptable value for $U$ in the current case is 0.9.

## 5.2 Experiment 2

The aim of Experiment 2 was to prove the assumption that applying the proposed stability criterion increases the stability of schedule execution however it reduces schedule efficiency. As a second scope of the experiment the effect of schedule stability factor on performance measurements was analysed.

In this experiment $U = 0.9$ and $p = 140$ with a triangle distribution {140, 1, 300}, then the mean of $b$ equals 160. Three rescheduling interval were considered 500, 2000 and 3500 to have results from a wide range of RI. The second group of input parameters was SF, set to 0, 0.25, 0.5, 0.75 and 1.

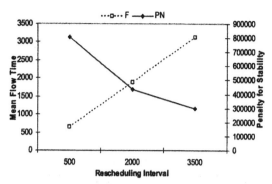

Figure 3. Effect of rescheduling interval on mean flow time and penalty values, in case SF=0

As we assumed, the lengthening of the rescheduling interval increases stability but decreases the efficiency of the system. Figure 3 shows the illustrative results where SF was set to 0. Efficiency measurement $\bar{F}$ is represented by the linear increasing dotted line, while the penalty values of the stability measurement are represented by the continuous line having a negative steepness. The penalty values decreased, because a higher number of modification made in the schedule at RI=500 with lower $\overline{PN}$ values resulted a greater product than the same parameters at RI=3500.

The effect of the parameter SF on penalty values given for stability and efficiency measurement $\bar{F}$ at different rescheduling intervals are shown in Figure 4 and Figure 5.

Figure 4 shows for all RI curves, that the values increase in a monotonic way, i.e. increasing SF decreases system performance (increasing $\bar{F}$) in each case. Comparing the results to SF=0, in case SF was set to 1, the outcome of the simulation showed an 8% increase of the performance measurement $\bar{F}$, in case RI was set to 500. Analyzing the other two cases, when RI was equal to 2000 and 3500, the performance of the system worsened only a few percent. Using these results it can be stated, that the negative effect of a higher SF level on $\bar{F}$ decreases as the length of the rescheduling interval is growing.

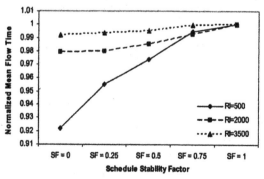

Figure 4. Effect of SF on normalized mean flow time at different rescheduling intervals

29

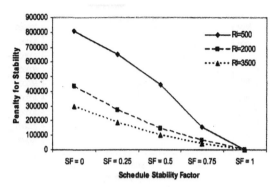

Figure 5. Effect of SF on penalty values at different rescheduling intervals

On the other hand, penalty values decreased significantly at each RI (see Figure 5), because the higher SF values reduced the total $\overline{PN}$ values, i.e. enabled less modification in the schedule.

Comparing $\overline{PN}$ values at different SF and RI parameter settings, it is interesting, that a penalty value given for SF=0 and RI=3500 is less than a penalty value for SF=0.5 and RI=500, while the efficiency is much better for RI=500.

Applying a limit for penalty values, e.g. let total $\overline{PN}$ be ab. $2*10^6$, then the optimal SF values can be selected for the given rescheduling intervals RI=500, 2000 and 3500. These values from Figure 5 are 0.7, 0.4 and 0.25 respectively.

## 6. CONCLUSIONS

The paper discussed the job shop scheduling problem and schedule measurement techniques, especially outlining the methods that can be applied in a dynamic environment. The results of the simulation study based on the proposed architecture showed that both rescheduling interval and the newly introduced variable schedule stability factor have a significant effect on schedule quality as well as stability. In case applying limitations for stability, then for the given rescheduling intervals the optimal SF values can be determined. This significantly improves stability measurements but inconsiderably reduces system performance.

## 7. FUTURE WORK

We would like to extend this experiment to a multi machine job shop system, using the results on stability gathered in this study. We propose a hybrid rescheduling strategy in a dynamic job shop environment defining two types of rescheduling events. The first type is done periodically (e.g. daily or weekly) using RI, releases new orders and involves tasks associated with order release. The second type is done when a disturbance occurs. It does not release new orders but instead reassigns work to off-load a down machine or utilize a newly-available one.

We assume that finding the appropriate schedule stability factor for each given rescheduling situation may results a compromise between the stabile schedule execution and schedule quality.

## REFERENCES

Baker, K. R. (1974). *Introduction to sequencing and scheduling*, John Wiley & Sons, USA.

Baker, A. D. (1998). A Survey of Factory Control Algorithms That Can Be Implemented in a Multi-Agent Heterarchy: Dispatching, Scheduling, and Pull. *Journal of Manufacturing Systems*, Vol. 17, 297-320.

Bidot, J., P. Laborie, J. C. Beck and T. Vidal (2003). Using simulation for execution monitoring and on-line rescheduling with uncertain durations. *Proceedings of the ICAPS'03 Workshop on Plan Execution*, Trento, Italy.

Church, L. K. and R. Uzsoy (1992). Analysis of periodic and event-driven rescheduling policies in dynamic shops. *International Journal of Computer Integrated Manufacturing*, Vol 5(3), 153-163.

Cowling, P. and M. Johansson (2002). Using real time information for effective dynamic scheduling. *European Journal of Operational Research*, Vol 139, 230-244.

Kádár, B. (2002). Intelligent approaches to manage changes and disturbances in manufacturing systems. PhD thesis. Technical University of Budapest, Hungary.

Kádár, B., A. Pfeiffer and L. Monostori (2004). Discrete event simulation for supporting production planning and scheduling decisions in digital factories. *Proceedings of the 37th CIRP International Seminar on Manufacturing Systems; Digital enterprises, production networks*, Budapest, Hungary, 444-448.

Kempf, K., R. Uzsoy, S. Smith and K. Gary (2000). Evaluation and comparison of production schedules. *Computers in Industry*, Vol 42, 203-220.

Law, A. and D. Kelton (2000). *Simulation modelling and analysis*, pp. 669-672, McGraw-Hill.

Rangsaritratsamee R., W. G. Ferrell Jr. and M. B. Kurz (2004). Dynamic rescheduling that simultaneously considers efficiency and stability. *Computers & Industrial Engineering*, Vol 46(1), 1-15.

Szelke, E. and R.M. Kerr (1994). Knowledge based reactive scheduling state-of-the-art. *Int. Journal of Production Planning and Control*, Vol 5 (March-April), 124–145.

Szelke, E. and L. Monostori (1999). *Modeling Manufacturing Systems*, Chap., Reactive scheduling in real-time production control, pp. 65-113. Springer, Berlin, Heidelberg, New York.

Vieira G. E., J. W. Herrmann and E. Lin (2000). Predicting the performance of rescheduling strategies for parallel machines systems. *Journal of Manufacturing Systems*, Vol 19(4), 256-266.

Williamson, D. P., L.A Hall, J.A. Hoogeveen, C.A. Hurkens, J.K. Lenstra, S.V. Sevastjanov and D.B. Shmoys (1997). Short Shop Schedules. *Operations Research*, Vol. 45, 288-294.

ELSEVIER
IFAC
PUBLICATIONS
www.elsevier.com/locate/ifac

# AGILE DESIGN AND MANUFACTURING IN COLLABORATIVE NETWORKS FOR THE DEFENCE INDUSTRY

Paul Maropoulos
Nikolaos Armoutis
David Bramall
Chris Lomas
Peter Chapman
Brian Rogers

*Institute of Agility and Digital Enterprise Technology,
School of Engineering, University of Durham, DH1 3LE, UK
Tel: +44 191 3342389, Email: p.g.maropoulos@durham.ac.uk*

Abstract: Designing products that satisfy ever-complex design criteria, planning effective processes and finding appropriate suppliers are of key importance for the prosperity of the defence industry. Within the Agile Design and Manufacturing concept establishing and coordinating dynamic supplier bases becomes a key challenge. Focusing on the agile notion, this paper introduces a framework to identify suitable manufacturing Small-Medium Enterprise (SME) and suggest dynamically viable process plans and product designs. *Copyright © 2004 IFAC*

Keywords: Agile Manufacturing, Concurrent Engineering, Design, Networks, Planning

## 1. INTRODUCTION

The increased complexity of engineering systems in sectors such as defence and aerospace, make it increasingly difficult for prime contractors to develop the entire final products themselves. They are increasingly placing a larger burden on the supply chain. This requires prime contractors to search for suppliers that can meet, in the best possible way, their needs across a wide range of products and geographic regions (globally). Small-Medium Enterprises (SMEs), many of which have limited resources, need to demonstrate competence in response to prime contractors and first tier supplier needs. They need to exhibit appropriate knowledge of technology and processes to develop new enhanced products and give customers confidence that they will meet their demands and will allow them to obtain work and sustain competitiveness. The issue is how SMEs with limited funds can develop this capability.

In contrast with mergers or takeovers, collaboration could allow the aggregation of competences analogous to those of a larger scale supplier without sacrificing the flexibility of the small enterprise. With the emergence of Information and Communication Technologies (ICT), collaboration is now even more attractive. Internet based tools such as virtual teaming kits are now available, which are claimed to be low-cost and easy to use. These types of tools provide the mechanisms for virtual organisations and allow SMEs to build effective agile networks without loosing their identity. However, an effective solution needs also to be able to identify and demonstrate the competences of engineering companies, as well as to match competences and based on these matches suggest dynamic, flexible, and adaptable partnerships with consideration to rapid product and process realization.

To address these requirements, an innovative framework is proposed in the following paragraphs. Although the framework is in its embryonic stage pending results of its testing, its constituent elements have been already independently validated through various industrial applications creating optimism for the results to come.

## 2. AGILE DESIGN AND MANUFACTURING

### 2.1 A definition

While the words Agile and Design have been linked together for some time, particularly in the software engineering environment, a clear definition of Agile Design for manufacturing has yet to be agreed upon. This section seeks to define the terms agile and design in order to lay down a definition of Agile Design which will be explored further in the subsequent sections.

Agile manufacturing is a well established and researched methodology which has grown from lean manufacturing. Goldman et al. (1994) suggest agility is a comprehensive response to the challenges posed by a business environment dominated by change and uncertainty. The concept of agility being responsiveness to a changing environment is also supported by the Design and Manufacturing Technical Network (D&ME) (2003), who define Agile Manufacturing as the ability to thrive and prosper in an environment of constant and unpredictable change. It is certain that one underpinning principle of Agility is responsiveness to a changing environment.

In defining Agile Design from a Design perspective, there are three clear routes the term could take. Firstly, there is design for Agile Manufacturing, ensuring that a product is designed with the manufacturing capabilities at the forefront of the design constraints. In this way a product can be manufactured with the least effort time and cost (Corbett, J. et al., 1991).

Secondly, we can define Agile Design as any product which has agile functionality. This concept of a product which once manufactured can have multiple functions, or even have its function changed entirely, has been defined as Adaptable Design by Gu et al. (2004) who propose a modular design process with functionally independent components or sub-assemblies. With simple interfaces between modules, upgrades are more feasible and therefore more likely. This concept of modularity is also proposed by Suh (2001) as the Axiomatic Design Independence Axiom. A secondary benefit to this methodology is the re-usability of modules, reducing waste and recycling.

Thirdly and finally is the definition of Agile Design proposed by the authors, as a flexible, scaleable, adaptive, responsive Design Process, encompassing not just design, but all processes which impact upon that design, from definition of requirements through to end-of-life.

Inherent to this definition are many of the modern design methodologies such as Axiomatic Design & Adaptable Design (in particular the independency of functional modules), Design for Manufacture, Assembly and the Environment, Concurrent product and process design and so on. Recent advances in software technology and management frameworks such as Digital Enterprise Technology (Maropoulos et al, 2003) have also done much to integrate design, manufacture and logistics.

## 2.2 Collaboration in Design

The modern manufacturing world is constantly changing, and there now exists a situation where 65% - 70% of the value added to a product is done so by companies other than the final assembler/manufacturer (MIRA, 1997). Large companies are forming partnerships with smaller more focussed and specialised partners, in some instances for long-term alliances, and in other cases for short term projects where specialist expertise is required. This phenomenon has given birth to what has become known as the Virtual Enterprise (Yagdev, 2001).

Clearly, the adoption of a modular design process as suggested by Suh (2001) and Gu (2004) lends itself to this Virtual Enterprise framework, facilitating the use of specialists for particular parts of a product design. For large complex systems such as those in the defence industry, this is of particular relevance.

Many software systems have been developed in recent times to address the need for collaboration amongst companies. This began initially in-house, before the large CAD vendors began to integrate collaborative working into their offerings. Concurrently to this development, resource planning software within (mainly large) enterprises began to flourish, in order to join previously (technologically and/or geographically) distributed departments. The two systems have both grown in their reach across the Product Life Cycle until converging at Product Lifecycle Management, at which point companies have been forced to integrate multiple systems or else tear out legacy systems in favour of a single instance replacement. In spite of these plentiful multi-million dollar resources, the ease of use of such systems by Small and Medium sized Enterprises (SMEs) has been largely neglected.

It is certain that for a successful implementation of Agile Design, such issues must be addressed to allow SMEs and large enterprises to interact seamlessly and quickly for the benefit of a reduced development period and optimal product design solution. Therefore, along with the design methodologies discussed earlier, collaboration techniques and tools will also be vital to Agile Design, and ways in which this can be effectively achieved should form the basis of further research.

## 2.3 Partner Identification

The need for the use of specialists for Agile Design has already been discussed, as has the need for an effective, flexible collaboration procedure between partner companies involved in any collaborative design project. It is therefore reasonable to suggest that with the need for experts, and an effective collaboration infrastructure, what remains is the ability to identify potential partners quickly and effectively, based on up-to-date, validated information about them. This information should include core competences, resource and human capabilities, and resource availability (Armoutis and Bal, 2003).

## 2.4 Process Planning

Concurrent design and process planning is established as a way of reducing product development time, while also impacting upon the design of a product, based on the capabilities of the resources available for manufacture (Maropoulos et al., 2003). In order to respond effectively to the environment, the impact of design changes must be minimal on the manufacturing planning undertaken, and therefore it is important to have a good understanding of the resources available for process planning, and to have the ability to identify potential process plans, or process limitations, at the earliest stage of design or re-design.

The combination of these 3 factors, effective collaboration within a Virtual Enterprise, rapid partner identification and process planning facilitate an Agile Design process. Changes in the external environment during a design project can be combated with the quick introduction of new partners, with appropriate expertise to change, or introduce new modules of a design. Process planning across the enterprise will ensure minimal disruption to existing plans by working at an early stage of design to search for optimal process plans from a wide search space based on the partner identification technology.

## 3. FRAMEWORK FOR AGILE DESIGN AND MANUFACTURING IN COLLABORATIVE NETWORKS

A framework (figure 1) has been developed to address the identified requirements of collaborative design, partner identification and process planning. The key aspects of it are described in the following paragraphs.

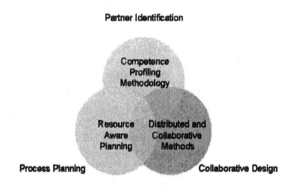

Fig. 1. The agile design and manufacturing framework for collaborative networks.

### 3.1 Resource-Aware Planning

Key elements of competitiveness for the defence industry are rapid product and process realization, early integration of design with manufacturing operations and the technical integration of the supply chain. Ideally then, integration efforts should be focussed on the early stages of product development, where the manufacturability and the majority of product lifecycle cost is determined. "Resource aware, aggregate planning" is a methodology for performing the technical evaluation (and optimisation) of product designs based on key manufacturability metrics (quality, cost and delivery) as early in the design cycle as possible. This is a Type II emergent synthesis problem (Ueda et al., 2001) because much of the detailed product specification is missing. The methods are capable of capturing the engineering capability of contractors, translating this into a number of possible production

scenarios and making this information automatically available to designers and planners in the form of a "rough-cut" process plan. The methods have been validated through the creation of a pilot process planning system called CAPABLE which has so far been tested with a limited number of industrial collaborators which supply the aerospace and defence industry.

Methodology. Figure 2 shows how the aggregate planning methodology is "driven" by the evaluation of feedback, mainly generated via the prioritisation mechanism of "capability analysis". The feedback enables the interactive evaluation of interim results that may be combined with the interactive specification of new options concerning the supply network's configuration. The key technology components employed for implementing the new pilot methods include; aggregate-level models for product, process and resource and a flexible planning engine supporting planning scenarios by using evolutionary computing for optimization and capability analysis techniques for planning feedback evaluation.

Product, Process and Resource Modelling. The specification of planning systems for use in early design required substantial development of the underpinning modelling, information management and knowledge retrieval, representation and classification technologies. The term "resource aware" was chosen to indicate the intention to create a much closer and dynamic inter-relationship between the main entities of the planning environment; the product model the generic, simplified process and assembly models and the resources, both humans and machines, available within a distributed enterprise.

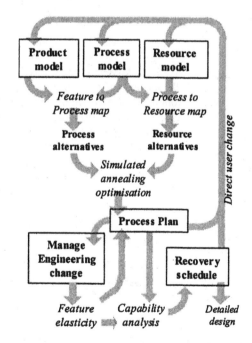

Fig. 2. Resource-Aware Planning Methodology.

The aggregate product model is a feature-based model which allows the effective representation of early product configurations at various levels of abstraction, beginning with structural elements only and culminating in a fully tolerance-enriched model.

A number of generic process models have been developed to model manufacturing operations. Central to the aggregate process model classes are a series of algorithms, derived from the simplification of detailed process models, which estimate quality, cost and delivery (QCD) performance based upon the feature characteristics of the product model and operating parameters from the resource. Finally, via a distributed architecture the chosen suppliers can create resource models of their own manufacturing facilities by supplying pre-defined operating parameters for specific classes of equipment. These resource models are then considered for the realization of particular processes by the planning engine. The ability of many supply network companies to independently and concurrently configure which of their resources can be made available for use by a remote planning optimization engine is at the heart of the "resource-aware" planning approach.

Optimization. The core functionality of aggregate planning is the rapid generation of a "rough-cut" process plan by the exploration of the large search space created by the feature-to-process and process-to-resource mappings. A Simulated Annealing (SA) algorithm is used to intelligently explore (via random selection, substitution and evaluation of an alternative within the plan) the search space for each feature by employing an energy function with user-defined weightings of QCD.

Implementation. CAPABLE's object-oriented class libraries and optimisation functions have been developed in Java. The product, process and resource data models are stored in a central SQL-compatible database, and are accessed across the web. The system uses the Java Remote Method Invocation API (RMI), in which the methods of remote Java objects can be invoked from other Java virtual machines, on different hosts. Access to the database is controlled via a series of private and public areas, CAPABLE. This distributed architecture allows the chosen supply companies to access the modelling tools in order to remotely create resource models of their own plants and supports the distributed, agile planning functions outlined in Section 5.2.

### 3.2 Distributed and Collaborative Methods for Agile Design

As well as novel methods for early planning, the defence industry can also benefit from recent advances in computer modelling, graphic visualization and distributed information and knowledge management, in order to positively impact product development and realisation and develop objective risk mitigation strategies. These technologies allow the development of new business processes to take advantage of the rapid formation of production networks and effectively manage workflow during all phases of a product's life cycle. 'Digital Enterprise Technology' (DET) is a new synthesis of technologies and systems for product and process development and life cycle management (Maropoulos et al, 2003). DET provides a coherent framework, acting in support of existing methods and systems such as resource-aware planning and digital manufacturing systems within an e-consortium. In the DET paradigm, CAD models are utilised to accurately simulate manufacturing and assembly processes to eliminate risk and shorten time to production. These simulations can be used to validate robot and tool programming, do ergonomic analysis of workstations and carry out material flow analysis for entire supply networks. The highly visual results can also be used to produce multimedia documentation and standard operating procedures for efficiently transferring processes and best practices with the network. The DET framework also maintains a bi-directional linkage between the physical product and the CAD model through the incorporation of methods for measurement planning and feedback using large-scale metrology.

### 3.3 Competence Profiling Methodology

A methodology has been developed to address the identified requirements of competence identification, structure, search, and matching. The methodology consists of four main stages:

I. Competence data collection. An on-line questionnaire has been development to assist in collecting and updating competence related information. Companies can load their competence information by registering on the 'Competence Profiling' web-site. The questionnaire aims at revealing the key skills and capabilities of engineering SMEs by focusing on both hard and soft factors, such as:
   - Company key indicators e.g. type of business and key markets
   - Awarded standards
   - Key manufacturing and management process capabilities
   - Core skills
   - Business philosophy

II. Normalizing. To ensure comparability among the various companies and correct use of the on-line tools, training and validation of the provided company data is provided. The 'normalising' stage also allows estimation of the quality of the identified competences. Since some of the information requires subjective judgement, the assistance of experts is considered essential within the normalisation process. Visits to the company premises by experts are organized and further-on validation and assistance is provided in capturing and assessing key competences.

III. Making competence information available for searching. In this stage competence information is stored in the relevant database which enables users to search for the appropriate set of business skills and capabilities to undertake a customer specific project. For this purpose the web-based Competence Search module has been developed (figure 3).

IV. Partnership formulation. In this final stage, companies with appropriate complementary competences are identified and matched to generate viable dynamic partnerships in response to a product need.

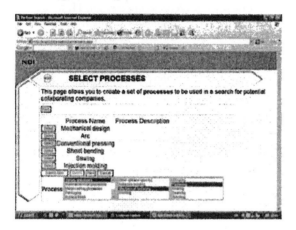

Fig. 3. Competence Search module

Several variations of the Competence Profiling Methodology (CPM) have already been tested and validated through European funded initiatives mainly within the automotive and general engineering sectors. This new and improved development will allow implementation and testing in an agile framework for the defence industry. The sector is technology driven and focused on demanding and rigorous quality standards. It is also characterized by intense international competition and high product complexity. Product development and manufacturing is many times a large-scale multi-partner task.

## 4. IMPLEMENTATION OF THE AGILE DESIGN AND MANUFACTURING FRAMEWORK FOR COLLABORATIVE NETWORKS

The ideas described are implemented and tested within the defence manufacturing sector. Initial testing has proved that the ideas proposed have been well perceived and accepted, leading to further on implementation and testing. Forty marine manufacturing SMEs have been invited to participate in a pilot implementation. At present stage companies complete the competence profiling process and populate the relevant database. It is anticipated that over 100 defence manufacturing companies will be competence profiled within the next two years and eventually fully benefit from this

framework. All participating companies are located in the North East of the UK.

## 5. BENEFITS OF THE FRAMEWORK

The implementation of the developed framework is proving to have several benefits. These are discussed in the following paragraphs.

### 5.1 Competence Profiling in Agile Design and Manufacturing

The Competence Profiling Methodology (CPM) promotes the companies' key skills and capabilities to fulfil customer requests. It has the ability to suggest rapidly a set of companies with appropriate complimentary competences to undertake a specific project. SMEs usually do not have the resources to undertake large and complex projects on their own. Being engaged in CPM they can be involved in larger and more complex projects via participation in dynamic collaborations. In addition, since information is normalized and validated via expert visits, CPM allows users of its system to search for, compare and confidently identify and match competent companies. Finally, through efficient communication with the other elements of the proposed framework, CPM enables speedy response in partner identification as a result of design or process planning changes. Such changes usually require new skill or capability which may not be available in the initial consortia.

### 5.2 Usage of CAPABLE in Agile Design and Manufacturing

In the context of agile design and manufacturing, CAPABLE is deployed, after the initial identification of potential suppliers by competence profiling. Its purpose is to carry out more specific technical evaluation of production options to facilitate agile planning and late decision making. This makes it particularly suitable for deployment in loose frameworks such as 'Engineering as Collaborative Negotiation' proposed by Jin and Lu (2004). As well as providing a designer with automatic feedback about the manufacturability of the design, the system also supports the concept of agile planning by maintaining a number of possible manufacturing scenarios to be carried forward to the detailed stage for evaluation. A number of ways to exploit this system have subsequently been identified, depending on whether the emphasis is to be placed on design improvement or overall manufacturing efficiency. For example, during the earliest stages of design the optimisation routines can use generic process models for evaluating the outsourcing of machining operations, selecting a number of potential suppliers based on cost criteria alone. However, during the more detailed design, the companies identified may be asked to connect the system and input more detailed information to enable a quality-based evaluation to take place at the level of individual machining centres. Ultimately, the aim of the system would be to allow the manufacturing company to use their specialist knowledge to critique the design and suggest changes to enable it to be made more effectively.

## 6. CONCLUSIONS

Trends such as the:
- intensification of the global competition
- fragmentation of markets into niche markets
- customer expectation of low volume and high quality custom products
- speedy development and production of products

mandate quick and effective application of agile principles in designing and manufacturing of products. Recognising also that dynamic collaboration would assist SMEs to respond within this intense environment and meet competition, a framework for agile design in collaborative networks has been proposed.

Competence Profiling when combined with CAPABLE and other proprietary "virtual manufacturing" software has demonstrated how web-centric co-development is influenced by the independent selection and deployment of resources within a global supply network.

Initial testing was very promising and future research will focus in integrating more closely the three components of the framework. In addition future research will concentrate on evaluating how a more unconstrained environment can be modelled and analysed and this will include the interactive evaluation of planning and designing functions developing the properties of self-coordination and co-evolution.

## 7. REFERENCES

Armoutis, N., and J. Bal, (2003). E-Business through Competence Profiling, In: *Building the Knowledge Economy: Issues Applications Case Studies* (Cunningham, P., Cunningham, M., and Fatelnig, P. (Ed)), 474-482, IOS Press, Amsterdam.

Corbett, J., J. Meleka, C. Pym, M. Dooner, (1991). *Design for manufacture: strategies, principles and techniques.* Addison-Wesley, Wokingham

D&ME (2003). Agile Manufacturing. *Pacific Northwest National Laboratory.* www.technet.pnl.gov/dme/agile/index.stm

Goldman, S.L., R.N. Nagel and K. Preiss, (1994). *Agile competitors and virtual organisations: strategies for enriching the customer.* Van Nostrand Reinhold, UK

Gu, P., M. Hashemian, A.Y.C. Nee, (2004). Adaptable Design. *Annals of the CIRP* **53(2)**.

Jin, Y. and Lu, S. C.-Y., 2004. Agent Based Negotiation for Collaborative Decision Making. Annals of the CIRP, **53(1)**, 121-124.

Maropoulos, P.G., D.G. Bramall, K.R. McKay, B Rogers, P. Chapman, (2003). An Aggregate Resource Model For The Provision Of Dynamic 'Resource-Aware' Planning. *Proceedings of the Institution of Mechanical Engineers Part B Journal of Engineering Manufacture* **217(10)**, 1471-1480.

Maropoulos, P. G., Rogers, B. C., Chapman, P., McKay, K. R. and Bramall, D. G., (2003). A Novel Digital Enterprise Technology Framework for the Distributed Development and Validation of Complex Products. Annals of the CIRP, **52(1)**, 389-392.

MIRA (1997). *Car Manufacturers of the World.* Automotive Information Centre – MIRA

Reich, Y., S. Konda, E. Subrahmanian, D. Cunningham, A. Dutoit, R. Patrick, M. Thomas, A.W. Westerberg, (1999). Building Agility for Developing Agile Design Information Systems. *Research in Engineering Design*, **11**, 67-83.

Suh, N.P., (2001). *Axiomatic Design: advances and applications.* Oxford University Press, New York

Ueda, K., Markus, A. Monostori, L., Kals, H.J.J., Arai, T., 2001, Emergent Synthesis Methodologies for Manufacturing, Annals of the CIRP, **50(2)**, 1-17.

Yagdev, H.S. and K.D. Thoben, (2001). Anatomy of Enterprise Collaborations. *Production Planning and Control*, **12(5)** 437-451.

ELSEVIER

IFAC
PUBLICATIONS
www.elsevier.com/locate/ifac

# NEW CONCEPTS OF MODELLING AND EVALUATING
# AUTONOMOUS LOGISTIC PROCESSES

**B. Scholz-Reiter, K. Windt**
**J. Kolditz, F. Böse, T. Hildebrandt, T. Philipp, H. Höhns**

*Department of Planning and Control of Production Systems,*
*University of Bremen, Bremen, Germany*

Abstract: Due to the existing dynamic and structural complexity of today's logistics systems, central planning and control of logistic processes becomes increasingly difficult. This intensifies the ongoing paradigm shift in logistic processes from centralised control of 'non-intelligent' items in hierarchical structures towards decentralised control of 'intelligent' items in heterarchical structures. The paper explains the observed paradigm shift in terms of new requirements related to modelling and evaluating methods of autonomy in logistic processes. The opportunities of autonomous logistic processes in contrast to conventionally managed logistic processes will be discussed on an exemplary scenario. *Copyright © 2004 IFAC*

Keywords: Adaptation, Agents, Autonomous and Adaptive Control, Evaluation, Manufacturing Processes, Methodology, Measuring Points, Modelling.

## 1. INTITIAL SITUATION
## AND CALL FOR ACTION

The field of production management and logistics is currently undergoing major changes, due to increasing structural (e.g. complexity and versions of products) and dynamic complexity of the production systems itself as well as to this production system is regarded as a sub-system of a logistics network.

This development is not new, and has been observed for the last years. Competition forces companies to develop new optimisation potentials. After nearly all internal possibilities of companies to improve their processes have been almost exhausted, some of the off-site concepts like Supply Chain Management (SCM) seem to be very promising to generate competitive advantages. On the one hand these changes are basically related to the fusion of several information- and web-technologies, which are technologically available and partly affordable like Radio Frequency Identification (RFID-) Technology and PDA- (Production Data Acquisition) within PPC-Systems, as well as on the other hand they are due to the deployment and fusion of a wide range of different methodologies of controlling and monitoring, for example from control theory and artificial intelligence. Coming from the field of control theory, new and further developed concepts have been discussed (Gassmann, 1998), for example related to adaptive controllers (Sastry and Bodson, 1989), or controllers using fuzzy-theory, as well as learning and knowledge based controllers (Luger, 2002). Even mixtures of these very different concepts have been sketched out, in the sense of so called hybrid approaches (Viharos and Monostori, 2001; Tsakonas and Dounias,

2002), sometimes integrating simulation models, as well.

This in turn will and has to lead to new ways of approaching complex scenarios of processes in production logistics. It may be pursuit for example in terms of locally modelling and linking, autonomous controlling and decision entities, e.g. multi-agent systems, which comprise several autonomous and heterogeneous agents acting together as a loosely coupled network to cooperatively solve given problems in an information-rich environment. These and similar understandings of the notion autonomy are currently under investigation within the Collaborative Research Centre (CRC) 637: "Autonomous Cooperating Logistic Processes - A Paradigm Shift and its Limitations" at the University of Bremen. Approaching these complex processes with different concepts of closed-loop control modelling e.g. based on cybernetics systems theory, as the basis for modelling, developing and building robust and reliable architectures for monitoring and controlling information systems, seems not only very promising, but rather inevitable.

In contrast to these aspects today's conventional control of an internal production chain or an external supply chain mainly pursue sequential, top-down planning approaches supported by different MRP (Material Requirements Planning) or ERP (Enterprise Resource Planning) concepts and information technologies in order to coordinate the supply flows within and between the different companies, which very often causes time lags. Sudden disturbances within the internal production chain or in an external supply chain basically ripple all the way through it and therefore easily make the complex and inherently local, distributed planning processes invalid. Expensive re-planning sessions concerning the quantity to produce or to deliver, the delivery times and in relation to this choices of new suppliers are the most likely consequences. The majority of today's conventional production planning and control systems is based on a collection of the following premises (Adam, 1992):

- predictable throughput times,
- no production bottlenecks,
- fix operation times per order,
- short downtimes of machinery.

But in some cases these premises are only able to support the mass production of more or less standard products with few different versions. Within other production situations, like for example sketched out in (Scholz-Reiter et. al. 2004), which considers a very customer specific job shop production of industrial pump sets, these traditional PPC-systems mostly do not lead to very useful results. According to (Rohloff, 1995) the following major weak points of PPC-systems in this context can be identified.

*The built-in feedback loops and coupling between the different subtasks and -processes are not sufficient or missing.* The main planning process pursues a single, sequential run. The two-way dependencies are basically not considered at all. Observed mistakes are regarded as mistakes originating from the preceding planning steps. This basically requires the already mentioned plan revision, which often is not supported by the conventional PPC-systems.

*The construction of a global model is often not possible.* Traditional PPC-systems assume that during the planning phase all matters-of-facts and timing cohesions between the most important decision criteria are fully known and fix. The production planning and scheduling processes are usually carried out weeks before the real start of production. At the point of time where these plans are activated, the considered boundary conditions of the planning phase are not valid any more, which again often leads to plan variants.

*The centralised planning approach is rather unsuitable.* The top-down and centralised MRP approach, which is being conducted by just one organisational unit, is closely connected with the assumption of being able to build up a global planning model. As a result the assigned production units are provided with predetermined and precisely defined tasks and sub-tasks (e.g. processing times, work content) without any freedom for local decision making and therefore a further use of personal know-how.

*Rigid and inflexible planning processes.* The rigidness and inflexibility of traditional PPC-Systems can be clearly and best identified, because they hardly do consider any enterprise specialties at all (Kurbel and Endres, 1995), although the requirements concerning the design and configuration of PPC-systems for the different types of producing enterprises and job shops may differ fundamentally. The decoupled production planning process for an anonymous mass customer market is much easier than a pure customer driven processing of orders, which normally imply the absence of large lot sizes but include much more complex products.

*Missing real time planning and control.* The different centralised planning steps of the traditional ERP respectively MRP based PPC-Systems are run sequentially, therefore the adaptation to changing boundary conditions (e.g. planning data) is only possible within quite long time intervals. This means that changes of the job shop situation cannot be considered immediately, but the next planning run at the earliest. As a result the current planning is based on old data and the needed adaptation measures cannot be performed in time for a proper reaction of the discrepancy between the planned and the current situation.

To summarize, these principle entrapments and constructional flaws strongly support the idea to basically redesign the deployed PPC-systems. In this context, within approximately the last ten to fifteen years a collection of decentralised concepts for the field of production planning and control – each of them emphasising different aspects – have been developed. Two of the maybe most relevant within the context of autonomy or self control respectively, are going to be sketched out shortly and delineated from the concept of "Autonomous Logistic processes" within the following chapter. Chapter three will introduce an exemplary scenario to discuss some opportunities of autonomously controlled logistic processes within production systems. This discussion will be complemented by the chapters four and five through outlining of some of the most necessary modelling requirements, as well as measuring and evaluation criteria of such a new process paradigm.

## 2. AUTONOMOUS LOGISTIC PROCESSES WITHIN THE CONTEXT OF OTHER KNOWN APPROACHES OF AUTONOMY AND AUTOMATIC CONTROL

The questions raised concerning the design and implementation of autonomous logistic processes are manifold and therefore multidisciplinary. First of all the question has to be answered satisfactorily, what the characteristics of logistic processes are. According to (Schönsleben, 2000), it can be basically distinguished between the following sub-areas relevant to planning and control of logistic processes, each of which considers different relevant flows of physical goods (logistic or business objects) and related flows of information:

- sourcing and procurement logistics,
- research and development processes,
- production logistics,
- distribution logistics,
- disposal or redistribution logistics.

The already mentioned CRC 637 "Autonomous Co-operating Logistic Processes - A Paradigm Shift and its Limitations" is presently mainly focussing on the aspects of distribution and production logistics and its different requirements and specialties. Aiming at more intelligent modelling of complex production logistics as well as distribution logistics systems and at its autonomously performed logistic processes it is of major importance to first of all identify the relevant logistics objects (e.g. trucks, machines). Moreover it is very important to precisely identify the locally relevant methods, in the sense of an appropriate procedure, business rules (e.g. decision rules), or basic principles of procedures in terms of defining appropriate control strategies (including economic goals like costs and locally added value). Upon these foundations the modelling and design of the autonomous controllers still is a major challenge in order to improve the internal reliability and robustness (e.g. high variances of the throughput time) within an enterprise, for example regarding its manufacturing processes. Furthermore at least an initial set of different possible local states of the participating intelligent business entities respectively logistic objects needs to be specified as well as the notion of events (e.g. distortions) has to be defined. One problem, which may occur and currently is under discussion within the CRC, with the notion autonomous logistic processes, is that the understanding of autonomy in close relation to heterarchy (= co-subordination), originally founded by McCulloch (Goldammer, 2003), which for example originates from biology and the theory of living systems (Goldammer and Paul, 1995), is based on the so-called *closure thesis*. This means that every autonomous system is organisationally closed and rejects the traditional *input-output-system* approach. Furthermore this leads to the fundamentally raised question of an entirely different *system/enviroment-relationship*, for example in terms of an adaptive state observer and its environment (Scholz-Reiter et. al. 2004), which needs to be reflected in a new systems analysis and design approach. This basically has to lead to a more or less bottom-up analysis and modelling approach (e.g. deploying distributed problem solving approaches), by modelling the local goals, start set-up of decision rules regarding the different intelligent monitoring and controlling entities, while finally observing and judging the overall system behaviour after bringing them all together. Similar research approaches have already been discussed for quite a while, but they can be distinguished from the research approach of the CRC 637, by taking for example the activities like *holonic manufacturing* or *fractal factory* into account.

The notion *holon* basically refers to the philosopher Arthur Koesteler (Koestler, 1968). It describes a strictly hierarchical open (social) system by deploying a *whole/part-systematics*. This idea was used during the late 80-ties and the early 90-ties to design so called *holonic manufacturing systems* (HMS), built from more or less modular cooperative Information and Communication Technology (ICT) components (e.g. products, resources). As a whole, these systems could be regarded as technical multi-agent systems (Lüth, 1998), in order to implement and scale machinery faster and more reliably. During the

HMS-Project a now available conceptual framework was developed (Bongearts, et. al. 1997; Langer and Bilberg 1998; Bochmann et al. 2000). According to Langer (Langer, 1999) the *holon* in its latest version is defined as an autonomous and cooperative basic building block of a manufacturing system for the transformation, the transport as well as the storage and/or validation of the information and physical product. The *holon* comprises a part capable of information processing and a part capable of the physical transformation of the produced good. Therefore a *holon* can be and often is a part of another *holon*.

The concept of the *fractal factory* first introduced by Warnecke (Warnecke, 1993) is basically focussing on the organisational units – fractals – and a principally discovered respectively assumed self-similarity between different analysed organisational units. The units operate autonomously with their own set of goals and their own exactly definable input and output parameters. The fractals conduct self-organisation and self-optimisation under consideration of their local goals either on the strategic or on the operative level. The process level is, if at all addressed just very indirectly and without considering any certain methodology. The core concept of the *fractal factory* still is a classical *input-output-system* approach, which leads to the same questions as already mentioned above.

To summarize, this already displays the fundamental difference to the approach choosen for the CRC 637, which basically assumes a heterarchical system setup and a more or less non-determined process flow. Nevertheless, the adaptation and further development of the multi-agent systems paradigm, whose characteristics can be considered important (Jennings et. al., 1998) especially for the requirements of autonomous logistic processes seems to be very promising:

- Multi-agent systems always emerge when several more or less autonomous and heterogeneous agents act together as a loosely coupled network to cooperatively solve a given problem.
- Each agent has incomplete information or capabilities for solving the problem, thus each agent has a limited view.
- There is no global system control.
- Data is decentralised.
- Computation is asynchronous.

Finally, all definitions of agent technology can be summarized by the following statements, which can easily act as guidelines for the development and application of suitable agents (Jennings and Wooldridge 1998; Wooldridge and Jennings 1999):

- Agents are a powerful, natural metaphor for conceptualising, designing, and implementing many complex, distributed applications.
- Agent systems typically use AI techniques – in this sense, they are an application of AI technology – but their "intelligent" capabilities are limited by the AI's state of the art.
- The development of any agent system, however trivial, is essentially a process of experimentation. Unfortunately, the experimental process encour-

ages developers to forget that they are actually developing software.

- It has to be considered as common to all agent-based applications that they are no overall system controllers and have no global perspective, already by definition.

This means that the deployed agents are situated (situatedness) in and experiencing (embodiment) a *system/environment-relationship*, according to the complexity of the conducted tasks (universal versus specialised agent) and due to the complexity of the environment (low versus high) (Kordic et. al., 2001). As a result the broad range of requirements on how to derive and design a "complete monitoring and controlling entity" are discussed under the focus of autonomous logistic processes in one of the following chapters of this paper. To summarise, the modelling and design of industrial, agent-based and autonomous controlling systems still is very challenging and up to now not solved satisfactorily. This is mainly due to the lack of systematic methodology for the systems analysts and designers as well as the lack of widely available industrial-strength multi-agent system toolkits (Jennings et. al., 1998; Wooldridge and Jennings 1999).

## 3. EXEMPLARY SCENARIO OF AN AUTONOMOUSLY CONTROLLED PRODUCTION SYSTEM

Applications of autonomous cooperating logistic processes are manifold and possible over the entire supply chain. In detail it is necessary to analyse, in which scale and in which logistics domain (procurement, production, distribution and disposal) it will be efficient to establish autonomous logistic processes and to find their limitations.

Because of the high complexity of this research project here it seems reasonable to focus on a concrete object of investigation in the form of a specific, exemplarily scenario. Based on this scenario of logistics in an autonomously controlled logistic production system the changes compared to conventionally controlled processes will be explained as well as arising benefits pointed out. By means of these changes new modelling and evaluation requirements of autonomous logistic processes will be deduced in the following chapters.

### 3.1 Basic Scenario.

In the context of this paper an exemplarily scenario of production logistics is examined. **Figure 1** gives an overview of a scenario of a two-stage job shop production. The material flow layer on the lowest layer shows the material flow net of the manufacturing system. The process layer is based on the material flow layer and describes the lead and performance processes of the manufacturing system. The layer of production controlling lastly assigns activities of the process layer to measurement points. With

theses measurement points relevant logistics metrics and performance figures are deduced.

*Material flow layer.* This level of abstraction describes the material flow of a two-stage job shop production. The first production stage contains the manufacturing of a part on two alternative machines ($M_{ij}$). The raw materials that are needed for production are provided by the source (So). In the second production stage the assembly of the parts that have been made in the first stage is executed alternatively on two machines ($A_{ij}$). The manufactured items leave the material flow net at the sink (Si). Every machine of this scenario has an input buffer ($I_{ij}$) and an output buffer ($O_{ij}$), in which the raw materials or parts are stored temporarily.

*Process layer.* The process layer represents the lead and performance processes of the job shop production scenario. These processes are assigned to the underlying material flow net. Lead processes can be defined as planning and control or coordination processes (Krüger, 1993). In this case the production planning and control processes are lead processes. Performance processes can be characterised as production or service processes, which are directly involved in adding value (Krüger, 1993). So the production logistic processes in-house transport, stocking, manufacturing and assembly belong to performance processes.

*Layer of production controlling.* The layer of production controlling is based on the process layer. Several measurement points can be defined between the processes. These measurement points allow determining diverse logistics metrics. For example throughput times of manufacturing orders can be developed from adequate measurement points. The throughput time of a manufacturing order is composed of operation time, consisting of setup time and processing time and of transit time, which is divided into waiting time before and after handling and transportation time (Wiendahl, 1997).

### 3.2 Scenario of an autonomously controlled production system.

In this chapter the approach of autonomous logistic processes in production logistics will be explained on the basis of the adapted exemplarily scenario of the two-stage job shop production displayed in figure 2. In detail it will be pointed out, how the weak points of traditional production planning and control systems, described in chapter 1, can be eliminated by establishing autonomous logistic processes. Furthermore expected changes concerning production controlling will be introduced.

A precondition to autonomy of logistic processes in the considered scenario is that the logistic objects of the material flow net (machines, buffers, parts etc.) have their own intelligence. For example the logistic objects could be equipped with RFID chips, which

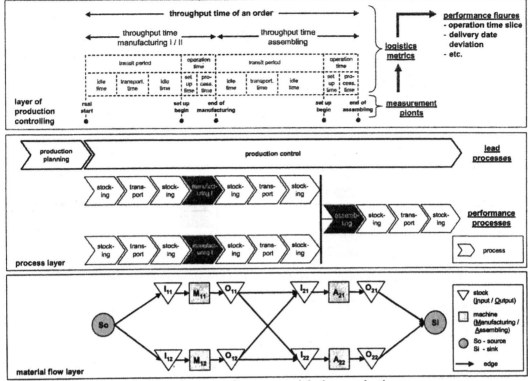

Fig. 1. Scenario of a two-stage job shop production.

feature processing and data-storage capacity. The existence of a communication infrastructure as well as appropriate software is preconditioned, too. So the logistic objects are able to interact with each other and to make decisions, which is meant with intelligence. In consideration of these technological requirements there are several changes to the material flow layer, the process layer and the layer of production controlling, which will be described below.

*Material flow layer.* The sequence of the production steps of this exemplary scenario persists independently of the production planning and control method. The production of goods in this scenario requires at first a stage of production and afterwards a stage of assembling. That means that the sequence of production steps in this scenario is technologically predetermined, but not the material flow. For example intelligent parts can autonomously choose one of the alternative assembling stations. Consequently establishing autonomous logistic processes excerts influence on the – not predetermined – material flow.

*Process layer.* Compared to traditional production planning and control systems the PPC processes in this scenario are partially linked to the performance processes. Some production planning processes do not take place in a centralised manner a long time before manufacturing, but in a decentralised one while producing. For example every machine can continuously plan and adapt its own allocation by communicating and negotiating with intelligent parts. Also a conveyor could negotiate with a commodity its transport from one machine to another. Other production planning processes like master pro-

duction scheduling or rough-cut planning are still part of the centralised production planning and control. In addition to planning of production processes the logistic objects of the material flow net, e.g. machines or orders, assume the production control. Machines autonomously initiate the release of self-planned production orders, monitor their production processes and react immediately to a possible breakdown during the manufacturing process.

As a result autonomous production systems are characterised by distributed production planning and control. Some PPC functions are still part of the centralised production planning and control system, some functions belong to decentralised PPC systems of several logistic objects. The limitations or transitions between conventional and autonomous planning and control will be investigated in the CRC 637.

*Layer of production controlling.* Conventional performance measurement systems are based on a set of logistics metrics, which are determined by several measurement points. Some of these logistics metrics and possibly dedicated performance figures of conventionally managed production systems were described above. Autonomously controlled production systems offer new potentials.

As explicated in this scenario logistic objects are able to store, process and exchange data at any time and any place. So the amount of measurement points is no longer limited. Logistic objects can provide information regarding the status of the current processes at every time.

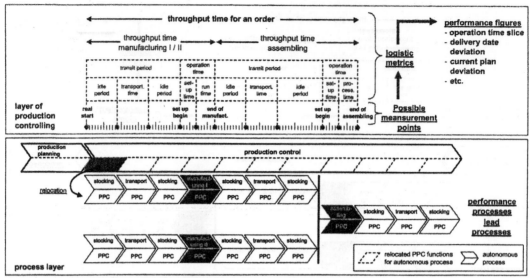

Fig. 2. Scenario of an autonomously controlled two-stage job shop production.

This high amount of measurement points allows the development of new, previously not possible logistic metrics and performance figures and thus a higher transparency of manufacturing processes within the scope of production controlling. For example an appearing delay of a machine can be identified during the assembly process by a decentralised monitoring system. The monitoring system permanently collects data of the current status of the assembly process and compares them with the data of the planned status. This discrepancy between current and planned process status can be represented by new performance figures, for example current plan deviation. The performance figure could be used as an indicator for an early detection of manufacturing delay. On the basis of this new performance figure the machine is able to recognise its delay and to initiate necessary steps immediately (e.g. capacity adjustment in form of extra shifts).

By means of the description of the autonomous production system scenario it becomes apparent that the weak points of traditional PPC systems described in chapter 1 can partially be reduced. Some planning processes do not pursue at one time and sequential run any more, but basically happen during the production process. In case of disturbance the logistic objects autonomously detect the problem, for example by a sensor system, and immediately initiate appropriate steps. One essential step is the adaptation of its own production planning. The dependencies of the current planning step from preceding planning steps can be reduced because of the ability of the logistic objects to adapt their decentralised planning at every time. Therefore the need for construction of a global model is no longer existent. In fact there are a lot of local models that show a high quality because of the short planning horizon and the lower complexity than in hierarchical systems. The hope is to show during the CRC project time that planning deviation in the form of discrepancy between current and target plans will be avoided.

As one result of the decentralised planning the know-how of the organisational units which are assigned to the several production processes can be involved.

## 4. REQUIREMENTS RELATED TO MODELLING METHODS OF AUTONOMOUS LOGISTIC PROCESSES

On one hand there are general requirements to modelling of autonomous logistic processes that can be formulated based on general considerations related to modelling of conventionally managed logistic processes. On the other hand there are specific requirements that have to be fulfilled, which call for extensions of existing modelling methods and development of new ones respectively. Figure 3 gives an overview of these requirements, which are described in detail in the following sections.

### 4.1 General Requirements

An important contribution to the formulation of general requirements to modelling methods is due to (v. Uthmann, 1999) with the formulation of their "Guidelines of Modelling" (GoM). There they postulate correctness, clarity, comparability, relevance, systematic design and economic efficiency as general requirements in order to achieve a high model quality. Therefore a method for modelling autonomous logistic processes should meet these guidelines.

Some of the requirements given in the following sections partly overlap with these guidelines or can be assigned to them. But, as they have a special meaning in the context considered, they will nevertheless be discussed here explicitly.

First of all it is necessary that a process expert, i. e. someone who is used to the processes to be modelled on a daily basis, should be able to use the method with no or only little learning effort. This calls for a graphical modelling method, based on a common and standardised notation. Examples of such notations

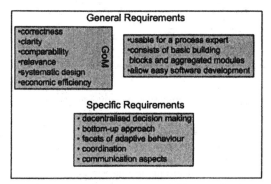

General Requirements

- correctness
- clarity
- comparability
- relevance
- systematic design
- economic efficiency

GoM

- usable for a process expert
- consists of basic building blocks and aggregated modules
- allow easy software development

Specific Requirements

- decentralised decision making
- bottom-up approach
- facets of adaptive behaviour
- coordination
- communication aspects

Fig. 3. Requirements to modelling methods.

are Event-driven Process Chains (EPC) or the Unified Modelling Language (UML). Furthermore in some cases it seems more adequate to pass on some generality of the modelling method in order to reduce complexity. In these cases focussing on the domain of logistics should be possible, though resulting in a loss of generality.

The modelling method should offer basic building blocks but also allow for aggregated modules on a higher level of abstraction. It should be possible to combine theses modules into libraries, easing re-use and thus reducing efforts of later modelling projects.

A further challenge is to ensure a good support for subsequent software development based on the model built of the process under consideration. On one hand support of logistic processes with the help of information technology plays an ever increasing role, so their implementation has to be considered during modelling and designing the processes. On the other hand, as already mentioned in chapter 2, realising autonomous logistic processes requires even further use of and support by information technology. Software implementation thus plays an even more important role there. In this regard it is important not to reduce understandability for process experts which would result from a modelling method that focuses too much on software development. In some sense the increasing importance of software implementation already is a requirement specific to modelling autonomous logistic processes.

### 4.2 Specific Requirements

Considering the above scenario of an autonomously controlled manufacturing system, several specific requirements to modelling can be derived.

In case of a machine breakdown for instance, logistic objects have to be able to decide on a further course of action. First of all for modelling this means that this decentralised decision making in heterarchical organizational structures, i. e. lacking central control, has to be representable. This proceeding suggests a bottom-up approach to modelling to allow to begin building a model, starting from the autonomous logistic objects available. This bears the danger of neglecting the global view on the processes which could lead to a model lacking consistency. Furthermore it becomes more difficult to design processes reaching global objectives this way. Thus the best

way seems to be a combination of bottom-up- and top-down-approaches. The difficulty is to find the right balance.

Autonomous logistic objects adapt to a changing environment by themselves. So machine failure can be faced by reactively executing predetermined actions, like IF-THEN-rules. This means that in our exemplary scenario jobs have to have alternative ways of action based on the current state of damaged and alternative functioning machines and other relevant environmental influences. This corresponds to classic approaches to business process modelling.

Beyond this reactive action there are higher levels of decentralised decision making that would allow adaptive behaviour of jobs or other autonomous logistic objects. The logistic objects could adjust to a new state of their environment, e. g. machine breakdown, without using alternatives that are predetermined in detail. One possibility to obtain such behaviour would be the use of a planning method from artificial intelligence (Russel, Norvig, 2003). Other possibilities are learning on one hand and evolution on the other. Both cases rely on feedback of the consequences of actions executed. Learning requires an entity to be able to draw conclusions from the consequences of its own behaviour or the behaviour of other entities in response to its own actions. Again evolution means that current behaviour affects future generations, which can be seen as one kind of feedback, too. In both cases a modelling method has to support a way of representing these different kinds of feedback.

Furthermore one should be able to explicitly model knowledge required for decision making by decentralised autonomous units, maybe in the sense of a more cognitive modelling of mental states, decision processes and functions (Schmid and Kindsmüller, 1996). This local knowledge can be distinguished into knowledge the unit has about itself and the knowledge it has of its environment as well as connections between unit and environment. Knowledge about itself for instance means for a machine to know working operations possible on itself in conjunction with associated processing times, maintenance intervals or which parts it is able to process. Environmental knowledge e. g. consists of knowledge about other machines and times required to transport goods to them. In this case jobs know about their due dates and the production steps required to manufacture them. These examples describe mostly static knowledge. In addition there is dynamic knowledge, i. e. knowledge that is likely to change with time, for instance which jobs are currently scheduled to be processed on a machine.

To enable processing and completion of jobs it is necessary to connect the knowledge of several autonomous units and their coordination. To coordinate the behaviour of participating logistic objects a decentralised planning method should be used instead of centralised planning, allowing decentralized autonomous decision making.

To determine information that is not available locally, but required to execute decentralised distributed decision making as described in the exemplary scenario above, information exchange between autonomous logistic objects is necessary. Thus it should at least be possible to represent the following three aspects of communication; first of all which logistic objects are involved in the communication, second which information is exchanged and finally the temporal order of information exchange.

# 5. REQUIREMENTS ON EVALUATION SYSTEMS

In the CRC 637 fourteen subprojects are involved with different fields of investigation. Representatives from each subproject are joined together in a research group "scenarios". Within this research group CRC-wide scenarios will be developed in order to compare the different approaches to autonomous cooperating logistic processes. To evaluate these specific approaches an evaluation system valid for all CRC-subprojects is needed. The idea is to evaluate this with a mutual basis of various performance figures for logistic processes.

The intention of implementing autonomous processes in logistics systems is to increase logistic performance and to satisfy the new demands of logistic systems. The aim of a new or enlarged evaluation system for autonomous logistic processes is to cope with new demands due to autonomy. To measure the performance of logistic systems performance figures are needed to determine the target achievement. These specific logistic performance figures are combined in a performance measurement system. Due to new demands this performance measurement system has to be individually tailored to autonomous logistic processes and to conventionally managed processes as well. The specific system allocates the performance figures to a system of objectives. This performance measurement system is necessary to make a logistic system observable by increasing the transparency of the system. In figure 4 different levels of objectives are shown to specify the various logistic targets (Luczak *et al.* 2004). The arrows next to objectives indicate whether to increase or to decrease this value in order to get a *high logistic efficiency*. The objective *high logistic efficiency* describes the uppermost level of the system. In a second level this objective is divided into two further objectives. A *high logistic efficiency* is achieved by *high logistic performance* on the one hand and *low logistic costs* on the other hand. On the third level these targets are again divided into more detailed objectives. The logistic performance is divided in *high availability, high productivity, short throughput time* and *high delivery service* while the logistic costs are divided into *low inventory costs* and *low process costs*.

By implementing autonomous units with the ability to measure their current state at any time the specific characteristic values are determined near real-time.

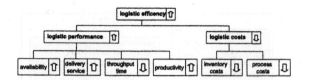

Fig. 4. Logistic objective system (Luczak *et al.* 2004)

Thus the transparency of the system will increase significantly. The intelligent objects (e.g. orders, machines, parts) are able to automatically perform a computer-aided production data acquisition, which is necessary for monitoring of the concerned system.

In the scenario described in chapter 3 different customer orders are manufactured in an exemplary production system. The customer order is defined by products and their structures, an amount of these products per order and a due date. In the following some exemplary performance figures are explained, which can be detected in this scenario. Depending on the product the order has to complete $n$ operations on different machines. Thus the *total operation time* is defined as

$$t_{o,total} = \sum_{i=1}^{n} t_i ,\qquad (1)$$

where $t_i$ is the individual processing time for each operation on the different machines in the described scenario. The due date of each operation is appended to the order concerning these operations. At any time the order is comparing actual versus estimated times so that deviations are recognized in real-time and necessary actions like rescheduling can be performed. The appropriate performance figure in this case is defined as *current plan deviation*

$$PD = t_{act} - t_{est} \qquad (2)$$

with $t_{act}$ as the actual time and $t_{est}$ as the estimated time of the current stage of the order given by the order scheduling.

In the described scenario each order has to complete one manufacturing and one assembly stage before it is completed. All orders in the system have to coordinate their machine scheduling in order to get the best result for each order with regard to the due date. While competing for the limited resources (two manufacturing and two assembly stations in this scenario) it is possible to use priority rules to decide the sequence of order processing.

In case of a breakdown of one of the assembly stations the information about this breakdown is transmitted near real-time to the order in such a way as to enable the order to reschedule itself. At the same time the manufacturing and assembly stations adapt their capacity planning. While monitoring the performance measure *current plan deviation* (*PD*) the time lack between noticing a breakdown and reacting to this breakdown is minimised. The hope is to reduce the effects of unexpected events like a breakdown in comparison to conventionally controlled production systems.

By knowing all specific logistic metrics, orders are able to determine their own performance figures like the *operation time slice* (*OTS*) which is defined as

$$OTS = \frac{t_{o,total}}{t_{through}} \qquad (3)$$

with $t_{through}$ as the throughput time of the specific order. In terms of the level of autonomy it has to be pointed out that some of the logistic metrics have to be divided into conventionally managed parts and autonomously controlled parts. With regard to the throughput time this value is given as

$$t_{through} = t_{through,a} + t_{through,c} \qquad (4)$$

while $t_{through,a}$ is the time slice of autonomously controlled operations and $t_{through,c}$ the one for conventionally managed. These performance figures which are divided into conventionally managed and autonomous processes will allow specifying effects by changes of the system. The effects by these specific changes will thereby be assigned to autonomously and conventionally managed subsystems. To detect the border of autonomy a criteria catalogue for autonomous processes will be developed within the CRC 637. By means of this catalogue autonomous processes will be identified for different types of manufacturing systems, so that it will be possible to identify $t_a$ and $t_c$.

By implementing intelligent objects (e.g. intelligent orders) with the capability to measure their specific indicators at any time these performance figures will no longer be determined at a fixed time or in a fixed period of time but will be determined continuously. By finishing an order the performance figure *delivery date reliability* (*DR*) for the entire system is updated automatically by the order. It is defined by the ratio of outgoing products delivered in time $n_{out,it}$ by the total number of outgoing products $n_{out}$

$$DR = \frac{n_{out,it}}{n_{out}}. \qquad (5)$$

Thus the system transparency is always up to date because of meaningful performance figures. Beside these exemplary performance figures the new evaluation system is able to provide a wide range of different performance figures specifying the level of logistic efficiency.

Logistic systems with intelligent objects and autonomous cooperating processes also generate new requirements to performance measurement systems because of their dynamic target system. While in conventionally managed processes the system objectives are clearly defined, this is different in the case of autonomous systems. A customer's order for example may have the objective *high delivery reliability* while suborders like the manufacturing order of the different parts may have the objective *short throughput time*. The other objects in this system have different objectives, like the machines trying to increase their utilisation ratio or the buffers trying to minimize their stock. In fact the different types of objects may have different logistic objectives like the ones shown in figure 4. Changes in the different logistic objectives over the time, which will also lead

to a dynamic objective system, are possible as well. A set of different objectives with different priorities is also possible. The new performance measurement system has to cope with emergent behaviour of autonomous subsystems, which means that a global behaviour of the whole system is not explainable with its subsystems. A positive effect of emergence obviously is a faster and higher achievement of the global objectives.

As the intelligent objects know their current state and are able to communicate at any time the evaluation system has to manage this amount of different information from different types of objects like machines, orders, parts etc. The task is to filter this large amount of information and generate meaningful performance figures. These significant performance figures must be available for the concerned objects in order to support their decision making process for the next steps. In addition to the preparation of information the performance measurement system has to evaluate the subsystems as well as the overall production system. Thus it has to measure the performance of different autonomous subsystems and maybe different conventionally managed subsystems on the one hand and has to reach a conclusion of the global objective achievement on the other hand. Conventional performance measurement systems are so far not able to cope with these requirements so that there is a need of new concepts of evaluation systems and performance figures, which will be developed in the CRC 637.

## 6. CONCLUSION AND FUTURE OF RESEARCH

The upper mentioned CRC 637: "Autonomous Co-operating Logistic Processes - A Paradigm Shift and its Limitations" at the University of Bremen, with its several sub-projects, is for example strongly motivated by the still existing broad range of lacks and unfulfilled requirements concerning the systems' analysis and design of such a new process paradigm. Some very interesting fundamental ideas of autonomy and self-control, originating from natural (e.g. biology, physics) or cognitive science, up to now just cannot be expressed satisfactorily with some of the common notation for business process modelling like UML or EPC. These new ideas of autonomy will lead to emergent behaviour of the global system which cannot be evaluated by conventional performance measurement systems so far and thus new or extended ones have to be developed to close this gap.

Therefore concerning for example the future application and dissemination of agent technology as industrial-strength autonomous controlling entities, it will be of major importance to provide these more comprehensible modelling methods respectively some generally extended foundations (e.g. basic building blocks). They must clearly address the actual problems of the business processes in order to derive the needed services as skills or functions provided by the software agent. Up to now the available methods are clearly developer driven, and basically the aspect of system integration (e.g. RFID, PDA, Legacy Systems) is often not considered.

## ACKNOWLEDGMENTS

This research is funded by the German Research Foundation (DFG) as the Collaborative Research Centre 637 "Autonomous Cooperating Logistic Processes - A Paradigm Shift and its Limitations" (SFB 637).

## REFERENCES

Adam, D. (1992). *Fertigungssteuerung*. Gabler, Wiesbaden.

Bochmann, O., Valckenaers, P. and H. Van Brussel (2000). Negotiation-based manufacturing control in Holonic Manufacturing Systems. In: *Proceedings of the ASI'2000 Conference*. Bordeaux, France, September 18th - 20th.

Bongaerts, L. , Valckenaers, P., van Brussel, H. and P. Peeters (1997). Schedule Execution in holonic Manufacturing Systems. In: *Proceedings of 29th CIRP International Seminar on Manufacturing Systems*, May 11th - 13th, Osaka University, Japan.

Gassman, H. (1998). *Theorie der Regelungstechnik*. Verlag Harri Deutsch, Frankfurt (a. M.).

Goldammer, E. von (2003). *Heterarchie – Hierarchie: Zwei komplementäre Beschreibungskategorien*. Download at 02.07.2004 from: http://www.vordenker.de/heterarchy/a_heterarchie.pdf.

Goldammer, E. von and J. Paul (1995). Autonomie in Biologie und Technik: Kognitive Netzwerke – Artificial Life – Robotik. In: *Selbstorganisation – Jahrbuch für Komplexität in Natur-, Sozial- und Geisteswissenschaften (*Ziemke, A. and R. Kaehr, Ed.), **6**, 277-298, Duncker & Humblot, Berlin.

Jennings, N. R. and M. J. Wooldridge (1998). Applications of Intelligent Agents. In: *Agent Technology: Foundations, Applications, and Markets* (Jennings N. R. and M. J. Wooldridge, Ed.), 3-28, Springer Verlag, Berlin, Heidelberg.

Jennings, N. R., Sycara, K. and M. J. Wooldridge (1998). A Roadmap to Agent Research and Development. *Autonomous Agents and Multi-Agent Systems*, **1**(1), 7-38.

Koestler, A. (1968). *The Ghost in the Machine*. Macmillan, New York.

Krüger, W. (1993). *Organisation der Unternehmung*. Kohlhammer, Stuttgart et al .

Kurbel, K. and A. Endres (1995). *Produktionsplanung und –steuerung*. Oldenbourg, München et al.

Langer, G. and A. Bilberg (1998). The architectural foundations for agent-based shop floor control. *8th International Conference on Flexible Automation and Intelligent Manufacturing (FAIM98)*, July 1st – 3rd, Portland, USA.

Langer, G., Sørensen, C., et al (1999). Research contributions to the modelling and design of Intelligent Manufacturing Systems. In: *Proceedings of the second International Workshop on Intelligent Manufacturing Systems*, September, Leuven, Belgium.

Luczak, H., Weber, J. and H.-P. Wiendahl (2004). *Logistik-Benchmarking*. Springer Verlag, Berlin, Heidelberg.

Luger, G. F. (2002). *Artificial Intelligence: Structures and Strategies for Complex Problem Solving*. 4th edition, Pearson Education Limited, Boston.

Lüth, T. (1998). *Technische Multi-Agenten-Systeme*. Hanser, München, Wien.

Rohloff, M. (1995). *Produktionsmanagement in modularen Organisationsstrukturen: Reorganisation der Produktion und objektorientierte Informationssysteme für verteilte Planungssegmente*. Oldenbourg, München et al .

Russell, S. J., P. Norvig (2003). *Artificial Intelligence: A Modern Approach,* 2nd edition, Pearson Education International, London et al.

Sastry, S. and M. Bodson (1989). *Adaptive Control: Stability, Convergence and Robustness*. Prentice Hall, Englewood Cliffs.

Schmid, U. and M. Ch. Kindsmüller (1996). *Kognitive Modellierung: Eine Einführung in die logischen und algorithmischen Grundlagen*. Spektrum Akademischer Verlag, Heidelberg et al.

Scholz-Reiter, B., Höhns, H. and T. Hamann (2004). Adaptive Control of Supply Chains: Building blocks and tools of an agent-based simulation framework. In: *Annals of the CIRP*, STC-O2, Kraków, Poland.

Schönsleben, P. (2000). *Integrales Logistikmanagement: Planung und Steuerung von umfassenden Geschäftsprozessen*. 2. Auflg., Springer Verlag, Berlin, Heidelberg.

Tsakonas, A. and G. Dounias (2002). Hybrid Computational Intelligence Schemes in Complex Domains – An extended Review. In: *Methods and Applications of Artificial Intelligence* (Vlahavas, I. P. and C. D. Spyropoulos, Ed.), 494-511, Springer Verlag, Berlin, Heidelberg.

Uthmann, C. von and J. Becker (1999). Guidelines of Modeling (GoM) for Business Process Simulation. In: *Process Modelling* (B. Scholz-Reiter, H.-D. Stahlmann, E. Nethe, Ed.), 100-116, Springer Verlag, Berlin et al.

Viharos, Z. J. and L. Monostori (2001). Optimisation of Process Chains and Production Plants by using a hybrid-, AI-, and Simulation-based Approach. In: *Engineering of Intelligent Systems* (Monostori, L., Váncaza, J. and M. Ali, Ed.), 827-835. Springer Verlag, Berlin, Heidelberg.

Warnecke, H.-J. (1993). *Revolution der Unternehmenskultur: Das Fraktale Unternehmen*. 2. Auflg., Springer Verlag, Berlin, Heidelberg.

Wiendahl, H.-P. (1997). *Betriebsorganisation für Ingenieure*. 4. Auflg., Carl Hanser Verlag, München et al.

Wooldridge, M. J. and N. R. Jennings (1999). Software Engineering with Agents: Pitfalls and Pratfalls. *IEEE Internet Computing*, **3**(3), 20-27.

ELSEVIER
IFAC
PUBLICATIONS
www.elsevier.com/locate/ifac

# AUTONOMOUS CONTROL OF SHOP FLOOR LOGISTICS

Bernd Scholz-Reiter* Karsten Peters*
Christoph de Beer*


*Department of Planning and Control of Production
Systems, University of Bremen, Germany*

Abstract: We investigate methods for implementing highly distributed control of production systems. As a first step we investigate the differences between conventional architectures and fully distributed systems. As a conventional system we understand here a system with a central control entity that plans and optimizes in advance the distribution of work among different work stations. In contrast we consider models of manufacturing systems with a flexible structure that allows parts when they are released to the shop floor to control themselves by means of autonomous shop floor control. The decision layer is transferred to the parts that are being processed. The parts themselves decide which resource they allocate for their processing. Their decisions are based on only local and actual available information and follow simple rules. Exemplary models for autonomous control of a production logistic scenario without any global planning or optimization in advance are presented. Simulation results make these control techniques comparable in terms of performance, throughput time and WIP to conventional systems. The results may stimulate a further exploration of distributed autonomous manufacturing

Keywords: Autonomous control, Shop floor oriented systems, self-adaptive control

## 1. INTRODUCTION

In todays manufacturing systems the production planning and scheduling tasks are of increasing complexity and are aggravated by the necessities to be flexible enough to adapt to outside changes. Due to the dynamic and structural complexity of modern manufacturing systems and networks, traditional central planning and control, that is efficient under steady operating conditions becomes increasingly difficult and is insufficient in the face of dynamic changes, as reported by numerous authors as Zweben and Fox (1994), Baker (1998), Bongaerts et al. (2000), Scholz-Reiter et al. (2004b). To cope with these challenges there is an ongoing paradigm shift from centralized control of non-intelligent items in hierarchical structures towards decentralized control of intelligent items in complex and flexible manufacturing systems (Scholz-Reiter et al., 2004a).
First attempts to overcome the problems of centralized architectures through heterarchical control of production systems arose, when the in-creasing complexity of computer integrated manufacturing was recognized (Havany, 1985), (Duffie and Piper, 1986). During the last decade a number of far reaching new conceptual frameworks for distributed manufacturing control have been proposed, presenting similar concepts with different origins (Tharumarajah et al., 1996). This includes the fractal factory concept (Warneke, 1993), which focuses mainly on an assumed self-similarity of organizational units, the Bionic Manufacturing System (BMS) (Okino, 1993), (Ueda and Vaario, 1998) which emphasizes biological evolution principles for dealing with dynamic changes, or the holonic manufacturing system (HMS) (Bongaerts, 1998), (Gou et al., 1998). The latter approach identifies key elements of manufacturing, such as machines, factories, parts, products or operators etc. with so called holons, which should have autonomous and cooperative properties. With this identification it compares closely to the main ideas of distributed artificial intelligence (DAI), especially the multi agent systems (MAS) paradigm (Jennings, 1998), (Parunak, 1994).

The mentioned concepts often address the whole spectra of manufacturing planning and control systems, ranging from product and process design to purchase and procurement and represent mainly a systems engineering approach to the development of manufacturing control infrastructure, rather than a solution or evaluation for solving individual manufacturing control problems.

In this paper we focus specifically on the distribution and routing of parts between workstations on the shop floor. The basic idea is, to equip each workpiece with some sort of intelligence, that enables it, to make the routing decisions by itself. This approach has no centralized instance for the allocation of manufacturing resources based on optimization procedures in advance. Instead, local heuristics are to be applied for the items, which also receive actual information on their environment, to decide which workstation should be used for the next production step. The main advantages of such an autonomous control are an added robustness against unforeseen changes in production times and machinery breakdown and an increased flexibility of the production system. Furthermore, our approach is comparatively simple to implement, since we consider only simple decision making by using decision rules. The challenge in distributed approaches to shop floor control is in achieving optimized global performance. Due to the self-organizing properties of autonomously controlled systems it is at the actual state of the art often impossible to verify these systems by formal methods. We therefore use simulations for evaluating the global performance of exemplary model systems.

The rest of this paper is organized as follows: At first we briefly review the main requirements and challenges for flexible manufacturing systems (Sec. 2). Then the advantages of autonomous control with respect to these requirements are discussed in more detail in Sec. 3. Whereas the discussion is on a more general level we will present simulation results for exemplary production scenarios with decentralized control algorithms in the second part of this paper. For the sake of completeness the used methodologies are introduced in Sec. 4. Then simulation results for two basic layouts are discussed in detail in Sec. 5 and Sec. 6. Finally we draw some conclusions for distributed autonomous control mechanisms of shop floor logistics and line out directions of future research.

## 2. REQUIREMENTS AND CHALLENGES

The majority of currently used production planning and control systems is based on centralized optimization procedures. Their successful application requires therefore besides a fixed set of orders, an (in principle) complete knowledge about the state of the production facility in advance, which is hardly to obtain in realistic systems. Furthermore such a system can not cope with unforeseen downtime of machinery or other outside disturbances and uncertainties.

To adapt to such changing boundary conditions or even for adaption to future alternative use of the same system because of product mix or design changes over time, the planning and control system has to be flexible and convertible. Another point is the robustness of the system which means that it should continue to work even in unpredicted cases of disturbance. To afford these tasks the systems structure has to be convertible and scalable to facilitate the possibility to adapt to different demands.

### 2.1 Routing Alternatives

A robust manufacturing system requires machines with overlapping capacities. It should be possible to perform every manufacturing step at always more than one machine. In case of breakdown or repair needs this type of redundancy provides the possibility to perform the tasks of the unused machine at an other one. In any manufacturing system a greater number of alternative paths a part can take to be finished indicates therefore a higher degree of flexibility and robustness. However, this flexibility can only be beneficial, if the routing mechanism itself is flexible enough to adapt to changes in the capacities. This is just, what an autonomous control can ensure in a natural way by its design.

## 3. SELF-ORGANIZATION OF MATERIAL FLOW BY AUTONOMOUS CONTROL

Every part that has to be produced in a manufacturing system is characterized by a sequence of production steps that must be applied until the production is finished.

Our approach is strictly based on spatial and timely local information and decision making. It is assumed that a part knows, when it is released to the production system or a previous production step is finished, which workstations are in principal able to perform the next production step. I.e. a list of routing alternatives is given for each autonomous part. Then in a communication process, that is assumed to be as simple as possible the actual workload of the potential next stations (the sizes of input buffers and the workstations state) are requested. Based on these informations a decision at which workstation it queues is rendered by the product itself, based on simple heuristics called *(decision) rules* in the following. Thus the routing of a workpiece is organized through a sequence of decisions made by the piece at the end of production steps.

Now we shall consider the behavior on system level that results by this approach. If a workstation breaks down, no further parts visit this station, i.e. the material flow is immediately diverted to alternative machines. Even if a workstation has a lower production rate than other alternative stations the work in process is automatically balanced between all workstations that can perform a certain production step. In summary the actual flow adjusts itself to the current situation, even if the product mix or other boundary conditions change.

Thus, the topology of material flows is not predetermined. On the one hand it emerges dynamically from the products and their types respectively, and on the other hand the available production capacities and their actual load influence the material flow structure. The system may therefore regarded to be self-organizing.

The main advantages of such a self-organization by simplified local feedback mechanisms are the comparatively easy implementation as well as the robustness, which is ensured by the distribution of decisions in many simple and decentralized autonomous units.

The question is, which rule performs best and how to evaluate the performance of such heuristics.

## 4. NUMERICAL EXPERIMENTS

Generally manufacturing systems can be approached by discrete event simulation (Yao, 1994). The systems dynamics is modeled as a series of events representing the start and end of specific actions. If an action is started, the time of its duration is determined randomly according to a given distribution of inter-event times. For demonstration purposes we use Poisson-processes and a uniform distribution ($U_1$) of inter-event-times in the interval $[t_{min}, t_{max}]$ with $t_{min} = 0.8 t_{max}$, respectively. Whereas the $U_1$ distribution is used as one example for scenarios, where the standard deviation of order times or rather order sizes is very low, the Poisson process serves here as an stochastic process which represents wide scattering order or production times. The Poisson process is frequently used in queuing theory as a 'completely random process' due to the Markov property (the usual Kendall Notation of queuing processes therefore uses M as abbreviation for Poisson processes).

A Poisson process implies a exponential distribution of inter event times, i.e. $P(\Delta\tau \le t) = 1 - e^{\lambda t}$ (where $\lambda$ is a 'event'- rate) and a standard deviation $\sigma_{\Delta\tau} = \sqrt{Var(\Delta\tau)} = 1/\lambda$ which is as large

as the mean inter-event time (or the expectation of $\Delta\tau$).

Since both small and large standard deviations of inter event times occur in practical realizations of production systems, we investigate both situations in the following simulations and discuss the consequences of different distributions in detail. Thus, uncertainty and disturbances of production are included in the simulation. Because the resulting systems are strongly nondeterministic (whereas the applied rules are) it is necessary to obtain averaged values for performance evaluations.

To any manufacturing system (as a whole as well

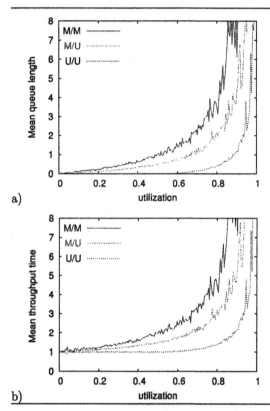

Fig. 1. Curves of performance measures vs. utilization $\rho$ for single server queues for different combinations of inter-arrival time and service time distributions. The production rate is in all cases $\mu = 1$.
a) Shows the mean queue length vs. $\rho$
b) Shows the mean throughput time vs. $\rho$
In both plots the dashed curve represents a M/M/1 system, the solid line a M/U1/1 system and the dotted curve stands for a U1/U1/1 system as indicated inside the figures.

as to any workstation in the system) we can assign a production rate (service rate) $\mu$. This is just the rate at which the system can produce new parts, given that in every moment a new incoming part is available. For new work, that is released to the system we can also find an arrival rate $\lambda$, which gives the rate at which new orders (material) arrives for production. With these rates, the utilization is simply:

$$\rho = \frac{\lambda}{\mu} \qquad (1)$$

$\rho \approx 1$ means a highly utilized manufacturing system working at the border of its capacity, whereas $\rho \approx 0$ implies a manufacturing system that has most of the time nothing to do.

As performance measures we apply herein the *mean queue length* which is similar to the work in process, and the *mean throughput time*, i.e. the time a piece of work needs from the moment it is released to the production system to the moment, where its production is finished. Thus a short throughput time gives a good base for due date performance.

Usually it is of little validity to obtain results for a single realization of a (at least partially) stochastic production process or to consider a configuration just for a certain workload situation. Therefore we use as a comprehensive visualization a plot of the investigated *performance measures* vs. *total utilization* of the manufacturing system. Fig. 1 gives some examples of these also as logistic operating curves known plots (Nyhuis, 1994), (Nyhuis and Wiendahl, 1999) for a single workstation with a buffer and a machine to produce the part ( called single server queue in queueing theory) under different distributions of inter arrival and service times.

## 5. DISTRIBUTION OF WORK IN ONE PRODUCTION STAGE BETWEEN PARALLEL WORKSTATIONS

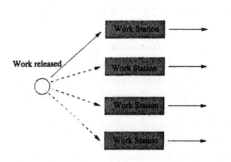

Fig. 2. Visualization of the model used in the first example. Once the workpiece is released, it has to choose one of the workstations to receive its processing.

As a first example for an autonomous control we use a system with four parallel workstations, as depicted in Fig. 2. The processing is done in a single stage step, and all products are of one class. Once a workpiece is released to the production, it should decide which workstation it chooses for processing. We investigate three possible rules for that decision:

*Rule A*

Just choose the cyclic next workstation   (2)
compared to the foregoing part.

That rule serves as an example for an inflexible, non autonomous approach, since it does not regard any information of the actual system state. The performance of a system operated with this rule should be compared with systems, where autonomous controlled products are used. These items use the following decision rules , which consider the actual input buffer size of the parallel workstations:

*Rule B*

Choose the first server with the smallest
buffer, (i.e. if 2 and 3 have empty buffers   (3)
use server 2).

*Rule C*

Choose the first buffer, unless its queue
lenght exeeds a certain threshold, then
go to the second. If the second also
exeeds the threshold then choose the   (4)
next etc. If also the last buffer contains
more then the threshold value increase
this value. (*In the following we used 5
as increment*)

For the latter rules we imply that a workstation that does not respond to a request for buffer size or indicates that it is down, due to some failure for instance, is not included in the evaluation process. We consider two scenarios: First we assume that all workstations have equal properties (i.e. the same machinery etc.: case I). Secondly we study the situation, that the production at different workstations requires different times, i.e. different production rates are offered by the workstations (case II). The product itself needs no information about this rates, since this information is provided also in the actual buffer sizes in a better way. Typical results of performance evaluation for different distributions of inter event times and these rules applied to the model system are shown in Fig. 3.

### 5.1 Discussion

In Fig. 3 the mean throughput times for the whole system vs the utilization of the whole system are shown, i.e. every job passing the system, regardless which workstation it uses, is considered for the evaluation.

At first we shall discuss the results obtained by using *rule C*, since here the differences to *rule A* and *B* are obvious. If *rule C* is used, then even

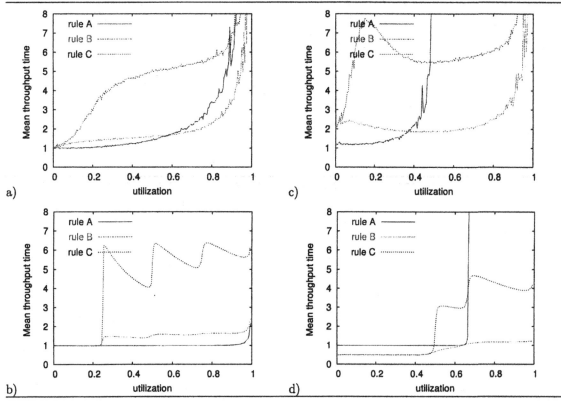

Fig. 3. Typical results of simulations for the model system from Fig.2. The curves show the mean throughput time vs. utilization $\rho$ if decision rules A,B or C (2)-(4) are applied for different systems parameters. the first column contains diagrams for systems (case I) where all servers have a equal production rate $\mu = \mu_i = 1$ whereas the second column depicts the situation for a system (case II) with $\mu_1 = (0.5)^{-1}, \mu_2 = (1.5)^{-1}$, $\mu_3 = (0.75)^{-1}$, $\mu_4 = (1.25)^{-1}$

a)$t_{through}$ vs. $\rho$, inter-release times and service times are exponentially distributed, all workstations have equal production rates.

b)$t_{through}$ vs. $\rho$, inter-release times and service times are $U_1$ distributed, all workstations have equal production rates.

c)$t_{through}$ vs. $\rho$, both inter-release times and service times are exponentially distributed, the workstations provide different production rates (case II).

d)$t_{through}$ vs. $\rho$, both inter-release times and service times are $U_1$ distributed, the workstations provide different production rates (case II).

In all plots the rules that result in the different curves are indicated.

for small utilization a relatively large throughput time is observed. This is caused by the specialty of rule C to imply a type of capacity control. At first the first workstation is highly utilized, whereas all other workstations remain unused. This leads to a large queue for workstation one and therefore to long throughput times. The second workstation comes into processing if the utilization of the first exceeds a certain threshold. Then we have more or less a two parallel server system and with increasing utilization of the whole system further servers are used. Therefore the mean throughput time remains more or less constant for a wide range of utilization. The described behavior is obvious, if we consider Fig. 4, where the successive inset of the buffers with increasing utilization becomes visible. In a nearly deterministic system these effects are much more distinctive, as the strange

behavior of the curve in Fig. 3b demonstrates. If the preferred first workstation offers a relatively low production rate, it is possible that for low utilization the throughput times are larger than for a more utilized system, where more workstations are in use, according to rule C (see Fig. 3b).

If we consider *rule A* and *rule B* the principial differences of a decision made according to local information and a 'blind' routing becomes visible. The information on buffer sizes used for decision in rule B does not contain any information if the workstation is actually processes a product or has finished and can start the production of a new part instantaneously. Thus, in cases where workstations are low utilized and therefore often a production of a new product can start immediately after release, the unintelligent rule A can lead to lower mean throughput times than rule

Work released

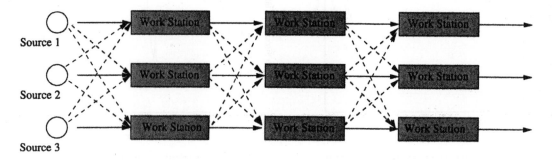

Fig. 5. A simple model of a flexible multistage production system for different classes of products. Depicted is a system where 3 classes of products ( labeled with $\alpha \in \{1,2,3\}$) are released by source 1-3 to the production system (with rates $\lambda_1,\lambda_2,\lambda_3$ respectively) consisting of 3x3 workstations providing production rates $\mu_{ij}(\alpha)$ for products of class $\alpha$ . Whereas the solid arrows show the default routes for three separated production lines (one line for each product class) the dashed arrows indicate all routing possibilities in a flexible system that can be obtained by an autonomous self-controlled part.

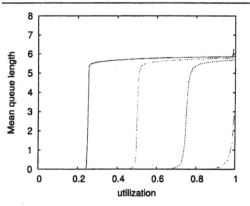

Fig. 4. The figure shows the behavior of the four buffers of the model system from Fig.2 depending on the utilization for rule C (4), if both inter-release time and service time are have a $U_1$ distribution and $\mu = \mu_i = 1$ . Rule C systematically makes a capacity control feasible since the successive buffers become used only with increasing workload. For the nearly deterministic system the transitions are very clear (compare Fig. 3b).

B (see Fig. 3). However, if the utilization is increased and queues are unavoidable due to the uncertainty in production times, rule B clearly leads to a better performance. This is possible because temporarily 'faster' workstations with small buffers can be used for production. The critical utilization, where the mentioned advantage of rule B provides a better overall performance, strongly depends on the systems parameters. We conclude, that for high utilization, the autonomy in routing

gives a competitive edge even if the information base (here: actual buffer sizes) is not optimal. Furthermore, in case II, where the offered production rates are different, a 'blind' equal distribution of work for all workstations (rule A) leads to an overloaded system, where the autonomous routing can find the alternative workstations. This behavior is shown in Fig. 3c,d where the throughput times for rule A reaches unacceptable values for moderate utilization whereas the autonomous controlled system provides a good performance.

## 6. DISTRIBUTION OF WORK IN A MULTI-STEP MULTI-CLASS PRODUCTION SYSTEM

As a second model system we use the production system depicted in Fig. 5. It serves as an example for a multistage production of different product types in a system, where in every stage three workstations are able to perform the processing for all products assigned to that stage. Thus for every released part there are $3^3$ possible routes to complete the processing. For this system we have investigated two situations. The simple example (case III) is a system, where only one product type is manufactured and all workstations have the same production rate. In this case only the way the routing decisions are made is relevant for the global performance.

Of more interest is the situation (case IV), where three different product classes are manufactured. Each class has a 'line' where the processing of this class products requires a unit time in all production stages. However, it is useful if a workstation of an other 'line' can take over a product from the preferred line if the corresponding workstation for the stage is overloaded. In that case we penalize the line change by a lower production rate for the part, that can be caused for instance by additional set-up times. The redundancy included in the system is beneficial, if the product mix is variable. Especially if a separated line for one product would be overloaded whereas an other product type does not utilize its preferred workstations the redundancy allowes a flexible utilization of the capacity resources. In that case not only the routing method is a relevant parameter for the overall performance but also the product mix has to be considered. It shall be shown in the following that autonomous control is able to cope with unbalanced product mixes.

We consider again three possible decision rules for control:

*Rule A'* (5)
Just choose the next server in line.

This rule is the example for an inflexible, non autonomous approach, since the lines are separated and no routing alternatives are available. Also no actual system information is used. The performance of a system operated with this rule should be compared with systems where the routing is organized by autonomously controlled products that use the following decision rules, which consider the actual input buffer size of the alternative workstations in each production stage:

*Rule B'*
Choose the first of the two following servers with the smallest buffer. (i.e. the (6) rule has a slight preference for the in-line buffer)

*Rule C'*
Choose the following buffer in line, unless its queue lenght exeeds a certain threshold, then go to the other line. If the other buffer exeeds the treshold go to the next alternative etc. If all alternative buffers exeed the threshold, then increase the threshold value by a fixed amount. (*In the following we used 5 as increment*). (7)

For the rules B', C' we imply that a workstation that does not respond to a request of buffer size or indicates that it is down, due to some failure for instance, is not included in the evaluation process. Typical results of performance evaluation for different distributions of production and release times and these rules applied to the model system are shown in Fig.6.

*6.1 Discussion*

In Fig.6 the mean throughput times for the whole system vs. the utilization of the whole system are shown, i.e. every product passing the system, regardless which specific path through workstation it uses, is considered for the evaluation. For the calculation of the total utilization the largest possible production rate (reached if every product is processed only at its designated line) is used.

For small utilization all three decision rules imply that the three lines are decoupled. Only the distribution of work to the line-heads implies a weak coupling for rules B' and C'. Therefore the differences in performance are small if the system is not utilized.

Furthermore, if we consider a more or less *deterministic processing* (see Fig. 6b), the different rules of exchanging work between lines have practically no impact for case III, where only one product class is produced. The reason is simply, that the line head workstations completely buffer the stochasticity out, and in the following line the system becomes in fact deterministic, which implies vanishing queues, and therefore no reason for a product to change its line.

Contrastingly, even for case III the rules make a difference if the system load is increased and the processing times in the workstations scatter more randomly (Fig. 6a). Then the autonomous control can use unforeseen gaps at alternative workstations for production and provides a better overall performance in terms of mean throughput times (and WIP respectively). Also here rule C' tends to produce larger buffers (and therefore a larger $t_{through}$) if compared to rule B'. Notheless the situation is different to the example for rule C discussed in the previous section. Here generally products use all alternative workstations since all 'lines' are fed by the parallel sources and thus no stepwise utilization as in Fig. 4 is observed.

Now we shall discuss the situation if case IV applies. Typical simulation results for this scenario are shown in Fig. 6c,d. The most obvious result is, that in the case of an unflexible system according to rule A' the systems performance gets worse if one product class overloads its line. In the example the release rate for products of type 1 is two times the rate the 'line' is dimensioned for, and thus the system breaks down for an overall utilization of not more than 50%. In contrast the performance for routing according to rule B' and C' is much better, because the not so frequently visited parallel workstations in every production stage borrow their capacities to the other product class. This happens due to the self-organizing features of the autonomously controlled material flow. Only for heavy utilization the payoff of increased production times (if a product is not processed at the designated workstation) leads to a higher mean

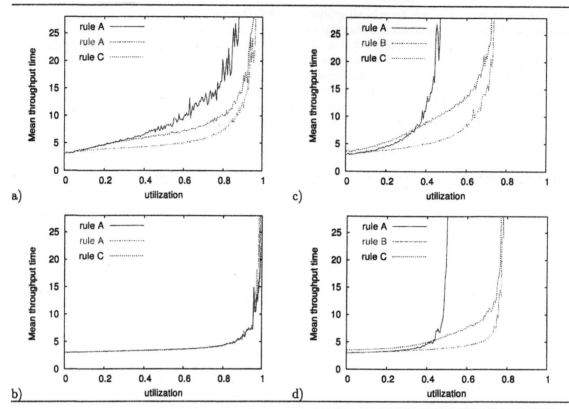

Fig. 6. Typical results of simulations, showing the mean throughput times vs. total utilization for the multistage production system from Fig. 5 with different parameter sets.

The first column shows curves for situations, where only one class of products has to pass the system and all production rates $\mu_{ij} = \mu = 1$ of the workstations are equal $\mu_{ij} = \mu = 1$ for all parts (case III). The second column (case IV) shows results for a system used for the production of three classes of products, released with $\lambda_1 : \lambda_2 : \lambda_3 = 2.0 : 0.1 : 0.9$ (i.e. for the case of separated lines the first line is overloaded, the second in all cases relatively weak utilized and the utilization of third products line is similar to the total utilization). Furthermore in this case different production rates are offered to parts depending if they are in their preferred line or not: $\mu_{ij}(\alpha = i) = 1.0$ and $\mu_{ij}(\alpha \neq i) = 1.5$ otherwise.

a)$t_{through}$ vs. $\rho$, inter-release times and service times are exponentially distributed, parameters according to case III, only one product class.

b)$t_{through}$ vs. $\rho$, inter-release times and service times are $U_1$ distributed, parameters as in a).

c)$t_{through}$ vs. $\rho$, both inter-release times and service times are exponentially distributed, parameters according to case IV.

d)$t_{through}$ vs. $\rho$, both inter-release times and service times are $U_1$ distributed, parameters as in c)

In all plots the rules that result in the different curves are indicated.

throughput time for the whole system than it would show, if no product has a release rate larger than 1. It is even worth to note that in this scenario (case IV) the scattering of production times is not the most relevant factor (Fig.6 c and d reflect qualitatively a very similar behavior). It is even more important how the system can cope with the asymmetric load of different product classes.

We further remark that the behaviour of the system in scenario IV is for other relations for the different product classes very similar to the herein discussed special parameters. This clearly indicates the overwhelming advantages of autonomous control of shop floor logistics. A system where the

products schedule themselves by rules like B or C instantaneously adapts itself - without any further changes to any infrastructure or system design - to a changed product mix. It is considerably robust against unforeseen and widely scattering production times and provides an overall performance that is for high utilization better than for a system with separated lines for each product type.

## 7. CONCLUSIONS

We have analyzed highly distributed control in form of autonomous routing of intelligent products by means of simulation. By using very simple decision mechanisms we enable individual parts released to the production system to choose a workstation themselves. Different strategies are compared; their benefit in terms of throughput time and buffer sizes for the local assignment of jobs to the next production step under a number of distinct stochastic distributions of production times was evaluated. It turns out, that autonomous controls can be more effective than traditional approaches for logistics control and provides additional robustness and flexibility for a production system. The introduced type of local feedback mechanism needs no fixed central planning component, which often is a bottleneck in other systems. The more alternatives for the next production step exist, i.e. the higher the chance to find an un-utilized workstation that can be equipped for the desired production in short time, the better distributed control works. This behaviour is most favorable if large (stochastic) variations of manufacturing times make it impossible to have a good planning and scheduling in advance.

However the design of appropriate systems in detail remains a challenging problem since an evaluation by formal methods provides certain problems. Here the development of analytical models is urgently needed. First approaches for systems with similar properties as discussed here have been recently presented by Dachkovski *et al.* (2004).

Such models will also help to analyze the dynamics of autonomous distributed control systems. This is of interest, since distributed systems can expose a very complicated dynamic behavior, that also may cause performance losses (Diaz-Rivera *et al.*, 2000), (Peters and Parlitz, 2003), (Avrutin and Schanz, 2000). Thus the investigation of decentralized and autonomous control theories is a highly interesting, ambitious and promising research area even in a theoretical point of view.

Future research has also to tackle the question, which benefits a more sophisticated design of local decision making algorithms can provide. This is important due to the fact that such mechanisms will need additional communication and information processing capacities and are not easily to implement. Thus comparative studies of such systems under realistic preconditions will provide informations of great practical relevance.

## 8. ACKNOWLEDGMENTS

This research is founded by the DFG (German Research Foundation) within the CRC 637 'Autonomous cooperating logistic processes - A Paradigm Shift and its Limitations'. The authors wish to thank Prof. N.A. Duffie for interesting and valuable discussions.

## REFERENCES

Avrutin, V. and M. Schanz (2000). On the scaling properties of the period increment scenario in dynamical systems. *Chaos, Solitons and Fractals* 11, 1949–1955.

Baker, A. (1998). A survey of factory control algorithms which can be implemented in a multi-agent heterarchy: dispatching, scheduling, and pull. *Journ. of Manufacturing Systems* 17(4), 297 – 320.

Bongaerts, L. (1998). Integration of Scheduling and Control in Holonic Manufacturing Systems. PhD thesis. Katholieke Universiteit Leuven, Belgium.

Bongaerts, L.L., D. Monostori and B. Kada (2000). Hierarchy in distributed shop-floor control. *Computers in Industry* 43(2), 123 – 137.

Dachkovski, S., F. Wirth and J. Jagalski (2004). Autonomous control of shop floor logistics: Analytic models. In: *Proc. IFAC Conference on Manufacturing, Modelling, Management and Control, MiM 04*. Athens, Greece.

Diaz-Rivera, I., D. Armbruster and T. Taylor (2000). Periodic orbits in a class of re-entrant manufacturing systems. *Mathematics of Operations Research* 25, 708–725.

Duffie, N.A. and R.S. Piper (1986). Nonhierarchical control of manufacturing systems. 5(2), 127–129.

Gou, L., P.B. Luh and Y. Kyoya (1998). Holonic manufacturing scheduling: architecture, cooperation mechanism and implementation. *Computers in Industry* 37, 213 – 231.

Havany, J. (1985). Intelligence and cooperation in heterarchial manufacturing systems. *Robotics and Computer Integrated Manufacturing* 2(2), 101–104.

Jennings, N.R. (1998). Applications of intelligent agents. In: *Agent Technology: Foundations, Applications, and Markets* (N.R. Jennings M.J. Wooldrige, Ed.). Springer Berlin Heidelberg New York.

Nyhuis, P. (1994). Logistic operating curves - a comprehensive method of rating logistic potentials. In: *EUROXIII/OR36, University of Strhycle, Glasgow*.

Nyhuis, P. and H.-P. Wiendahl (1999). *Logistische Kennlinien*. Springer Berlin Heidelberg New York.

Okino, N. (1993). Bionic manufacturing system. In: *Flexible Manufacturing Systems, Past-Present-Future* (J. Peklenik, Ed.). pp. 73–95. CIRP.

Parunak, H.V.D. (1994). The aaria agent architektur: From manufacturing requierements to agent-based system design. In: *Workshop on Agent-Based Manufacturing ICAA '98, Minaeapolis* (H.V.D. Parunak A.D. Baker S.J. Clark, Ed.).

Peters, K. and U. Parlitz (2003). Hybrid systems forming strange billiards. *Int. Journ Bif. Chaos* **13**(9), 2575 –2588.

Scholz-Reiter, B., K. Wind, J. Kolditz, F. Böse, T. Hildebrandt, T. Philipp and H. Höhns (2004*a*). New concepts of modeling and evaluating autonomous logistic processes. In: *Proc. IFAC Conference on Manufacturing, Modelling, Management and Control, MiM 04*. Athens, Greece.

Scholz-Reiter, B., K. Windt and M. Freitag (2004*b*). Autonomous logistic processes: New demands and first approaches. *Proceedings of the 37th CIRP International Seminar on Manufacturing Systems* pp. 357–362.

Tharumarajah, A., A. Wells and L. Nemes (1996). Comparison of the bionic, fractal and holonic manufacturing systems concepts. *Int. Journ. of Computer Integrated Manufacturing* **9**(3), 217–226.

Ueda, K. and J. Vaario (1998). The biological manufacturing system: Adaption to growing complexity and dynamics in manufacturing enviroment. *Manufacturing Systems* **27**(1), 41–46.

Warneke, H.J. (1993). *The Fractal Company*. Springer.

Yao, D.D. (1994). *Stochastic Modeling and analysis of manufacturing systems*. Springer.

Zweben, M. and M. Fox (1994). *Intelligent Scheduling*. Morgan Kaufman.

ELSEVIER
IFAC
PUBLICATIONS
www.elsevier.com/locate/ifac

# MODELLING OF DIE SHAPE IN DRAWING PROCESS SIMULATION BY FEM

A.M. Camacho[1], E.M. Rubio[1], M. Marcos[2], C. González[1]

[1]National Distance University of Spain (UNED). c/ Juan del Rosal, 12, 28040-Madrid, Spain
E-mail: amcamacho@bec.uned.es
[2]Univesity of Cadiz. c/ Chile s/n, 11003-Cadiz, Spain

Abstract: Drawing is a metalforming process widely used in the industry. The finite element method has become a very useful tool for studying drawing processes more deeply than with other conventional methods. In the present work a drawing process has been analysed taking into account different die geometries, under plane strain conditions. The evaluation of the influence of this parameter over global aspects of the process and also over local phenomena is studied. The main objective of this study is the optimization of the drawing process by means of improving the design of dies. *Copyright © 2004 IFAC*

Keywords: Manufacturing processes, Finite element method, Mechanical stress, Coulomb friction, Simulation, CAM.

## 1. INTRODUCTION

According to the definition of Wagoner and Chenot (2001) "Drawing process consists of pulling a preform with a constant section through a die to give a precise profile to the workpiece". Drawing process is used for obtaining plates, wires, bars, and other geometric sections in one or several operations. The main features of the final product are a good surface finish and dimensional accuracy of the parts. In this work, sheet drawing is studied and plane strain conditions are assumed.

The main variables of the process are: the die semiangle, $\alpha$, the reduction ratio in cross-sectional area, $r$, and the Coulomb friction coefficient at the die-workpiece interface, $\mu$. These variables are close related to the energy necessary to carry out the process. Nevertheless, other factors as die geometry affect to the mechanics of the process and to the properties of the part. In this way, three are the geometries that have been submitted to study: the wedge shaped, the wedge shaped with calibration zone and the profile according to circumference arch and in all of them, a constant reduction has been considered.

For the mentioned cases, different aspects are analysed; it is necessary to emphasize the drawing mean stress, the contact pressure distributions between the die and the workpiece, and the stress fronts at the entrance and exit of the die. The evaluation of the results allows to choose the most appropriate geometry from different points of view: from an overall improvement of the process and, besides, from a local point of view, taking into account variables that can predict some phenomena of interest for us.

To develop this study different drawing processes have been simulated using the ABAQUS finite element code (Hibbitt, *et al.*, 2003). Unlike to other methods that are commonly used for studying this sort of process ( Rubio, *et al.*, 2003; Sánchez and Sebastián, 1983), the Finite Element Method (FEM) allows to analyse the local phenomena starting from the calculation of stress or strain distributions at areas as die-workpiece interface. The main objective of the study of this kind of phenomena is the improvement of the process with a better design of the dies.

## 2. ANALYSIS PROCEDURE

This study has been carried out using the ABAQUS/Standard finite element code. Its main characteristic is that for solving the equations generated by this analysis method an implicit methodology is employed. This calculation tool is an appropriate general purpose software for simulating the different cases that have been considered in this work.

Table 1 Material properties

| $\rho$ (kg/m$^3$) | $E$ (Pa) | $\nu$ | $Y$ (Pa) |
|---|---|---|---|
| 2700 | $7 \cdot 10^{10}$ | 0,33 | $2,8 \cdot 10^7$ |

The workpiece has been meshed with the CPE4R element type, and its properties are the following: it is a continuum, plane strain, lineal interpolation and reduced integration element. These properties are appropriate for this element to be used in this kind of analysis, where large deformations and contact non linearities are assumed. In relation to the material, the workpiece has been modeled by means of an aluminum alloy, whose main mechanical properties are shown in the table 1.

Three are the geometries that have been chosen. The first one is named as W and consits of a commonly used geometry in the drawing process simulation (Fig. 1.a). It is a wedge shaped geometry with an angle, $\alpha$, that is the process parameter known as die semiangle. The second geometry, C, is a modification of the first one, since a calibration zone at the exit of the die has been included as it can be seen in the figure 1.b. Several studies that have been developed with steel wire (Godfrey, et al, 2000) indicate that wire residual stresses are reduced if a calibration zone is included, and also the dimensional accuracy of the part is improved (Ruiz, et al, 2003).

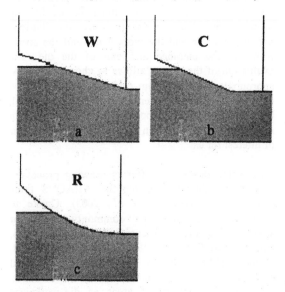

Fig. 1. Three die shapes considered.

Finally, a third geometry, R, has been modeled using a circumference arch tangent to the workpiece at the exit area, as can be seen in figure 1.c.

The technological conditions that have been considered are plane strain and a reduction ratio in cross-sectional area, $r = 0,3$, without friction and using a Coulomb friction coefficient of $\mu = 0,3$ at the die-workpiece interface. Different aspects of the mechanics of the process are analysed. At first, the necessary power to carry out the process is evaluated because this variable is taken into account in the calculation of the process efficiency. In this way, an adimensional variable, $\sigma_{zf}/2k$, has been defined. And so, the results can be compared easily.

The variable $\sigma_{zf}$ is known as the drawing mean stress and it presents dimensions of specific energy, while $k$ is the shear yield stress. For obtaining this adimensional variable, is necessary to calculate before the drawing force, $F_f$. The ABAQUS output variable NFORC is a measure of the nodal force due to the common element stresses. Therefore, the sum of the contributions for all the nodes at the exit section, $A_f$, allows to obtain the drawing force. Finally, the drawing mean stress, $\sigma_{zf}$, is the result of the quotient $F_f/A_f$. Likewise, the minimum equivalent stress, $s_{eq}$, and the maximum equivalent plastic strain, PEEQ, have been calculated comparing the results in the three cases. Both definitions (Hibbitt, et al., 2003) are shown in equations (1) and (2).

$$s_{eq} = \sqrt{\frac{3}{2} s_{ij} s_{ij}} \qquad (1)$$

$$PEEQ = \dot{\bar{\varepsilon}}^{pl} = \sqrt{\frac{2}{3} \dot{\varepsilon}^{pl}_{ij} \dot{\varepsilon}^{pl}_{ij}} \qquad (2)$$

On the other hand, the distributions of the contact pressure between the workpiece and the die have been obtained taking into account the nodal output variable CPRESS. Also, stress distributions along several sections (Fig. 2) have been represented in a graphic, at the entrance and the exit of the die. As it can be seen in the figure 2, the section at the entrance is named as EE', and it is an unique element separated from the die entrance.

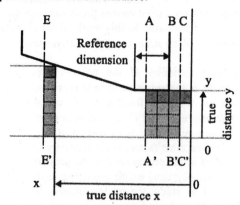

Fig. 2. Cross sections at entrance and exit of the die and references for the true distance x and y.

At the exit, three sections have been defined: AA', BB' and CC', and their location is shown in the figure 2. They allow to see the stress evolution at the exit.

# 3. APPLICATIONS

## 3.1 Frictionless application

The first case is going to be approached supposing that there is not friction at the die-workpiece interface. The drawing mean stress, minimum equivalent stress and maximum equivalent plastic strain, for each geometry, are shown in the table 2.

### Table 2 Results for $\mu = 0$; $r = 0,3$

| Tipo G. | $\sigma_{zf}/2k$ | $s_{eq(min)}$ | $PEEQ_{(max)}$ |
|---------|------------------|---------------|----------------|
| W | 0,380 | $2,621 \cdot 10^3$ | 0,1181 |
| C | 0,380 | $3,947 \cdot 10^3$ | 0,1122 |
| R | 0,395 | $1,798 \cdot 10^3$ | 0,1580 |

The results are similar for all the geometries, although it is possible to establish some behaviour trends. In relation to the contact pressure between the die and the workpiece, the distributions of the figure 3 have been obtained.

From now, the same legend is supposed in all the figures, unless another legend appears in a specific graphic.

Subsequently, the equivalent stress and x direction stress distributions at the die entrance are presented (Fig. 4). To complete the analysis, the stresses at the exit die are shown for the three sections AA', BB' and CC' (Fig.5).

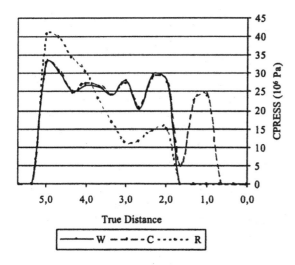

Fig. 3. Contact pressures for $\mu = 0$; $r = 0,3$.

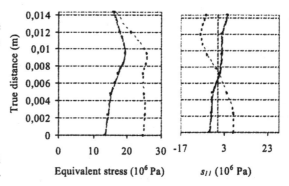

Fig. 4. Stress distributions along the entrance section EE' for $\mu = 0$; $r = 0,3$.

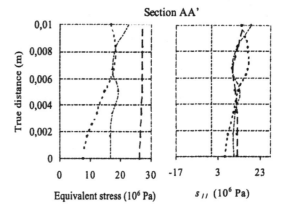

Section AA'

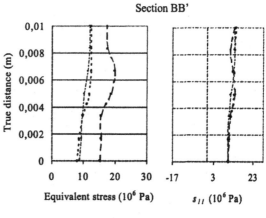

Section BB'

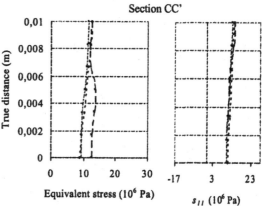

Section CC'

Fig. 5. Stress distributions along several sections at the exit for $\mu = 0$; $r = 0,3$.

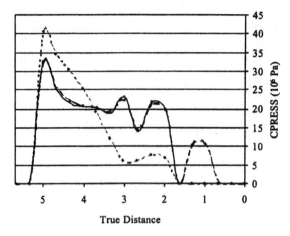

Fig. 6. Contact pressures for $\mu = 0{,}3$; $r = 0{,}3$.

### 3.2 Application with friction

The same analysis is developed with friction at the die-workpiece interface and a Coulomb friction coefficient of 0,3 is supposed. In this case, the three first parameters that have been evaluated previously are given by the table 3. As in the first application, the results are similar, although they present different behaviour.

Table 3 Results for $\mu = 0{,}3$; $r = 0{,}3$

| Tipo G. | $\sigma_{zf}/2k$ | $s_{eq(min)}$ | $PEEQ_{(max)}$ |
|---------|--------|--------|--------|
| W | 0,734 | $4{,}839 \cdot 10^3$ | 0,1109 |
| C | 0,774 | $4{,}916 \cdot 10^3$ | 0,1055 |
| R | 0,712 | $2{,}923 \cdot 10^3$ | 0,1654 |

On the other hand, contact pressures (Fig.6) present a similar trend in both cases, although the lowest values are obtained in the case with friction. Also stress distributions for the sections at the entrance and the exit of the die can be seen in the figures 7 and 8, respectively.

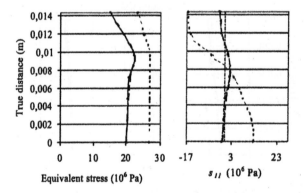

Fig. 7. Stress distributions along the entrance section EE' for $\mu = 0{,}3$; $r = 0{,}3$.

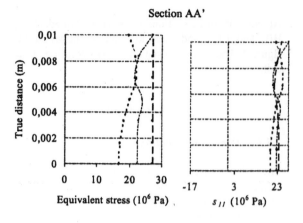

Section AA'

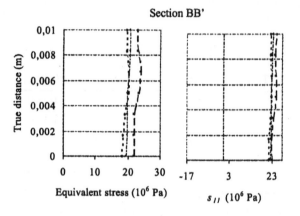

Section BB'

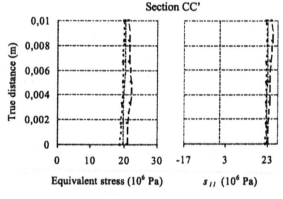

Section CC'

Fig. 8. Stress distributions along several sections at the exit for $\mu = 0{,}3$; $r = 0{,}3$.

### 4. RESULTS

At first, the drawing mean stress, minimum equivalent stress and maximum equivalent plastic strain values are compared. In both cases ($\mu = 0$ and $\mu = 0{,}3$) two kinds of table have been elaborated.

The first one is related to the best behaviour observed for each variable (lower values of $\sigma_{zf}/2k$, $s_{eq(min)}$ and $PEEQ_{(max)}$), while the second one shows the geometric situations with the worst results (higher values of $\sigma_{zf}/2k$, $s_{eq(min)}$ and $PEEQ_{(max)}$).

Table 4 Best properties for $\mu = 0$; $r = 0.3$

| Tipo G. | $\sigma_{zf}/2k$ | $s_{eq(min)}$ | $PEEQ_{(max)}$ | Total |
|---|---|---|---|---|
| W | × | | | 1 |
| C | × | | × | 2 |
| R | | × | | 1 |

Table 5 Worst properties for $\mu = 0$; $r = 0.3$

| Tipo G. | $\sigma_{zf}/2k$ | $s_{eq(min)}$ | $PEEQ_{(max)}$ | Total |
|---|---|---|---|---|
| W | | | | |
| C | | × | | 1 |
| R | × | | × | 2 |

As it can be seen, tables 4 and 5 show the frictionless values. The geometry C gives two of the best results and only one of the worst ones. That is the reason why geometry C can be considered the best one when friction effects are neglected. On the contrary, geometry R is the least adequate in this case.

Table 6 Best properties for $\mu = 0.3$; $r = 0.3$

| Tipo G. | $\sigma_{zf}/2k$ | $s_{eq(min)}$ | $PEEQ_{(max)}$ | Total |
|---|---|---|---|---|
| W | | | | |
| C | | | × | 1 |
| R | × | × | | 2 |

Table 7 Worst properties for $\mu = 0.3$; $r = 0.3$

| Tipo G. | $\sigma_{zf}/2k$ | $s_{eq(min)}$ | $PEEQ_{(max)}$ | Total |
|---|---|---|---|---|
| W | | | | |
| C | × | × | | 2 |
| R | | | × | 1 |

However, more realistic conditions should take into account the friction phenomenon. Figures 10 and 11 show equivalent stresses and plastic strain distributions in all the workpiece, for $\mu = 0.3$.

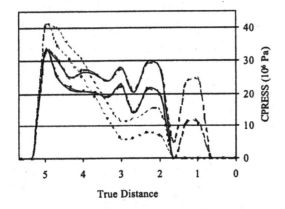

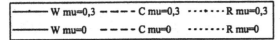

Fig. 9. Contact pressures comparison.

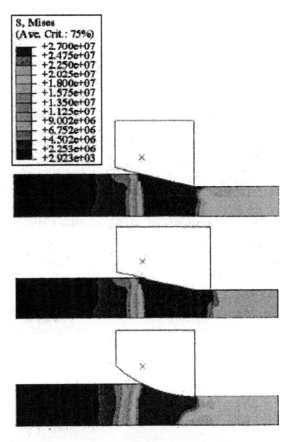

Fig. 10. Equivalent stress distributions, $\mu = 0.3$.

Tables 6 and 7 reflect the observed behaviour in this case. Geometry R has got the best properties, while C has got the worst ones because the calibration zone increases friction effects.

In figure 9 the obtained contact stress distributions have been drawn, so the results can be compared easily. As it was commented before, geometry R presents the highest pressure values at the initial contact area (even overtaking the material yield stress) and a progressive decrease is observed toward the exit area, where the lowest values are found. Otherwise, geometries W and C (which also overtake the material yield stress at the initial area) present a more homogeneous distribution, although a higher mean pressure than in the geometry R is observed. Both distributions are identical, except in the calibration zone where the geometry C presents a peak of pressure. In all cases, the contact pressures are lower when friction exists, as it was demonstrated by Johnson and Mellor (1983) through several tests. This behaviour could occur because friction helps the pressure decreases at the die interface for high reductions and friction coefficients, due to friction contributes to the metal deformation (Rowe, 1977).

The following valuations can be extracted from the analysis of the stress distributions. At first, it must be emphasize that geometries W and C show the same results at the die exit but not at the entrance.

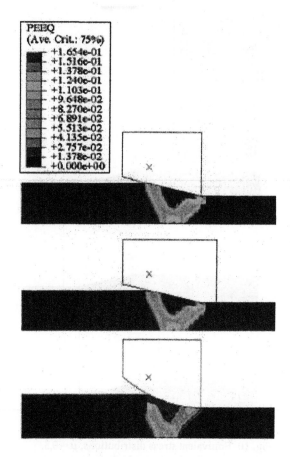

PEEQ
(Ave. Crit.: 75%)
+1.654e-01
+1.516e-01
+1.378e-01
+1.240e-01
+1.103e-01
+9.648e-02
+8.270e-02
+6.891e-02
+5.513e-02
+4.135e-02
+2.757e-02
+1.378e-02
+0.000e+00

Fig. 11. Equivalent plastic strain distributions, $\mu$=0,3.

If friction is neglected, the stresses at the die entrance are similar on the section surface for the used geometries. When friction is considered this behaviour is not observed. Otherwise, for geometry R, compressive stresses on the surface and the slight tensile stresses at the symmetry axis are higher in the case with friction. However, geometries W and C present slight tensile stresses on the surface in the frictionless case and change to compressive ones when this phenomenon is taken into account; while at the axis, the stresses are compressive in both cases, and the compression decreases when friction is included.

The stresses analysis at the exit confirms, as it was expected, that stresses are tensile for all geometries. Stresses $s_{eq}$ and $s_{11}$ are higher when friction exists. In section A, geometry R gives the lowest stress values on the surface and at the axis in all cases, while geometry C gives the highest ones. In section B, stresses $s_{11}$ are very similar. Geometries W and R present similar stresses $s_{eq}$ either on the surface or at the axis. Similar stresses appear in the center of the section when there is friction. Like in section AA', geometry C shows the highest values of $s_{eq}$. Finally, the results obtained for the section C are discussed. In this case, geometry R presents the lowest values for both kinds of stresses if friction is considered. If friction is neglected, the values of $s_{11}$ are almost coincident for the three geometries and also for the stresses $s_{eq}$ of

geometries W and R, being C the geometry with the highest values, like in the other cases.

## 5. CONCLUSIONS

The evaluation of the results related to $\sigma_{zf}/2k$, $s_{eq(min)}$, and $PEEQ_{(max)}$ allows to conclude that, when friction is neglected, the geometry that gives the best results is the wedge shaped die with calibration zone, while the worst values are obtained for the profile according to circumference arch. On the contrary, if friction is considered (a more realistic situation), the best geometry is R, and the worst is C.

Contact pressures at the interface die-workpiece show that the profile according to circumference arch has got the lowest mean pressure, and the contact pressure is lower when friction exists. The most homogeneous distribution at the die entrance corresponds to wedge shaped dies with and without calibration zone. The compressive forces registered on the surface for the three geometries could be the reason of the contact pressure peaks observed at the contact initial area. The distributions of stress at the exit show that a wedge shaped die with calibration zone gives a more stressed material. In general, a profile according to circumference arch gives the lowest stress values, overalls in the case with friction. At the die exit, the stress distributions for geometries W and R are quite homogeneous. When friction is neglected the stress distributions are more irregular.

## REFERENCES

Godfrey, H., F. Richards and S. Sason (2000). The benefits of using wiredrawing dies with smaller included angles and longer nibs. *Wire Journal International*, 102-113.

Hibbitt, D., B. Karlsson and P. Sorensen (2003). *ABAQUS v6.4 User's Manuals*. U.S.A.

Johnson, W. and P.B. Mellor (1973). *Engineering Plasticity*. Van Nostrand Reinhold, London.

Rowe, G.W. (1977). *Principles of Industrial-Metalworking Processes*. Arnold, London.

Rubio, E.M., M.A. Sebastián and A. Sanz (2003). Mechanical solutions for drawing processes under plane strain conditions by the upper bound method. *Journal of Materials Processing Technology*, **143-144**, 539-545.

Ruiz, J., J.M. Atienza and M. Elices (2003). Residual stresses in wires: influence of wire length. *Journal of Materials Engineering and Performance*, **12(4)**, 480-489.

Sánchez, A.M. and M.A. Sebastián (1983). Métodos analíticos en deformación metálica. Desarrollo histórico y actual. *Deformación Metálica*, **90-91**, 29-37.

Wagoner, R.H. and J.-L. Chenot (2001). *Metal forming analysis*. Cambridge University Press, Cambridge.

ELSEVIER

IFAC

PUBLICATIONS
www.elsevier.com/locate/ifac

# ANALYSIS OF A QUANTITATIVE APPROACH OF ROUTING FLEXIBILITY IN MANUFACTURING

**Rosario Domingo\*, Roque Calvo, Miguel Ángel Sebastián**

*Department of Manufacturing Engineering.*
*National University of Distance Education (UNED).*
*C/ Juan del Rosal, 12. E-28040 Madrid, Spain*
*\*Tel.: +34 92 398 64 55; fax: + 34 91 398 82 28; e-mail: rdomingo@ind.uned.es*

Abstract: A new approach to quantify the routing flexibility of an assembly system is suggested. It is focused on quantitative decision making through the value function. An assembly model is set and parameters influencing the cost, quality, and makespan are analysed. Thereby a heuristic algorithm is used and analysis of variance for orthogonal arrays experiments is carried out. System optimisation is achieved maximizing the value function, which contains an additive weighted influence of cost, quality and makespan. Routing flexibility evolution is evaluated and value change in a period represents the evolution of flexibility. *Copyright © 2004 IFAC*

Keywords: Assembly, Manufacturing systems, Simulation, Optimisation problems, Decision making, Flexible manufacturing systems,

## 1. INTRODUCTION

Flexibility is an essential factor to assess assembly systems in order to respond to different conditions under uncertainty, minimising impact in cost, time or quality. Literature review shows that flexibility quantification has been not consolidated as a group of measures widely accepted. Beach et al. (2000) in their literature review about manufacturing flexibility set that flexibility definitions include two common features: response to change and uncertainty. This is extended to routing flexibility, which it initially is identified by Brown et al. (1984), and after by Sethi and Sethi (1990) who point to that it is the ability to produce a part by alternative routes though the system.

In this paper a new approach to quantify the routing flexibility of an assembly system is discussed. It is focused on quantitative decision making through the function of value that can be introduced by the theory of utility. The objective of maximum value is pursued to get the better combination of cost, quality

and time. These factors present an interchange, which can avoid the production system optimisation (Chryssolouris, 1991).

Therefore the objectives of this research are the following: i) define an assembly model with capacity to allow decisions making, ii) analyse the evolution of value function respect to routing flexibility, like response to change.

Section 2 analyses the measurement of routing flexibility, section 3 define an assembly model. Analysis methodology is described in section 4, and finally results are discussed in section 5.

## 2. MEASUREMENT OF ROUTING FLEXIBILITY

Authors assume that production flexibility (F) is the temporal ratio of incremental value between two states (Calvo et al. 2003), equation (1)

$$F = \frac{\Delta value(D,Q,T)}{\Delta time}\bigg|_{state1}^{state2} = \frac{value_2 - value_1}{time_2 - time_1} \quad (1)$$

Value is function of the productive factors, which are function also of the system state parameters $x_1, x_2, \ldots, x_n$, thus we can express (2)

$$value = value(cost, quality, makespan) =$$
$$= value(x_1, x_2, \ldots, x_n) \quad (2)$$

State parameters can frequently take only integer values, e.g. product-to-line assignation, and discrete variations on manufacturing systems can be considered as continuous by taking average values in a period.

Routing flexibility is defined as the temporal ratio of incremental value between two states, given layout, process and products. Outstanding contributors to cost, quality and delivery come from the global performance of production, as the result of the assignment of product-to-line routes. This assignment allows the manufacturing of parts by alternatives routes, according Sethi and Sethi (1990). Those routes are inside the set of possible routes that the incidence matrix defines. Ordinary feasible routes are noted as 1 in the incidence matrix and incompatible routes as 0. Obviously, a higher number of possible routes increases the optimisation options. Some authors have suggested as a measure of routing flexibility the percentage of possible routes out of the potential total. This measure evaluates every route as equally important, which is a simplistic approach. A more detail approach results from the global optimisation of routes and its impact on cost, quality, delivery and finally on value. In addition to the incidence matrix (IM), other three different matrices can detail the assignation effect of every route. For every product on each line, the matrices product-to-line of cost (ID), quality (IQ) and delivery (IT) evaluate the production factors.

Moreover routing flexibility needs operator's flexibility to make different products in the same lines. Incidence matrix states whether product-to-line assignation is possible or not based on line design. Ordinary noted as 1 when the assignation is possible, 0 otherwise. The matrix index, $mi$, is defined (3) as the percentage of possible routes out of the total one (Nagarur and Azeem, 1999).

$$mi(\%) = \frac{\sum_{j}^{m}\sum_{i}^{n} b_{ij}}{m \cdot n} \times 100 \quad (3)$$

Hence, all the routes have not the same importance; it is more significant to use the matrix index, $mi$, in an incremental way. This is considering the increment of matrix index as the results of adding new routes to the previous reference matrix. Thus, routing flexibility is expressed by (4). Even so, $mi$

can quantify partially the complex among product-to-line assignation interactions.

$$RF = \frac{\Delta value(ID, IQ, IT)}{\Delta time}\bigg|_{\substack{Product \\ Lines \\ Process}} =$$
$$= \frac{\Delta value}{\Delta mi} \cdot \frac{\Delta mi}{\Delta time}\bigg|_{\substack{Product \\ Lines \\ Process}} \quad (4)$$

## 3. ASSEMBLY MODEL

Therefore an assembly model is set, which is non-linear. Assembly system is a flow shop model with semiautomatic parallel lines. A product portfolio, lines parameters and its incidence matrix define the state of reference. From this situation, routing flexibility evolution is evaluated. Routing flexibility is the temporal ratio of variation of value at constant product, lines, and process (Calvo et al. 2003). Incidence matrix determines whether product-to-line assignation is possible or not based on line design.

The model is used to simulate the response of a semiautomatic assembly system respect to routing flexibility. System optimisation is achieved maximizing the value function, which contains an additive weighted influence of cost, quality and makespan.

Some features of the manufacturing system built-in assembly lines are: a) Product is defined by manufacturing standard time per part and its daily demand; b) Lines are defined by its layout, machines, operators and processes established for the products they can process. Layout, machines or processes are established during line design. Therefore design, and the incidence matrix between products and lines estates initially determine the capability of a line. Significant line operative parameters include average number of operators, efficiency, availability and cycle time. Operators are determined depending upon the products finally assigned (planning) for production on those flexible lines; c) Production is managed make-to-order and zero stock, so every day planning is to produce the day after deliveries; and d) Optimised assignation product to line tries to maximize the objective function (set by management or decision-maker).

The model is the following:

Objective function. Maximization the value function

$$Value(cost, quality, makespan) =$$
$$= \frac{1}{v1 \cdot D\_r + v2 \cdot Q\_r + v3 \cdot T\_r} \quad (5)$$

Constraints:

$i$: index $n$ products, $i = 1, \ldots n$
$j$: index $m$ lines, $j = 1, \ldots m$

$$C_j \cdot e_j \cdot M_j - hh_i \cdot x_{ij} \geq 0, \quad \forall i, j \quad (6)$$

$$\left(avq(t)\cdot h_j - \sum_i^n st_j \cdot x_{ij}\right)\cdot e(t)_j \cdot M_j -$$
$$-\sum_i^n \left[pd(t)_i \cdot hh_i\right]\cdot x_{ij} \geq 0; \ \forall j \qquad (7)$$

$$M_j \leq \qquad M_j^{max} ; \quad \forall j \qquad (8)$$

$$C_J \geq Cmin_j \qquad , \forall j \qquad (9)$$

$$x_{ij} = \begin{cases} 1 & \text{if product } i \text{ is assigned to line } j \\ 0, & \text{otherwise} \end{cases} \qquad (10)$$

$$\sum_j^m x_{ij} = 1 \ , \ \forall i \qquad ; \qquad \sum_j^m \sum_i^n x_{ij} = n \qquad (11)$$

Data:

Incidence matrix:
$IM = [b_{ij}]$,      if $bij = 0$, then $x_{ij} = 0$    (12)
Lines: $Cmin_j, e_j, ava_j, h_j, st_j, M_j^{max}$
Products: $pd_i, hh_i$

The following notation has been defined. Thus, $v1$, $v2$, and $v3$ are the weighting factor given for each criterion. $Cj$ is the cycle time, the speed of the assembly line. $Cmin_j$ [hour/part] is the minimum cycle of the line $j$; $e_j$ the efficiency of the line, dimensionless, takes values between 0 and 1; $ava_j$ is the estimated availability (excluding planned downtime) of the line; $h_j$ [hour] is the daily labour time (e.g. 8 hour per shift); $st_j$ [hour] is the line set-up time; $M_j^{max}$ is the maximum allowed amount of operators in the line $j$; $pd_i$, $hh_i$ represent the daily demand [parts], and the standard time per part [operator-hours/part] respectively, both for product $i$. Moreover, efficiency, volume demand or availability might be considered a function of time.

Equations (6) show the limitation for a product to be made in a line, due to capacity per cycle (operator-hours per part) or headcounts presence. Equations (7) express the total capacity in terms of daily operator-hours of the line. The upper limit of operator's assignation to each line can be limited for training or labour constraints, expressed by (8). The shorter cycle of a line cannot be lower that the slowest machine in the line, for zero stock assembly system, as equates (9).

Equation (11) establishes that each product is assigned to a line. There is no sharing of production volume of each product among in different production lines.

Variable $M_j$ is the number of operators of the line $j$, $C_j$ [hour/part] is the cycle of the $j$ line and $x_{ij}$ is the assignation integer variable which takes value 1 whether $i$ product is assigned to line $j$, 0 otherwise, statement (10).

Incidence matrix (12) establishes assignment product-to-line limitations, due to process, layout, machinery or training possibilities on each line. So, $b_{ij}$ takes value 1 whether assignation is possible by design, and 0 otherwise.

The model does not contain process details. Description of internal operation is deliberated avoided in order to deal with a more general approach, useful independent on the process in the lines. Incidence matrix is a datum and expresses the process possibilities for each line to produce each product, previous to assignation optimisation through the objective function.

Finally the objective function completes the formulation of the model. This set of equations constitutes a mixed integer model. In general the model will be non-linear, depending on the objective function (5), unless they are considered simple objective functions including linear function of the variables. After establishing the objective function to maximize value, the program could be solved getting the optimal solution that permits to assess flexibility associated with system evolutions.

Therefore, objective function, according to decision-maker criteria (equation 5) can be expressed by equations (13), (14), and (15).

Cost:

$$D\_r = \frac{\sum_j^m M_j \cdot h_j}{\sum_i^n hh_i \cdot pd_i} \qquad (13)$$

Quality:

$$Q\_r = \left[ 1 + \frac{\sqrt{\sum_j^m \left( \frac{n_j}{n} \cdot \frac{S^2_{hh_{i\in Aj}}}{X^2_{hh_{i\in Aj}}} \right)}}{\frac{s\_hh}{x\_hh}} \right] \qquad (14)$$

where,

$$X_{hh\,i\in Aj} = \sum_{i=1}^n \frac{hh_i}{n_j} \cdot x_{ij}$$

$$S_{hh\,i\in Aj} = \sqrt{\left( \sum_{i=1}^n \frac{(hh_i - x_{hh\,i\in Aj})^2}{n_j} \cdot x_{ij} \right)}$$

$A_j$ is the set of products that are assigned to line $j$.
$n_j$ it is the number of different part types assigned to $j$.
$x\_hh$ is the mean of $hh_i$, and $s\_hh$ is the standard deviation of $hh_i$.

Makespan:

$$T\_r = \frac{\max_j \left( \sum_i^n (hh_i \cdot pd_i + st_j) \cdot x_{ij} \right)}{\frac{\sum_i^n (hh_i \cdot pd_i)}{m}} \qquad (15)$$

Equations (13), (14), and (15) express dimensionless values. Moreover the objective function considers main contribution to cost is the number of direct labour hours. Time is expressed by the makespan, the completion time. Homogeneity in product-to-line assignation is considered as an index of quality. The signal-to-noise (SN) weighted ratio (Taguchi, 1986) of the processing times on each line quantifies the homogeneity of assignation. High levels of SN ratio represent high level of response with little variability. Applied to our model, inverse ratio, noise-to-signal (NS) is used by an easier calculation.

These three criteria are represents by the numerator of D_r (cost in operators), Q_r (quality in NS), and T_r (makespan in hours), which are named Dmin, Qmax, and Tmin respectively.

## 4. METHOD OF ANALYSIS

A special branch and bound algorithm has been developed to solve the model. At each node of the tree, branching is done according to order products and lines. The exploration is carried out through calculating of $C_j$ and $M_j$. $C_j$ sequence is in the ascending order, and $M_j$ sequence is in the descending order. If a partial solution violas this condition, this solution and branch from this node are not considered. The complete resolution of algorithm is shown in Domingo et al. (2004).

The analysis of variance (ANOVA) for orthogonal arrays experiments is carried out to identify statistically significant factors. Parameters related to products, lines, and incidence matrix have been selected and the use of the matrix L12 allows including 11 parameters at 2 levels. Statistical analysis is conduced on relevant factors in order to test the three criteria. Moreover, percentage contribution has been calculated. It is the percentage of total sum of squares explained by the factor after an estimate of the error sum of squares (Taguchi, 1986). This analysis has been carried out according to criteria Dmin, Qmax, and Tmin.

Firstly values $mi$ are generated randomly. The results of simulation show that optimal values are between 70 and 100. Moreover when $mi$ takes values near to 100% (less routing constraints) the cost is lower, the quality is higher, and makespan is lower (Sebastián et al., 2000). Therefore a $mi = 100\%$ is taken in the statistical design.

Routing flexibility is analysed to determine its contribution to objective function by DSS Factory software. Value function is calculated. Value change in a period represents the evolution flexibility. It is necessary a dimensionless value, if the three criteria (cost, quality, and makespan) are additive.

## 5. STATISTICAL ANALYSIS AND DISCUSSION ON RESULTS

Table 1 shows the information for experiment. Moreover efficiency, availability, and set-up time are data. Efficiency and availability are 1 in each line, and set-up is zero. Experiment results are showed in table 2. Parameters considered are related to products, lines, and incidence matrix. Products parameters, for $n$ products, are the following:

x_pd: mean of $pd_i$, dimensionless
s_pd: standard deviation of $pd_i$; dimensionless
x_hh [hour-operator]: mean of $hh_i$
s_hh [hour-operator]: standard deviation of $hh_i$
hh_max [hour-operator]: maximum standard time

Lines parameters, for $m$ lines:

x_hM [hour–operator]: mean maximum capacity
x_h/Cm [part]: mean of $h_j/Cmin_j$
s_h/Cm [part]: standard deviation of $h_j/Cmin_j$

Incidence matrix parameters: $mi$

### Table.1 Problem data

| Line (j) | $h_j$ | $M_j^{max}$ | $Cmin_j$ |
|---|---|---|---|
| 1 | 8 | 67.5 | 0.02 |
| 2 | 8 | 103.5 | 0.01 |
| 3 | 8 | 121.5 | 0.01 |
| 4 | 8 | 157.5 | 0.01 |

### 5.1 Statistical Design

### Table 2. Orthogonal array L12

| | n | m | x_hM | x_h/Cm | s_h/Cm | x_hh | s_hh | x_pd | s_pd | hh_max |
|---|---|---|---|---|---|---|---|---|---|---|
| 1 | 10 | 3 | 900 | 600 | 150 | 0.8 | 0.2 | 80 | 30 | 1.3 |
| 2 | 10 | 3 | 900 | 600 | 150 | 1.2 | 0.4 | 120 | 40 | 1.5 |
| 3 | 10 | 3 | 1100 | 800 | 300 | 0.8 | 0.2 | 80 | 40 | 1.5 |
| 4 | 10 | 4 | 900 | 800 | 300 | 0.8 | 0.4 | 120 | 30 | 1.5 |
| 5 | 10 | 4 | 1100 | 600 | 300 | 1.2 | 0.2 | 120 | 30 | 1.3 |
| 6 | 10 | 4 | 1100 | 800 | 150 | 1.2 | 0.4 | 80 | 40 | 1.3 |
| 7 | 12 | 3 | 1100 | 800 | 150 | 0.8 | 0.4 | 120 | 30 | 1.3 |
| 8 | 12 | 3 | 1100 | 600 | 300 | 1.2 | 0.4 | 80 | 30 | 1.5 |
| 9 | 12 | 3 | 900 | 800 | 300 | 1.2 | 0.2 | 120 | 40 | 1.5 |
| 10 | 12 | 4 | 900 | 600 | 150 | 0.8 | 0.2 | 120 | 40 | 1.3 |
| 11 | 12 | 4 | 1100 | 800 | 150 | 1.2 | 0.2 | 80 | 30 | 1.5 |
| 12 | 12 | 4 | 900 | 600 | 300 | 0.8 | 0.4 | 80 | 40 | 1.3 |

According to Criterion Dmin, significant parameters are x_hh, and x_pd (see table 3). Thereby on a deterministic environment with flexibility and adequate capacity, the possibility optimisation depends of average standard time and average order volume. With criterion Qmax, ANOVA results points to that statistically significant parameters are s_h/Cm, x_hh, s_hh, x_pd, n, and m (table 3). But parameters influence is non linear or with strong interactions. Finally, criterion Tmin shows that significant parameters are $x\_hh$, $x\_pd$, and $m$ (table 3). Criterion Tmin have a relationship with intensive use of lines and production resources.

Therefore $x\_hh$ and $x\_pd$ are statistically significant in all the criteria. The percentage contribution of parameter $x\_hh$ is the highest for criteria Dmin and Tmin, while percentage contribution of $x\_pd$ is highest for Qmax (see figure 1).

Table 3. ANOVA table

| | Sum of squares (SS) | Variance ratio (F) | Significance | Contribution (%) |
|---|---|---|---|---|
| **Dmin** | | | | |
| n | 936.33 | 4.15 | | 3.75 |
| m | 0.00 | 0.00 | | 0 |
| x_hM | 147.00 | 0.65 | | 0.42 |
| x_h/Cm | 0.00 | 0.00 | | 0 |
| s_h/Cm | 261.33 | 1.16 | | 0.19 |
| x_pd | 5808.00 | 25.73 | * | 29.49 |
| s_pd | 432.00 | 1.91 | | 1.09 |
| x_hh | 10443.00 | 46.26 | * | 53.97 |
| s_hh | 0.00 | 0.00 | | 0 |
| hh_max | 0.00 | 0.00 | | 0 |
| error | 18027.67 | 1.00 | | 11.93 |
| **Qmax** | | | | |
| n | 0.00138 | 317.81 | * | 3.5 |
| m | 0.00515 | 1183.95 | * | 13.07 |
| x_hM | 0.00025 | 57.53 | | 0.62 |
| x_h/Cm | 0.00000 | 0.00 | | 0 |
| s_h/Cm | 0.00268 | 615.18 | * | 6.79 |
| x_pd | 0.00712 | 1637.89 | * | 18.09 |
| s_pd | 0.01194 | 2745.94 | * | 30.33 |
| x_hh | 0.00632 | 1452.97 | * | 16.04 |
| s_hh | 0.00043 | 98.21 | | 1.07 |
| hh_max | 0.00000 | 0.00 | | 0 |
| error | 0.04 | 1.00 | | 0.11 |
| **Tmin** | | | | |
| n | 5526.01 | 3.72 | | 2.99 |
| m | 26807.05 | 18.06 | * | 18.75 |
| x_hM | 0.00 | 0.00 | | 0.00 |
| x_h/Cm | 0.00 | 0.00 | | 0.00 |
| s_h/Cm | 2388.39 | 1.61 | | 0.67 |
| x_pd | 45772.97 | 30.83 | * | 32.79 |
| s_pd | 2010.02 | 1.35 | | 0.39 |
| x_hh | 46611.41 | 31.39 | * | 33.41 |
| s_hh | 0.00 | 0.00 | | 0.00 |
| hh_max | 0.00 | 0.00 | | 0.00 |
| error | 129115.84 | 1.00 | | 10.99 |

### 5.2 Calculation of value function

Establishing for instance a weight of $v1 = 33\%$ for cost, $v2 = 33\%$ for makespan, and $v3 = 33\%$ for quality, the value function to maximise can be expressed (16).

$$Value = \frac{1}{0.33 \cdot D\_r + 0.33 \cdot Q\_r + 0.33 \cdot T\_r} \quad (16)$$

An overcapacity has been considered. Likewise, it is possible to manufacture 30% more of parts, and to have 30% or 40% more of hours. Statistical analysis points that significant parameters are $x\_pd$ and $x\_hh$, and their percentage contribution is very high.

A weighed combination of productive factors allows improving cost, quality, and time, and finally the value (figures 2, 3, 4, and 5). A flexible line improves the value, if the $mi$ is increasing.

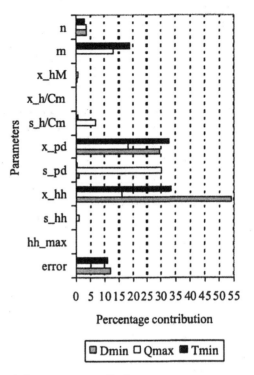

Fig. 1. Percentage contributions

For example, according to data from figure 5 and equation (4), whether the month is taken such as the unit of time. Thus, routing flexibility between $mi=0.51$ and $mi = 0.71$ is:

$$FR = \frac{0.5410 - 0.5121}{0.71 - 0.71} \cdot \frac{0.71 - 0.51}{1} = 0.02 \, \text{month}^{-1}$$

Obviously, between $mi=0.51$ and $mi = 0.61$, is zero.

This quantification allows decisions making, in short or long term, according to this expression. A positive value set that the response of system whether routes number is increased.

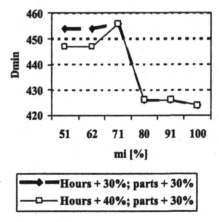

Fig. 2. Evolution of $Dmin$ [operators] respect to $mi$.

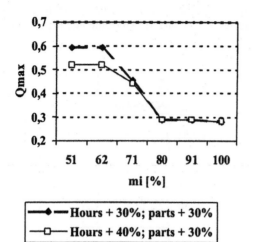

Fig. 3. Evolution of *Qmax* [dimensionless] respect to *mi*.

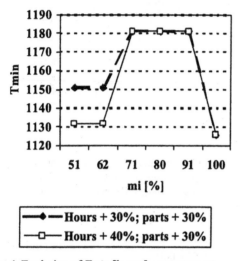

Fig. 4. Evolution of *Tmin* [hours] respect to *mi*.

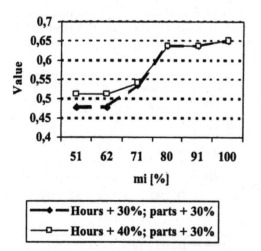

Fig. 5. Evolution of value [dimensionless] respect to *mi*.

## 6. CONCLUSIONS

With this paper an assembly lines model has been introduced and ten assessment parameters investigated.

A mathematical model has been development with the objectives related to minimization makespan, maximization quality, and minimization cost. Moreover a value function has been defined to get a combination of three objectives and analyse their evolution according to increase the routing flexibility. The model and the algorithm allow obtaining the value function dimensionless, and also the number of operator required, the hours required to complete a product, and the homogeneity of the assignation to lines.

With this study, two characteristics of routing flexibility are analysed: response to change and uncertainty. The response to change has been considered on its temporal dimension. The uncertainty has been suggested means the quantification of routing flexibility, which allows decisions making in a short or long term.

## REFERENCES

Beach, R., A.P. Muhlemann, D.H.R. Price, A. Paterson and J.A. Sharp (2000). A review of manufacturing flexibility. *European Journal of Operational Research*, **122**, 41-57.

Browne, J., D. Dubois, K. Rathmill, S.P. Sethi and K.E. Stecke (1984). Classification of flexible manufacturing systems. *The FMS Magazine*, **2 (2)**, 114-117.

Domingo, R., R. Calvo, C. González and E.M. Rubio (2004). Heuristic approach for assignment parts to assembly lines. *International Journal for Manufacturing Science and Production*, in press.

Calvo, R., R. Domingo and M.A. Sebastián (2003). An integrated assessment of manufacturing flexibility. *International Journal for Manufacturing Science and Production*, **5 (3)**, 151-161.

Chryssolouris, G. (1991). Manufacturing Systems: Theory and practice. Springer Verlang, New York.

Nagarur, N., A. Azeem (1999). Impact of commonality and flexibility on manufacturing performance: a simulation study. *International Journal of Production Economics*, **60-61**, 125-134.

Sebastián, M.A., R. Calvo, E.M. Rubio and P. Núñez (2000). Modelo para la asignación de tareas en líneas de montaje semiatomáticas. Anales de Ingeniería Mecánica, **3**, 2257-2269.

Sethi, A.K. and S.P. Sethi (1990). Flexibility in manufacturing: a survey. International Journal of Flexible Manufacturing Systems, **2**, 289-328.

Taguchi, G. (1986). Introduction to Quality Engineering: Designing Quality in Products and Processes. Asian Productivity Organization, Tokyo.

ELSEVIER

IFAC
PUBLICATIONS
www.elsevier.com/locate/ifac

# A MODELING FRAMEWORK FOR DISTRIBUTED MANUFACTURING SYSTEMS

**Viktor Zaletelj, Alojzij Sluga and Peter Butala**

*Department of Control and Manufacturing Systems, Faculty of Mechanical
Engineering, University of Ljubljana, Ljubljana, Slovenia*

Abstract: In the paper Adaptive Distributed Modeling Framework (ADMF) for
collaborative synthesis of dynamic distributed temporary manufacturing structures is
presented. Formation of such structures is triggered by emerging market
opportunities. Here, several partners are participating in a distributed environment.
The proposed modeling framework provides simultaneous synthesis of models over
the web. The modeling framework enables definition, model building, simulation
and instantiation of dynamic and adaptable temporary manufacturing structures.
ADMF is implemented as a prototype web application entitled wDME (web-based
Distributed Modeling Environment). A test case illustrates the approach.
*Copyright © 2004 IFAC*

Keywords: Systems design, decision support systems, distributed model, dynamic
modeling, IDEF

## 1. INTRODUCTION

Nowadays, manufacturing companies are forced to
adapt themselves dynamically to stochastic markets.
In traditional companies this requirement is beyond
their capabilities, due to structural rigidity, lack of
competencies, and time pressure.

New organizational forms and structures such as
production networks, clusters and virtual enterprises
are being developed in order to overcome the above
issue. These forms enable companies to form
business coalitions, to react jointly to business
opportunities, to gain the synergetic effects and the
competitive advantage by sharing of risks and
competencies, and to form flexible and goal-oriented
temporary dynamic cluster structures here referred as
Adaptive Distributed Manufacturing System (ADMS)
(Butala and Sluga, 2002). Building of ADMS opens
many questions about decision levers, collaboration,
communication and legal rights, which broaden the
aspects of customer-oriented production. These
issues as well as the high frequency of disturbances
and changes on all levels require new methodologies
and tools to support development of ADMS.

In the paper, a modeling framework entitled
Adaptive Distributed Modeling Framework - ADMF
- for distributed modeling of ADMS is presented.
The purpose of the framework is to enable definition,
model building, simulation, instantiation and
synthesis of ADMS. The developed models act as a
system ontology definition and are used for
simulation and performance evaluation, as well as
control roadmaps.

## 2. BACKGROUND AND MOTIVATION

*The complexity issue.* The source of complexity in
manufacturing lies in a large number of interacting
elements, where inter-dependencies are usually not
well known. Changing demands and stochastic
disturbances represent another source, which
increases the dynamics, uncertainty and scope of
elements.

Ueda (2001) introduced three types of problem
classes related to systems design and regarding the
completeness of information of a system
specification and environment description.
Considering ADMS, the third class problem appears
even on the lower levels of a model, where neither
the system specification nor the environment
description is complete. Ueda (2004) stated "To
solve this type of problems interactions between the
system and the human need to be established". This
is the reason, why tools for distributed model
creation, which enable real-time interactions of
human experts, are needed.

Peklenik (1995) argued the role of a human subject
in a manufacturing system. Here, the human subject
(expert) is a part of an elementary work system and
provides the competence needed to do a job.

Another major consideration regarding the role of a
human expert is given by Parunak (1995): knowing
that unlike custom manufacturing equipment, unique
skills cannot be reliable replaced on the open market
nor anticipated in advance, these characteristics of
the human capital can be critical to success.

*The role of a model.* One of the main mission of the model is to understand complex dynamics of the observed system and thus to be able to control it. In the field of manufacturing, analytic solutions for a system beyond a very modest level of complexity are unavailable. It is not enough to characterize a manufacturing system in terms of its formal dynamics. To make an unbiased business decision, the correlations between dynamical patterns and business performance have to be known.

When several distributed partners are involved in modeling, common understanding of the problem domain is the crucial issue. Objects identification and description, their formulation, identification of relationships between them, provide the ontology definition. Only this way the model consistency could be maintained.

Conceptually, an ADMS model is built dynamically in a distributed environment. Therefore, ADMS performance is not known in advance. The model inherently exhibits multi-dimensional properties; a one-dimensional optimization would not bring expected results. One way to solve the problem is to use simulation by exploring the effects of changing parameters. According to Parunak (1995), the reasoning can go in two directions. First, what can a given set of observed dynamics tell us about the scope that a given system can support? Second, if a proposed scope is given, can one estimate the resulting dynamics and thus determine whether the proposal can be met? Here the model plays the role of a simulation polygon.

The discussion until now is about how to gain an effective structure composition. Nevertheless, is that enough? What about the structure realization and the model role in this process? It should be reminded, that the structure instantiation problem is quite different from adoption of the existing one. In the last case, a basic infrastructure already exists, which can lead toward different integration problems.

As intensive efforts are needed to accomplish usable models of new manufacturing structures, it is argued that only a minor, additional effort is needed to turn models into execution roadmaps for operations. However, a prerequisite for this is a comprehensive modeling framework.

*Model versus reality.* A model should provide behavior patterns of the observed part of the designed system to be usable as a simulation and prediction tool. Here arises the question how deep should modeling go to capture the interesting behavior. It is obvious that the quantity of details does not necessary imply better results. On the other side, a too shallow model may not show special behavior patterns. In this contradiction, the qualitative aspect has an interesting role. At the same time, this is the area of human interaction especially in the case of the third class problems, and that is why experts are involved to synthesize the system architecture.

When using the model as an execution roadmap, the model itself becomes the logic pattern for the operational system. The main contribution of such approach is high and fast adaptability of the distributed system, as physical coding is minimized.

*Distributivity.* A distributed manufacturing system adds another aspect of complexity to our observation. A communication heterogeneity (misunderstanding), time and space dislocations and high business frequency presents new questions for control.

Traditionally, the model synthesis is performed in a single company. ADMS opens new requests to system modeling. Each partner of a temporary structure has to have the ability to model its own part of the model autonomously, exposing functionality interfaces toward the whole system. This way dislocated experts have an ability to define and express their knowledge and competence decisions simultaneously. When semi-automatic configuration of the dynamic structure is established (for example using an agent-based network), human experts are asked for specifying missing decision parameters.

*Modeling framework.* Modeling methodologies and tools, such as Integration Definition for Function Modeling (IDEF), Unified Modeling Language (UML), Architecture of Integrated Information Systems (ARIS), Petri-Nets, GRAI Integrated Methodology (Grai-GIM), are widely used, especially in the field of business process reengineering.

Considering building of new manufacturing structures, a need for a modeling framework is identified. Several approaches in this direction can be found in the literature. This research is mostly influenced by the work of Hongmei et al. (2003). It proposed a framework for virtual enterprise operation management.

Beside this, Wallace et al. (2001) introduced non-deterministic control of distributed simulation models. McLean and Riddick (2000) proposed the database structure to support distributed simulation. Shunk (2003) broadened the scope of IDEF modeling from a single to multi-enterprise view. Dynamic aspects of models are considered in (Mentink el al., 2003). Noel and Tichkiewitch (2004) introduced the concept of dynamic entities for simultaneous model building.

In the field of business modeling, the following initiatives are of interest: (1) Business Process Management Initiative developed BPML (Business Process Management Language), which provides an abstract model, grammar and XML syntax for expressing generic, executable business process models and supporting entities (BPMI, 2000); (2) Microsoft, IBM and BEA released BPEL4WS (Business Process Execution Language for Web Services) which provides similar functionalities (BPEL, 2003). Both initiatives are concentrated mainly on IT support of business processes.

## 3. ADMF – ADAPTIVE DISTRIBUTED MODELING FRAMEWORK

### 3.1 The ADMF framework

The idea of the adaptive distributed modeling framework (ADMF) is to support simultaneous building of ADMS structures in terms of model development, simulation, estimation and synthesis in a distributed environment. Typical partners in ADMS are manufacturing companies, especially small and medium sized, which identify opportunities to compete on the global market through collaboration in dynamic manufacturing clusters.

In order to realize this idea the framework should: (1) provide on-the-fly definition, integrated modeling, model distribution, simulation and execution; (2) enable sharing of functionalities, data and models; and (3) have features as openness and availability, platform independence, and adaptability.

The ADMF modeling framework is composed of: (1) distributed modeling environment, (2) palette of modeling methods, tools and generic building blocks, (3) simulation platform and (4) support for model building and operations. The framework structure is presented in Fig. 1. ADMF provides modeling, visualization, data storage and process execution services. Framework member enterprises (ME) expose their own services to the framework. Such a service is an autonomous work system (AWS), whose virtual part (VWS) is connected to ADMF over www. Framework core functionalities (FCF) are distributed as well as MEs. A parallel behavior is achieved through redundancy of AWS services. The framework also defines basic demands for plug-in components, which provide harmonized and cooperative environment.

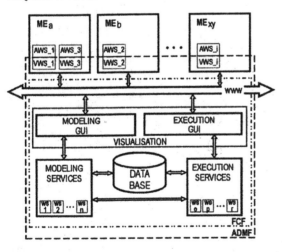

LEGEND: ADMF - adaptive distributed modeling framework
AWS - autonomous work system
FCF - framework core functionalities
GUI - graphic user interface
ME - framework member enterprise
VWS - virtual work system
ws - web services

Fig. 1: ADMF structure

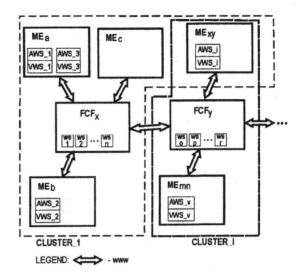

LEGEND: ⟷ - www

Fig. 2: Interrelations between ADMF and ME partners in temporal dynamic clusters.

Fig. 2 reveals an example of ADMS structures and interrelations between the framework and member enterprises.

### 3.2 Modeling support

ADMF provides four basic modeling methodologies: (1) xIDEF0, (2) UML class diagram, (3) manufacturing cybernetics, and (4) constraint logic programming (CLP). The use of these methodologies enables complementary definition and manipulation of modeling entities from different perspectives. It is essential that entities are defined run-time.

In ADMF the IDEF0 methodology (IDEF, 1993) is extended and named xIDEF0 (Zaletelj et al., 2004). The extension is applied on inputs, outputs, controls and mechanisms, which are substituted with so called flow-pipe objects of a certain type. Additional properties, i.e. push/pull, continuous/discrete, information/ material/human are defied. Each flow-pipe connects two activities and has an ability to perform a custom filter. The filter depends dynamically on the model state. The usage of filters enables creation of branching points (similar to IDEF3). Typifications of activities (transform, transport, store, and decide) enable auto control, suggestible building of a model and selective use or focus on the entities. All xIDEF0 activities are positioned on the same layer. To enable a hierarchical model, selected activities can be grouped inside high-level views, dedicated for studying of a specific topic. In Fig. 3 a generic xIDEF0 activity diagram is presented. It follows the basic specification of the IDEF0 notation and combines the advantages of the object-oriented programming, state transition diagram, and IDEF3 methodology. xIDEF0 covers the business level of entities flow description.

After identification of entities, UML is used for definition of additional entity properties, such as inheritance, life-cycle states, etc.

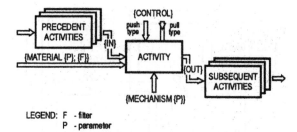

LEGEND: F - filter
        P - parameter

Fig. 3: Activity diagram in xIDEF0 methodology

For modeling of manufacturing systems the elementary work system (EWS) concept and notation (Peklenik, 1995) based on manufacturing cybernetics is adopted. The EWS diagram enables a detailed view on a manufacturing process, interacting structural elements, process implementation device, human subject, information and control flow, and their interrelations. Other business aspects, such as project objectives, subject role and control algorithm are defined as well.

CLP is used for defining relationships between entities. It enables causal definition of relations and it is much more powerful as the traditional relational databases.

### 3.3 Simulation support

As the optimal overall performance of the synthesized structure is searched, it is obvious that each problem can require its own optimal structure, not only different tuning. Thus several optimization cycles (structuring, simulation and evaluation) should be performed, supported by human experts.

Each modeling building block (entity, activity, EWS, etc.) has at least one simulation layer (Fig. 4). The layers are dedicated to particular model evaluation aspects and include a source code for the simulation performing (web service / parent object method). Free interchange of information between layers is enabled. Each layer can reach any of the entities (that pass by) working parameters and their current values, replicate real objects and use them inside its own simulation model. Additional simulation properties are also defined. Definition of the simulation scope is enabled trough a special simulation layer behavior. Outer requests trigger simulation of the observed element, which checks all necessary information. If some information is missing, the simulation layer responds with a request to nearby connected elements (or even to a human expert). This way the simulation is spreading through the model until the basic request is fulfilled. Depending on the simulation goal, the whole model can be simulated at once, or just its particular part. For the second case, a simulation distance (simD) should be provided. It defines how far away from the first requested element the simulation is to be performed. When reaching the boundary limited by simD, a virtual entities generator (simG) is activated, which provides input signals (random/function) to the model.

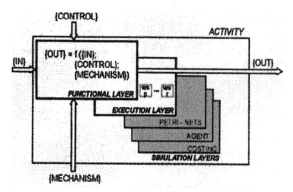

Fig. 4: Multi-layers of the modeling building block (example: xIDEF activity).

The high-level views can also possess the corresponding simulation layer, which abstracts the behavior patterns of all included activities. This way, simplifications of the simulated model can be realized.

An internal clock is incorporated in the framework to enable synchronization between independent simulation parts that can be executed remotely, parallel or redundant. It also enables identification between the entity instances in different time slots (from the past, present or simulation future).

### 3.4 Framework operational support

In order to make a built model operational, a number of aspects should be covered, which are not considered during the model simulation phase. Most of them arise when human-system interactions appear and physical world limitations have to be solved. Dealing with temporal, distributed and collaborative environment, all these features should be supported and coordinated to make the operational model feasible.

In Table 1 additional features of the framework are presented. Looking from an application developer point of view, these are services, which are always needed for a real-world operational system. Thus, they are provided as reusable components inside the ADMF. Some of them enable physical conditions for operability, some socio-business prerequisites (trust, security, etc.) and others framework management functionalities. They are needed to enable quick "time to model operability", as it is the crucial issue in temporal distributed systems.

Both modeling and operational functionalities require high interaction with a user. To support this process, graphical user interfaces (see Fig. 1) are established on the fly when entity definition is generated and later used for controlling properties. The views correspond to particular methodologies, but are based on the common information.

The very important part of the framework is the database system, which provides data storage, sharing, restoring, transaction behavior and definition

Table 1: The core framework functionalities, supporting operational phase of the model

**Table 1: The core framework functionalities, supporting operational phase of the model**

| Service name | Service description |
|---|---|
| version control | development tracking, proper entities version interaction inside emerging environment |
| life-cycle management | VERAM methodology |
| access control | visibility of entities – its state description, rights to change |
| action control | is function of (entity life-cycle state, current role(s) of an entity that interact); Actions: save, edit, promote, download, accept, deny |
| role management | entities role assignments, control of access and action rights |
| org. structure | organization, department, hierarchy |
| change control | opened, locked by, history backtracking (who, what, when, where) |
| asynchronous communication & execution | email, gsm-sms, web-services; used for notifications, long-lasting processes, long-lasting jobs exec. |
| synchronous communication | video, audio, chat, web-services; |
| relationship management | common for all entities; transparency and tracking |
| time-control | location and time dislocation for collaboration |
| multi-language control | personalized user interfaces; |
| classification | multiple; common roots; tree-like; |
| guard control | up-to-date information handling; obligatory checking |
| backup | safe copies |
| negotiation services | bidding |
| administration | modeling scope; users; roles; etc. |
| security | encryption, authentication; |

functionalities for the model building blocks, the model itself and the operational environment. This solution enables mixing of model data and operational data. Data are normalized. The data structure enables querying through various types of entities at the same time, thus abolish distinction toward de-normalized data forms. Additional performance gain is achieved by using an object-relational database (ORDBMS). Special attention is put on relationship management, where separation of entities definition and relationships is done, thus providing a more flexible data storage environment.

### 3.5 Framework implementation

Web technologies and tools, based on open standards HTTP, XML and its derivatives enable implementation of ADMF in a distributed environment over the web. On this basis, a prototype application entitled Web based Distributed Modeling Environment (wDME) is developed. .NET framework is used as a platform for building, deploying and running the distributed modeling framework application. .NET provides an adequate solution with the use of the web services (WS). The developed web-application runs on a MS-IIS server. With ASP.NET on-the-fly composing ability for preparation of web pages is enabled to those, who interact with the model. Cache ORDBMS acts as data storage system, partially connected with the P# constraint logic programming services in the field of relationship management.

*Model synthesis and execution.* The model building is performed interactively according to Fig. 5. The procedure consist of the following steps: (1) definition of entities and their simulation layers; (2) building of ADMS structure; (3) preparation of WSs, (4) deployment and exposition through framework resources, (5) simulation and analysis, and (6) improvement cycles until a desired behavior according to the criteria defined by the subject is reached.

## 4. TEST CASE

The test case model discussed here incorporates a complex production line for production of sandwich panels, a dislocated sub-suppliers assembly and a customer site inspection. They form the distributed temporal manufacturing system named TRIM. The objective was to realize an effective integration of the outside inspection for a particular production order. Dislocated experts designed the system using the common defined ontology. Each expert designed its part of the system. Production experts designed the model of the fabrication part of the system (using xIDEF0 and EWS methodologies), with a special focus on quality-related input-output product characteristics and time schedule. A third party inspector built a high-level model (xIDEF0) of the inspection process. In Fig. 6 the design user interface is shown.

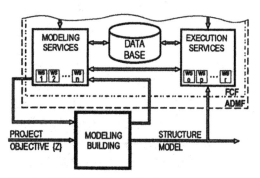

Fig. 5: ADMF model synthesis

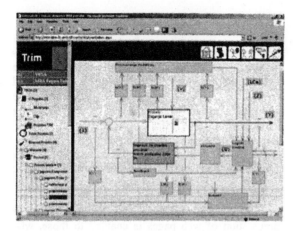

Fig. 6: Graphic user interface of the wDME

Partial models form logically one single model, which consistency is assured by wDME. The model effectiveness was checked with PN simulation. The modeling cycle was repeated until the desired conditions were fulfilled. After that, structural model was used as an information system execution template, which provided web interfaces to the corresponding decision makers and connected AWSs through automated information-flow. The involved partners experienced the following: (1) collaborative distributed environment, (2) working in a virtual team, (3) real-time visibility of the model, (4) distributed but still integrated manufacturing process.

## 5. CONCLUSION

In the paper, the ADMF modeling framework for design of distributed manufacturing systems is presented. It supports all modeling phases (ontology definition, design, simulation and analysis) as well as the model instantiation.

The ADMF modeling framework is composed of: (1) distributed modeling environment, (2) palette of modeling methods, tools and generic building blocks, (3) simulation platform and (4) support for model building and operations.

The structure elements of the ADMF are distributed in the environment. The most of the elements are redundant in order to achieve reliable operation. Simulation and execution layers are directly connected in such a way that the simulation considers the real system and the real state of the environment simultaneously.

ADMF behaves as an open system, where each layer can be customized and additional services can be added. It is implemented as a prototype web application wDME. It supports simultaneously and geographically dislocated modeling of the distributed manufacturing system or, in more general terms, dynamic manufacturing clusters, network organizations and virtual enterprises.

## ACKNOWLEDGEMENT

This work was partially supported by Ministry of Education, Science and Sport, RS Grand No.: S2-787-010/21454/2000.

## REFERENCES

BPMI, (2000). Business Process Management Initiative: *http://www.bpmi.org/index.esp*.

BPEL, (2003). Business Process Execution Language for Web Services Version 1.1, *http://www-106.ibm.com/developerworks/library/ws-bpel/*.

Butala, P. and A. Sluga, (2002). Dynamic structuring of distributed manufacturing systems. *Advanced engineering informatics*, **16**, no.2, pp. 127-133.

Hongmei, G., H. Biqing, L. Wenhuang and L. Xiu, (2003). A framework for virtual enterprise operation management. *Computers in Industry*, **50**, pp. 333-352.

IDEF, (1993). FIPS183 IDEF0, *www.itl.nist.gov/fipspubs/idef02.doc*.

McLean C. and F. Riddick, (2000). The IMS MISSION architecture for distributed manufacturing simulation. *Proceedings of the 2000 Winter Simulation Conference, Orlando, Florida*, pp. 1539-1548.

Mentink R., F.J.A.M.van Houten and H.J.J. Kals, (2003). Dynamic process management for engineering environments. *CIRP Annals*, **52**/1.

Noel F., S. Tichkiewitch, (2004). Shared dynamic entitites technology to support distant coordination in design activity. *CIRP Annals*, **53**/1.

Parunak, H.V.D, (1995). The Heartbeat of the factury. *Technical report, ERIM, Ann Arbor, http://www.erim.org/cec/publications.html*.

Peklenik, J. (1995). Complexity in manufacturing systems. *Proceedings of the CIRP seminars and Manufacturing Systems*, **24**, No.1, pp. 17-25.

Shunk, D.L., J-I. Kim and H.Y. Nam, (2003). The application of an integrated enterprise modeling methodology – FIDO – to supply chain integration modeling. *Computers & Industrial Engineering*, **45**, pp. 167-193.

Ueda, K., A. Markus, L. Monostori and H.J.J. Kals, T.Arai, (2001). Emergent Synthesis Methodologies for Manufacturing, *CIRP Annals*, **50**/2, pp.335-365.

Ueda K., A. Lengyel and I. Hatono, (2004). Emergent synthesis approaches to control and planning in Make to Order Manufacturing Environments. *CIRP Annals*, **53**/1.

Zaletelj, V., A. Sluga and P. Butala, (2004). A modeling framework for networked structures. *Proceedings of the 5th International Workshop on Emergent Synthesis IWES '04 (K. Ueda, L. Monostori, A. Markus (Ed))*, pp. 24-25.

Wallace,D., E. Yang and N. Senin, (2001). Integrated Simulation and Design Synthesis. *http://hdl.handle.net/1721.1/3802*

ELSEVIER

IFAC

PUBLICATIONS
www.elsevier.com/locate/ifac

# CUTTING TESTS DEFINITION FOR EFFECTIVE TOOL CONDITION MONITORING

## Eva M. Rubio[1] and Roberto Teti[2]

[1]*Dept. of Manufacturing Engineering, National Distance University of Spain (UNED)
Juan del Rosal 12, 28040-Madrid, Spain
e-mail: erubio@ind.uned.es*

[2] *Dept. of Materials and Production Engineering, University of Naples Federico II
Piazzale Tecchio 80, 80125-Naples, Italy
e-mail: tetiro@unina.it*

Abstract: In recent years, several tool condition monitoring (TCM) systems for machining processes have been developed. However, standard solutions have not been found yet, basically due to the complexity of the tool wear mechanism and the possibility to select among different alternatives in each step of their development. The main critical points associated with the cutting test definition have been analyzed in this paper. The study has revealed the need for a methodology of cutting test standardization. The minimum set of elements that it should contain has been established. Among them, a relevant role is held by the previously developed tool wear and the way of achieving it. *Copyright © 2004 IFAC*

Keywords: Machining, Tools, Cutting Tests, Sensors, Monitoring

## 1. INTRODUCTION

Turning, drilling, milling and grinding are material removal processes widely used in many industrial sectors. In some of them, such as the automotive or the aerospace, geometrical accuracy and surface finish requirements have been strongly increased in recent years. To reach these new needs, machine tools used in these processes have been replaced by numerical control machine tools (NCMT) trying, in this way, to eliminate the variability introduced by the operator and to obtain better quality parts (Byrne *et al.*, 1995; Li and Zhejun, 1998; Rubio *et al.*, 2003; Sánchez-Sola *et al.*, 2004).

Besides, in order to increase productivity and decrease manufacturing costs, the NCMT have been grouped along with other elements, mainly robots and computers, in flexible and computer-integrated manufacturing systems which are able to work automatically during long periods of time. Such advanced manufacturing systems demand an optimal performance in all machining stages. This includes the tool change just in time when the tool has reached a certain wear level. However, there is no previously determined moment to carry this out. Generally, to reduce costs, the utilization of the tool as long as possible would be recommended, provided the required geometrical and surface accuracy levels are respected. A tool condition monitoring (TCM) technique allowing for the on-line tool wear state evaluation and capable to act properly to counterbalance the loss of machine performance caused by worn out tools or unexpected catastrophic tool failure, can effectively provide for the optimization and improvement of the manufacturing process efficiency (Byrne *et al.*, 1995; Li and Zhejun, 1998; Rubio *et al.*, 2003; Waschies *et al.*, 1994; Hayashi *et al.*, 1988; Nayfeh *et al.*, 1993).

Development and implementation of tool condition monitoring systems have been the study subject for many researchers during the last years (Byrne et al.,

1995; Li and Zhejun, 1998; Teti, 1995; Leem and Dornfeld, 1996; Shawaky et al., 1998; Grabec et al., 1998; Inasaki, 1998;Lee et al., 1999; Silva et al., 2000; Pérez et al., 2000; Kopac and Sali, 2001; Xiaoqi et al., ,2001; Li, 2002; Haili et al., 2003). However, in spite of all these works, standard solutions for their industrial application have not been found yet. This is mainly due to the complexity of the tool wear mechanism and the possibility to select among different alternatives in each step of the tool wear monitoring system establishment (e.g. cutting test definition, type and location of sensors, monitoring test definition, signal processing method selection or process modeler selection), so that not only one solution is possible.

The analysis of the bibliographic review has revealed that, in order to obtain TCM systems applicable to industry, it is necessary to achieve, firstly, the development and implementation of an experimental TCM which allows to characterize, systematically, the tool wear process from different situations. This will allow to reach an adequate knowledge and control of all factors involved in this kind of systems by means of the individual variation of each one of them. In this way, when the system is applied to a real situation in a shop floor, it will be possible to detect variations in the results produced by other sources.

In this work, the definition of the cutting tests representing the basis for the establisment of a TCM procedure has been proposed as first step to improve the knowledge level about the experimental TCM system. Then, the main critical issues involved in the problem have been emphasized and analyzed. Finally, a methodology for systematizing the establishment of TCM systems in industry has been proposed.

## 2. FRICTION PHENOMENA, TOOL WEAR AND MONITORING SYSTEMS

During a cutting process, friction takes place at the workpiece-tool interface. This phenomenon is very complex and involves combined effects of adhesion, erosion-abrasion, fatigue-plastic, diffusion and corrosion (Carrilerro et al., 2002). The contribution of each of these effects to the global friction will depend, in each case, on different aspects such as the specific properties of the workpiece and tool materials, the workpiece surface roughness and the operating and environment conditions (Li and Zhejun, 1998; Grabec et al., 1998; Li, 2002; Carrilerro et al., 2002).

If the intensity of tool wear, $\delta$, is plotted as a function of cutting temperature, $T$, for the different wear mechanisms, it can be seen, as shown in figure 1, that the adhesion mechanism operates in the widest range of cutting temperatures (Carrilerro et al., 2002).

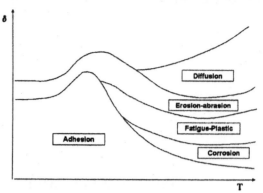

Fig. 1. Wear mechanisms versus temperature.

Friction produces a gradual wear of the cutting tool which has a negative influence on the quality of the generated surface and on the workpiece geometry and dimensions specified by the design. And, in the worst case, even the breakdown of the tool, the damage of the workpiece or the machine can occur (Lee et al., 1999; Kopac and Sali, 2001; Jemielniak and Otman, 1998).

If, as it has been said above, the machine works along with other machines and robots under the control of a computer, the breakdown of the machine involves serious problems for the whole manufacturing system. In order to prevent such problems, various methods for tool wear monitoring have been proposed, but none of these methods is universally accepted due to the complex nature of this type of phenomena (Byrne et al., 1995; Teti, 1995).

Tool wear monitoring methods have been classified traditionally into direct and indirect ones according to the sensors used in each case. In the direct method, the tool wear is measured directly using different types of sensors such as optical, radioactive, electrical resistance, etc. In the indirect one, the tool conditions are extracted from the value of certain parameters measured during the cutting operation such as forces and torque, motor current and effective power, vibrations, acoustic emission or audible sound (Byrne et al., 1995).

Besides, these methods can be performed on-line, while the cutting process is being carried out, or off-line, by interrupting the process to take measures either on the machine or away from it. These last ones are not very practical, particularly in automatical manufacturing systems.

The development and implementation of on-line direct methods are quite difficult to carry out since there is not a good vision of the tool during the cutting operation. Both, because of the tool movement (in the rotating tool case) and due to chips and cutting fluids. For these reasons, methodologies explored for the on-line direct tool condition monitoring have not received a good acceptation in shop floors.

The last trends in tool condition monitoring are focused on the development of on-line indirect methods. There are several developments for in-process indirect measurement systems based on different types of sensors, either stand alone or integrated. The most generally used sensors are vibration, acceleration, load, power, acoustic emission and force (Haili *et al.*, 2003; Yee *et al.*, 1986; Wilcos *et al.*, 1997; Peñalva and Fernández, 2000; Tolosa and Fernández, 1996). Among them, it is possible to remark the last one, as the most widely employed due to its acceptable results (Leem and Dornfeld, 1996; Tolosa and Fernández, 1996).

In general, extracting information about the tool condition by means of sensors, includes the following steps:

- Cutting test definition
- Tool condition monitoring test definition
- Signal processing method selection
- Process modeler selection
- Experimental layout
- Performance of the cutting and tool condition monitoring tests
- Signal processing
- Tool wear process modelling

However, the bibliography review shows that the information about some of these steps is generally incomplete and, therefore, it can happen that the whole process analysis (cutting/tool wear monitoring tests, signal processing and so on) does not give reliable results because some factors have not been taken into acount.

In this work, the main problems found in the first step, i.e. cutting test definition, are going to be exposed along with the solutions mostly adopted in practice or proposed herewith as best alternatives.

## 3. CUTTING TEST DEFINITION: PROBLEMS AND SOLUTIONS

The correct definition of the cutting tests can be considered as the first point to improve in order to develop and implement an adequate on-line sensor based TCM system. Although there are in the bibliography works that report a very good description of the cutting tests (Teti and Buonadonna, 1999), most of the authors do not give, or not with the desired detail, all the necessary data to be able to make a correct analysis of the results.

A good cutting test definition should include, at least, the following elements. The type of machine tool used: conventional or numerical control. The cutting operation: turning, milling, drilling or grinding. The specified cutting conditions: cutting speed, feed rate, depth of cut, use or not of cutting fluids (indicating in the first case: type, manufacturer and so on). The workpiece characteristics: material, dimensions, surface finish and other special properties which can help to define the experiment. The tools: material, coat material, dimensions and fresh/worn state.

The last item plays a very important role in the definiton of the cutting tests and, particularly, the way of achieving the wear of the tools used in the tests. Thus, the time of use, the used workpiece material or under what cutting conditions such wear has been achieved are considered critical issues since such data are needed to complete the information about the experiment and to obtain a better knowledge about the process which allows to extract more accurate conclusions.

For example, it is important to know the time of use in order to carry out a comparison of the obtained results when the cutting conditions, the workpiece materials or the type of tool are changed. Both if a comparison is made between tests carried out in the same laboratory and if it is made between different laboratories.

In addition, it would be good to have information about the workpiece material used for the obtainment of previous tool wear. In general, nothing is specified about the used workpiece material. For example, if it is the same as the one used for the current tests, if it is another one or if the tool wear is achieved working different types of worpiece materials. This is important since, depending on the used material and the cutting conditions, different variations of the tool geometry can occur. As it has been said above, the main type of wear comes from the adhesion mechanism. This kind of friction consists basically of transfering small particles from the tool to the chips (Boothroyd and Knight, 1989). But, sometimes, the incorporation of macroscopic fragments from the workpiece material into the tool surface can occur (Kendall, 1995). These fragments adhere on the rake face of the tool in two different forms: Built-up Edge (BUE) and Built-up Layer (BUL) as it is shown in figure 2 (Carrilero *et al.*, 2002; Trent, 1989).

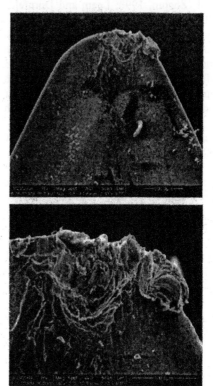

Fig. 2. BUE and BUL obtained by adhesion mechanism.

Both forms are mechanically unstable and can be removed from the tool surface by the action of the cutting forces. The existence of these small particles of material, which are going to partially or totally adhered or be detach during the machining process, can disturb the sensor signal obtained during the tests (Lee et al., 1999).

It is also important, to know the cutting conditions for developing the previous tool wear process. Such information is needed to identify if other types of wear, besides the tipical flank and crater ones, can be present in the tools at the moment of testing. For example, if the previous wear is developed under severe cutting conditions, it will be possible that microcracks appear inside the tool and can cause the catastrophic tool failure even when the tests take place under much litghter cutting conditions. This can produce erroneous conclusions from the analysis to the test results.

If information about time of use, material or cutting conditions for wear development is not available, then it should be good to characterize the tool wear state with other thecniques, as for example Scanning Electron Microscopy (SEM), Energy Dispersive Spectrometer (EDS) or Ultrasonic Analysis (UA) in order to verify the outside and inside alterations of the tool material.

Besides, used tool wear criterion (VB, $VB_{MAX}$, KT or any other one), the number of level for each one, the numerical value of each level should be indicated in order to complete the methodology. In the reviewed literature about tool wear monitoring, basically, only two parameters are considered for characterizing wear: flank wear (VB) and crater wear (KT). The first one has been more widely used (Li and Zhejun, 1998; Leem and Dornfeld, 1998; Li, 2002; Teti and Buonadonna, 1999) than the second one (Teti and Buonadonna, 1999). Besides, in most of cases, only two levels are taken into account: fresh and worn. In this sense, Leem and Dornfeld (1996) say that the fresh/worn dichotomy is impractical for the signal analysis and propose a finer distintion of wear levels. Particularly, they use three tool wear levels (fresh, medium and worn) measured by the main flank wear VB. Grabec (1998) and Li and Zhejun (1998) propose four tool wear levels based on flank wear, as well, to characterize the tool wear. This work philosophy has been applied by other research teams with the KT and the VB wear parameter simultanuously as shown in Teti's work (1999).

Summarizing, in this work a methodology for the establishment of the cutting tests like means to develop effective TCM systems has been proposed. The most important elements involved in the cutting process, such as the ones summarized in table 1, should be declared.

The systematizing and standardization gotten with such methodology will allow scientists to: have a more complete information on the system initially;

Table 1. Main elements to establish in the definition of the cutting tests.

| Element | Type/ Characteristics/Properties | | |
|---|---|---|---|
| Cutting operation | Turning Milling Drilling Grinding | | |
| Machine Tool | Conventional Numerical Control | | |
| Workpiece | Material Dimensions Surface finish Special characteristics or properties | | |
| Tools | Material Coat material Dimensions State | Fresh Worn | Time of use Cutting conditions Workpiece material(s) |
| Tool wear criterion | VB $VB_{MAX}$ KT Any other | | |
| Tool wear levels | Indicating the tool wear criterion Number of levels for tool wear state Numerical value of each tool wear parameter | | |
| Cutting conditions | Cutting speed Feed rate Depth of cut | | |
| Cutting fluids | No Yes | Type Manufacturer | |

characterize the signs better; analyze them in a more integral and more effective way and compare better among experimental work carried out in different conditions or in the same conditions but by different laboratories.

## 4. CONCLUSIONS

The main problems involved in the design of the cutting tests to perform tool wear monitoring tests based on sensors have been analyzed.

The study has revealed, first of all, the need of establishing a methodology which could help in the comparison of the results obtained from different monitoring tests and different laboratories.

As a first approach, the main elements and characteristics which should be contained in such methodology have been established. Particularly, the way of achieving the previous tool wear has been dicussed.

# REFERENCES

Boothroyd, G. and W..A Knight (1989). *Fundamentals of Machining and Machine Tools.* Marcel Dekker, New York.

Byrne, G., D. Dornfeld, I. Inasaki, G. Ketteles, W. König, and R. Teti (1995). Tool Condition Monitoring (TCM) – The Status of Research and Industrial Application. *Annals of the CIRP*, **44**, 541-567.

Carrilero, M.S., R. Bienvenido, J.M. Sánchez, M. Álvarez, A. González and M. Marcos (2002). A SEM and EDS insight into the BUL and BUE differences in the Turning Processes of AA2024 Al-Cu Alloy. *International Journal of Machine Tools and Manufacture*, **42**, 215-220.

Grabec, I., E. Govekar, E. Susic and B.Antolovic (1998). Monitoring manufacturing processes by utilizing empirical modeling. *Ultrasonics*, **36**, 263-271.

Haili, W., S. Hua, C. Ming and H. Dejing (2003). On-line tool breakage monitoring in turning. *Journal of Materials Processing Technology*, **139**, 237-242.

Hayashi, S.R., C.E. Thomas and D.G. Wildes (1988). Tool break detection by monitoring ultrasonic vibrations. *Annals of the CIRP*, **37**, 61-64.

Inasaki, I. (1998). Application of acoustic emission sensor for monitoring machining processes. Ultrasonics, **36**, 273-281.

Jemielniak, K. and O. Otman (1998). Catastrophic tool failure detection based on acoustic emission signal analysis. *Annals of the CIRP*, **47**, 31-34.

Kendall, L.A. (1995). Friction and wear of cutting tools and cutting tool material. In: *Friction, Lubrication and Wear*. ASM Metal Handbook Vol.**18**, ASM Int, Ohio.

Kopac, J. and S. Sali (2001). Tool wear monitoring during the turning process. *Journal of Materials Processing Technology*, **113**, 312-316.

Lee, K.S., K.H.W. Seah, Y.S.Wong and L.K.S. Lim (1999). In-process tool-failure detection of a coated grooved tool in turning. *Journal of Materials Processing Technology*, **89-90**, 287-291.

Lee, M., D.G. Wildes, S.R. Hayasi and B. Keramati (1988). Effects of tool geometry on acoustic emission intensity. *Annals of the CIRP*, **37**, 57-60.

Leem, C.S. and D.A. Dornfeld (1996). Design and implementation of sensor-based tool-wear monitoring systems. *Mechanical Systems and Signal Processing*, **10**, 328-347.

Li, X. and Y. Zhejun (1998). Tool wear monitoring with wavelet packet transform fuzzy clustering method. *Wear*, **219**, 145-154.

Li, X. (2002). A brief rewiev: acoustic emission method for tool wear monitoring during turning, International. *Journal of Machine Tools and Manufacture*, **42**, 157-165.

Nayfeh, T.H., O.K. Eyada and J.C. Duke (1993). An integrated ultrasonic sensor for monitoring gradual wear on-line during turning operations. *International Journal of Machine Tools and Manufacture*, **35**, 1385-1395.

Peñalva, M.L. and J. Fernández (2000). Caracterización del desgaste de la herramienta en procesos de torneado duro de acabado a través de la señal de emisión acústica. *Actas del Congreso de Máquinas-Herramienta y Tecnologías de Fabricación 2000*, 383-396.

Pérez, T., M. SanJuan and G. Paz (2000). Monitorización del estado de la herramienta en procesos de corte en máquinas-herramienta mediante aplicación de la emisión acústica. *Actas del Congreso de Máquinas-Herramienta y Tecnologías de Fabricación 2000*, 415-420.

Rubio, E.M., C. González, R. Domingo and A.M. Camacho (2003). Dimensional precision análisis of circular interpolation with CNC-Milling Machine. *Proceedings of the 7th International Research/Expert Conference. Trends in the development of Machinery and associated technology*, 405-408.

Sánchez-Sola, J.M., M.A. Sebastián, E.M. Rubio, M. Sánchez-Carrilero and M. Marcos (2004). Adhered film formed on coated tools during dry drilling processes of Titanium alloys. 3rd *International Conference on advances in production engineering* (APE 2004), 125-135.

Shawaky, A., T.Rosenberger and M. Elbestawi (1998). In process monitoring and control of thickness error in machining hollows shafts. *Mechatronics*, **8**, 301-322.

Silva, R.G., K.J. Baker and S.J. Wilcox (2000). The adaptability of a tool wear monitoring system under changing cutting conditions. *Mechanical Systems and Signal Processing*, **14**, 287-298.

Teti, R. (1995). A review of Tool Condition Monitoring Literatura Data Base. *Annals of the CIRP*, **44**, 659-666.

Teti, R. and P. Buonadonna (1999). Round Robin on Acoustic Emission Monitoring of Machining. *Annals of the CIRP*, **48**, 47-69.

Tolosa, I. and J. Fernández (1996). Carecterización de la fragmentación de la viruta en operaciones de torneado a partir de la señal de emisión acústica. *Actas del Congreso de Máquinas-Herramienta y Tecnologías de Fabricación 1996*, 1-15.

Trent, E.M. (1989). Metal Cutting, Butterworths, London.

Waschhies, E., C. Sklarczyk and E. Schneider (1994). Tool Wear Monitoring at Turning and Drilling. *Non Destructive Characterization of Materials*, **VI**, 215-222.

Wilcos, S.J., R.L. Reuben and P. Souquet (1997). The use of cutting force and acoustic emission signals for the monitoring the tool insert geometry during rough face milling. *International Journal of Machine Tools and Manufacture*, **32**, 481-494.

Xiaoqi, C., Z. Hao and D. Wildermuth (2001). In-process tool Monitoring through acoustic emission sensing, Automated Material Processing Group. *Automation Technology Division*, 1-8.

Yee, K.W., D.S. Blomquist, D.A. Dornfeld, and C.S.Pan (1986). An Acoustic emission Chip-Form Monitor for Single-Point Turning. *Proceedings of the 26th Machine Tool Design and Research Conference*, 305-312

ELSEVIER
IFAC
PUBLICATIONS
www.elsevier.com/locate/ifac

# EXPERIENCES IN VIRTUAL FACTORY PROTOTYPE: MODULAR PLANTS DESIGN AND SIMULATION

**M. Sacco, S. Mottura, L. Greci, G. Viganò, C. R. Boër**

*VME Group, Institute of Industrial Technologies and Automation (ITIA)*
*National Research Council (CNR), Milan, Italy*

Production systems are changing very fast according with the rapid development of the market, the customer needs and the requirements and the products themselves. Designers and Engineers are asking for supporting tools in order to follow this trend in the proper way; the factory layout should be redesigned (for improving throughput or producing new products) or the factory has to be designed for another location (moving in a more cost-effective countries). ITIA has conducted research on Virtual Reality and Discrete Event Simulation applied to factory design and process simulation. *Copyright © 2004 IFAC.*

Keywords: Virtual Reality, Simulators.

## 1. INTRODUCTION

By defining "product" a system that has to be conceived, designed, tested, realized and -after the use- recycled, the evolutionary trend of the product life-cycle is moving towards the digital life. The life-cycle is supported by SW tools for the logistics, the management of the information, the projects, the 3D modelling, the impact evaluation, the process simulation, the production simulation, etc. With all these factors organized together it can be close to the Digital Factory concept (Maropoulos, 2003), a SW system that leads the "product" from its conceiving to the realization. Today, the Digital/Virtual Factory is often intended as a software system (integrating process simulation and database resources) representing the factory system behaviour, and also as a 3D realtime immersive interactive world. ITIA has started some years ago a research in the context of the Virtual Reality (VR) and Discrete Events Simulation (SI) with the aim to create the Virtual Factory (VF). The VF is a factory in virtual reality where the production process is simulated, both in real-time or offline, by a discrete events simulator. The user is allowed to personally interact, in a immersive real-time 3D environment, with the factory process computed by the simulator. He can navigate and feel himself present on the shop floor. In the VF the user can also create the virtual layout of the production system and then create, almost in an automatic way, the discrete events simulation model. The idea to experience Virtual Factory-like environments has the aim of giving a new tool with which the user doesn't sees the production layout in 2D static view but he can build it, have it and navigate it in 3D and in real-

time. This give the VF the possibility to perform also spaces and environment impact evaluations of the displacement of the plant in the building.

## 2. THE VIRTUAL FACTORY

The research conducted at ITIA on VF has been divided in three main topics (Integration VR-SI and GUI, Object Library and Modularity, 3D simulation), each of them related to a different research project. The first one has been deeply studied in the Eureka project ManuFuturing (Sacco *et al.*, 2000; Boër and Jovane, 1996): in the project a prototypal application was realised in order to check the feasibility of the integration between VR and Discrete Events Simulation (SI) and the technology of the components of the VF system. The second step is represented by the Modular Plant Architecture (MPA) project (Schuh *et al.*, 2003) (www.mpa-online.net). The research activities were on improving the GUI both for the design of the factory (building and production layout) as well the navigation/interaction and for developing a common objects library. The common idea of these two case studies is to have an integration of the SI and the VR (Kelsick *et al.*, 2003) for giving the user a support tool in the production system design. On one side, the process simulation for the plant management, on the other side the VR as a immersive tool for the real-time visualization of the plant itself. The VR meant also as a interactive 3D environment where to work on the configuration of the production layout through a user friendly interface (Mesquita *et al.*, 2000). Last step was performed in the exploitation of the EUROShoE project (www.euro-shoe.net), where the resulting data of the simulation as been sent it back to the virtual environment for the 3D animation and visualisation of the production plant. Each of the three project has produced a prototype system.

### 2.1 The VF first prototype.

The first prototype developed is called VirtualFactorySimulation (VFSI), in the ManuFuturing project. With VFSI the user builds, in a immersive VR environment, the layout of a production system and simulates it with a discrete events simulator (based on Arena – Rockwell Software).

In this case the VR is utilized as a advanced and alternative interface for the layout editing. The research has been focused on integration and the draft of the user-virtual layout architecture. The activities of the user of the VR are summarized in various stages: production units creation, displacement and parameterisation, simulation of the production lines, structural fixing of the virtual layout after the results of the simulation. The mentioned activities are not all necessary performed by the user in a sequential way because the VF is an interactive 3D environment where the user is free to act as he wants at every time with a set of primitive operations. The layout simulation phase, performed by a discrete events simulation, is the only one to occur in a mandatory way in a second time, so the logical operative architecture of the VFSI can be shown in Fig. 1.

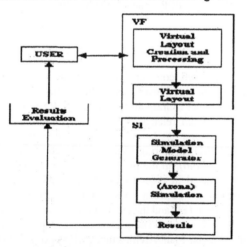

Fig. 1. The VFSI operation architecture.

VF is an immersive 3D environment so the user lives the scene in stereoscopic view on a power-wall wearing DataGloves and position trackers to handle the virtual objects. By using appropriate gestures the user can make appear 3D menus and select and grab from them the right tools for the units creations, like stores, machines, transports, etc. The placing of the entities is performed by grabbing and moving with the hand. In this manner the VF has the role of a man-machine interface for the creation of the layout to be simulated by the SI (Fig. 2). Such an interface, with the gestures language (to point, move, grab, etc.) aims to be an alternative with respect to the traditional

setup process of the simulation model so that it is semi-automatically created from the design layout for the SI that becomes merely a simulation engine.

The user interface of the VF is a 3D menu that can be triggered with a gesture by the user, from which he selects the items by pointing with the index finger or by extracting parts of the main menu as submenus or virtual tools for the creation and editing of the resources of the virtual layout (Fig. 3).

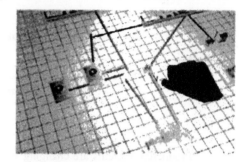

Fig. 2. The VF. The user assembles the shop floor.

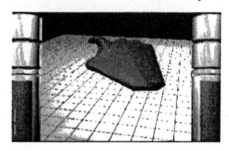

Fig. 3. The user points and grabs the operations from the 3D console.

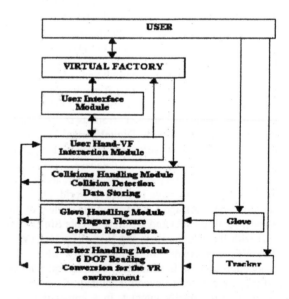

Fig. 4. The relations among the VF SW modules.

The user can create, edit and destroy entities (represented by textured boxes). An important operation is the saving of the layout for the SI on the PC side for automatic simulation generation.

The VF is an application structured in various related modules (Fig. 4), everyone of them is involved in a specific management activity: user interface, user-VF interaction, sensors, DataGloves, collisions. The user interface module deals with the user-VF interaction module in order to know the user-driven DataGloves activities. The interaction model has been designed as a simple model to avoid a particular, extended set of gestures that prevent the user is concentrated on the editing activity -difficult to perform and/or to remember- instead of the factory model/process. Once the creation and editing stage has been closed, the layout is saved to a file in a common proper data format and sent to the PC with the SI module running. The SI module is the SW part that manages the discrete events simulation. It is a Visual Basic application, based on special libraries dealing with the Arena kernel in order to access its functionalities directly from code and not from graphical user interface. This application reads and decodes the file from the VF side. It is an ASCII file subdivided in sections, everyone of them describes the data related to a resource class for the simulation so, for example, there is a section for the raw materials storage, a section for the machines and so on. Every section of the file, depending on the type of described resource, is composed by text strings with the entity data, like name, code, links to external transports lines and so on. The SI application creates the data structures inside Arena kernel that represent the simulation model. In reality, not always the correspondence between virtual layout and simulation model is 1 to 1: some resources created with VF are represented in the SI by structures more complex than the boxes of the VF. This is due to the fact that VF works with the concepts so the user edits the layout by placing the desired resources, the real representation in Arena is hidden for him.

The VF side of the VFSI application has been developed in C++ on a Silicon Graphics (SGI) Onyx2, with Vega (Paradigm Simulation) and Performer (SGI) real-time rendering libraries. The HW for the interaction is a 5th Dimension Technologies DataGloves, Polhemus position trackers, Barco power-wall and Crystal Eyes stereo-glasses.

*2.2 Towards the modularity.*

The first prototype, VFSI, allowed just the feasibility of such a complex system. In the MPA project , the modular design of the VF resources has been deeper evaluated. A object-oriented architecture of a resources database, called Standard Facility Library (SFL) has been designed. The SFL is a data structure common to both the VF side and the SI side and it deals with the VF with a Ethernet LAN and TCP/IP sockets. The VF runs on a SGI workstation as a client side, the SFL runs on a PC, as a server side (Fig. 5).

Based on generic concept and object-oriented technology, a factory is decomposed into the following modules as "building blocks" for factory design and simulation, and as objects in the view of object-oriented technology:
1. Production units that are the machines and equipments, including: processing units (lathe, grinder, punch, ...), transport units (tram,

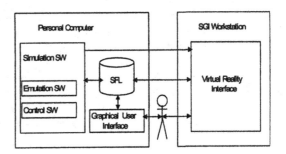

Fig. 5. New VF (VR and SI) architecture.

conveyor, AGV, ...), storage units (sink, container, buffer, ...).
2. Environment units, including: building units (office, meeting room, cafeteria, ...), service units (power supply, air conditioning, lighting, ...), external units (parking area, external stocking area, ...).
3. Production modules, made by production units.
4. Environment modules, made by environment units.
5. Module layouts which place production units and environment units.
6. Factory, made by production and environment modules.
7. Factory layout which places production modules and environment modules.
8. Parts produced by factory.
9. Path which describes the route of transport units.

For each factory resource object, the structured information (attribute) and the behaviour (method to manipulate the resource object) have been defined and stored in the SFL. The attributes cover:
1. Spatial description and layout information for virtual reality, such as position, connection, distance, etc.
2. The emulation/control capabilities of the resource object, in terms of: events generated by its emulation model (in order to notify the occurrence of a given event to the control software); supported commands (for actuating the control software decisions); possible states (defining the internal behaviour of the resource).
3. Other parameters for virtual reality and simulation, like speed, production capacity, for example.

The main purpose is to provide a friendly 3D interface for factory building or shop floor design under virtual reality environment. The user will select the desired resource objects from the SFL, which have been stored previously, and will put them on the virtual factory floor, connecting outputs to inputs and verifying the hypothetical layout against the existing spatial constraints.

In the case of a shop floor design, the layout of production resources will be arranged well with consideration of factory environmental constraints. For each production resource, which will be used by the emulation and control software, an interface is implemented to prompt user to enter the proper parameter values if absent (i.e. cycle time, resource capacity, etc.). The values are stored in the SFL. In the case of factory building design, the main purpose of the virtual reality interface is that of allowing the user to navigate in the building and obtain an immediate and realistic feedback of design choices and change accordingly.

In this new Virtual Factory "version" the different aspects of the factory, infrastructures and machines, has been classified and modularized in the SFL. The application can ask SFL for data modules, with their own private attributes.

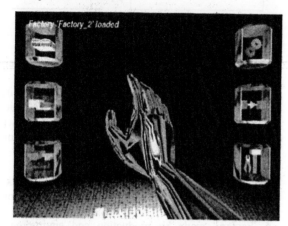

Fig. 6. The new 3D GUI.

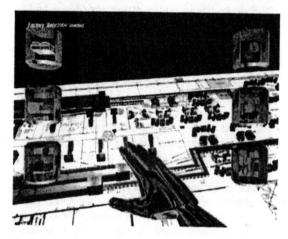

Fig. 7. Virtual shop floor.

The VR user interface has been enhanced for a better usability. The menus are still 3D consoles that expand in inner levels but can better support options and items lists (as 3D objects) more extended. The virtual layout editing capabilities have been enhanced with replication, resizing, snap to grid operations and so on. Also this Virtual Factory application has been developed in C++ on SGI Onyx2, with Vega and Performer libraries and the same HW as the first one.

*2.3 Integrated Pilot Plant.*

On the base of the previous steps ITIA is developing a virtual factory for the customized shoe production, within the EUROShoE project, where has been created in Vigevano (Italy) an automatic and modular plant for the customized shoe production. The same production line can build different kinds of shoes by reconfiguring the description of the production process. In this context the Virtual Reality Integrated Production Plant (VR_IPP) has been developed.

VR_IPP is a VR environment (with stereoscopic desktop and/or power-wall visualization) where the building and the shoe production plant are both modelled. It can be navigated in real-time and it can be linked with the simulator, SI, (Arena) for the execution of 3D simulations.

Actually the system is configured as follows: the layout of the plant is loaded into SI module. The simulation is performed and the operations of the simulated entities and resources is saved on a file so we can have the history of the whole production process (Fig. 8). In the VR_IPP, the user loads and runs the simulation and consequently he can see the production cycle of the shoes, from the initial store to the finished product at the end of the line. During the simulation, obviously, the user can navigate and also utilize some static point of view on the shop floor for a better observation.

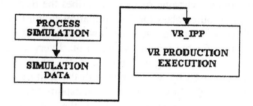

Fig. 8. The simulated production is inputted into the VF.

To do the simulation in the VR_IPP, the existing layout on the SI side has been coupled with custom Visual Basic application code so that every entity is tracked during the simulation time. Then, in the SI simulation time, all the operations performed by the components are written to a file.

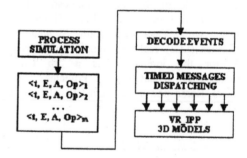

Fig. 9. The simulated events are dispatched to the virtual shop floor components.

Generally speaking, those operations are saved as time events in a quadruple $<t, E, A, Op>$: at the time instant $t$, the entity E does an operation A on a set of operands Op, with $Op=\{op_1, op_2, ..., op_n\}$. On the VR_IPP side, a SW module reads the file and translates the quadruples in statements to be scheduled to the 3D entities by following the right time elapsing (Fig. 9).

Since the real IPP plant is, as stated above, modular and re-configurable, in the VR_IPP the user can choose the parts of the plant to use and in this way he defines a customized production path. The path is saved in order to be simulated on the SI side, for testing that customized product. As a inspection utility, during the simulation in VR_IPP the alphanumerical codes of the entities are shown in 3D, near the related 3D models. Also the history of operations that are performed can be visualized in a real-time 2D-overlayed mode as a enhancement of the information about the virtual production process.

The 3D models of VR_IPP are the original ones of the CAD projects of the physical parts both of the building and of the shop floor. The models have been processed to reduce the geometrical complexity for the real-time rendering and to enhance their visual aspect with diffusive, specular, light-map and bump textures. Then they have been managed in hierarchies where every node is a part of the geometry related to a particular local reference system. With this design of the 3D environment, the SW module for the quadruples decoding and for the statements dispatching for the real-time animations, has been developed.

The user interface of VR_IPP is not composed by 3D menus, as in the previous Virtual Factories, but it is a overlay 2D menu on the 3D scene (Fig.10). The user plays with the mouse to activate, utilize, and deactivate the menu. The menu is not a Windows GUI-based, it is composed by polygons and textures. The hierarchical configuration of the menu items and the actions related are created in XML with a simple custom editor of ITIA. The VR_IPP application loads the XML, creates the menu and manages the GUI user interaction by using the data stored in the XML. For this type of application the overlayed GUI has been more simple to use because it avoids the collision detection precision problems, proper of the 3D virtual hand and the 3D menu.

Fig. 10. The 2D overlay menu.

With VR_IPP ITIA has disengaged itself from the commercial technologies. For the previous VF prototypes the VR environment was running on SGI workstation and the real-time 3D SW application was layered on commercial libraries. This kind of resources are enough cost expensive and are not friendly with an hypothetical end-user.

The SW application of VR_IPP is based on a custom library of ITIA, based directly on OpenGL, thus it is free of costs for the runtime. It also runs on a personal computer since the graphical HW performances, today, are competitive and the PCs are at a very lower cost rather the workstations. These factors can help in dealing with end-users, for example the small and medium enterprise, usually not easily ready for expensive investments.

Fig. 11. The VR_IPP environment.

## 3. CONCLUSIONS AND FURTHER DEVELOPMENTS

During the design and development of the aspects of interface, 3D models, simulation, ITIA has spent in the first approaches about 2 man-years, and for the last one (see 2.3) 6 man-months. By enhancing the tools the target is to reach 3 man-months.

The three steps approach has allowed ITIA to verify the feasibility of the VF idea and to create the main components. Today, the user through the VF system is able to design a complete plant (both the building as well the production plant), create the discrete events simulation model, almost in an automatic way, and finally to see the results both in the 2D simulation environment (mainly for statistical data) as well in a 3D virtual environment. The system, a SW prototype, requires now an engineering phase and software tuning in such a way it will become in the next future a good product for supporting production plant design and simulation.

## REFERENCES

Boër, C. R. and F. Jovane (1996). Towards a new model of sustainable production: ManuFuturing. *CIRP Annals*, **STC O 45/1/1996**, p. 415.

Kelsick, J., J. M. Vance, L. Buhr and C. Moller (2003). Discrete Events Simulation Implemented in a Virtual Environment. *Journal of Mechanical Design*, **Vol. 125, Issue 3**, pp. 428-433.

Maropoulos, P. G. (2003). Digital enterprise technology – defining perspectives and research priorities. *International Journal of Computer Integrated Manufacturing*. **Vol. 16, No. 7-8**, 2003, pp.467-468.

Mesquita, R., J. Dionisio, P. Cunha, E. Henriques and B. Janz (2000). Usability Engineering: VR interface for the next generation of production planning and training systems. *Computer Graphics Topics*, **Issue 5**, pp. 17-18.

Sacco, M., S. Mottura, G. Viganò, A. Avai and C.R. Boër (2000). Tools for the innovation: Virtual Reality and Discrete Events Simulation to build the 2000 Factory. *Proceedings of AMSMA 2000*, Guangzhou, P.R. China, pp. 458-462.

Schuh, G., H. Van Brussel, C. Boër, P. Valckenaers, M. Sacco, M. Bergholz and J. Harre (2003). A Model-Based Approach to Design Modular Plant Architectures. *Proceedings of the 36th CIRP International Seminar on Manufacturing Systems*, Saarbrücken, pp. 369-373.

BIBLIOGRAPHY

Jones, K. C., M. W. Cygnus, R. L. Storch and K. D. Farnsworth (1993). Virtual Reality for Manufacturing Simulation. *Proceedings of 25th Conference on Winter Simulation*, Los Angeles, California, USA, pp. 882-887.

Mecklenburg, K. (2001). Seamless Integration of Layout and Simulation. *Proceedings of 33nd Conference on Winter Simulation*, Arlington, Virginia, USA, pp. 1487-1494.

Quick, J. M., C. Zhu, W. Haibin, M. Song and Dr. W. Mullër-Witting (2004). Building a Virtual Factory. *Proceedings of 2nd International Conference on Computer Graphics and Interactive Techniques*, Singapore, China, pp. 199-203.

Rohrer, M. W. (2000). Seeing is Believing: the Importance of Visualization in Manufacturing Simulaiton. *Proceedings of 32nd Conference on Winter Simulation*, Orlando, Florida, USA, pp. 1211-1216.

Son, Y. J., A. T. Jones and R. A. Wysk (2000). Automatic Generation of Simulation Models from Neutral Libraries: an Example. *Proceedings of 32nd Winter Simulation Conference*, Orlando, Florida, USA, pp. 1558-1567.

Zhai, W., Fan Xiumin, Yan Juanqui and Zhu Pengsheng (2002). An Integrated Simulation Method to Support Virtual Factory Engineering. *International Journal of CAD/CAM*. **Vol. 2, No. 1**, pp. 39-44.

ELSEVIER
IFAC
PUBLICATIONS
www.elsevier.com/locate/ifac

# A METHOD FOR LIFECYCLE ORIENTED DESIGN OF PRODUCTION SYSTEMS

**Günther Schuh, Nils Wemhöner, Achim Kampker**

*RWTH Aachen University*
*Laboratory for Machine Tools and Production Engineering (WZL)*
*{g.schuh, n.wemhoener, a.kampker}@wzl.rwth-aachen.de*

In a world of increasingly dynamic markets, production systems have to offer the optimal degree of flexibility in order to achieve optimal performance over their complete lifecycles. Current production systems, however, are often inflexible, partly because it is still hard to quantify flexibility and its benefits in practice. Against this background, the authors present a method for lifecycle oriented design of production systems based on a simplified systems engineering approach. The method makes it possible to explicitly take into account risk and uncertainty when designing production systems and to therefore take informed decisions to implement the optimal degree of flexibility. *Copyright © 2004 IFAC*

Keywords: Design, evaluation, lifecycle, manufacturing systems, modelling, productivity

## 1. INTRODUCTION

Product technologies change rapidly. New innovations are introduced ever more often. The length of product lifecycles has decreased by about 30 % during the last ten years. Not only are more completely new products being pushed into the market more often; also revisions become more frequent. The number of variants in production increases. Additionally, proceeding globalisation leads to more volatile demands. Market dynamics are high – and keep increasing.

New technologies in production are introduced more and more often. These technologies keep becoming more complex and more expensive; capital investments per worker have tripled over the last three decades.

Increasing – not only local, but worldwide – linking of different manufacturing enterprises helps, on the one hand, to achieve flexibility in production; on the other hand, it leads to an increase in uncertainty and thereby in the complexity of the planning process.

This environment leads to higher risks in medium and long term planning of production systems. The ability to adapt to changing requirements quickly enough represents an increasingly important competitive requirement, particularly for innovative and therefore especially dynamic industry branches (Schuh, *et al.*, 2000; Wittmann, *et al.*, 2000). Enterprises have recognised this and have been trying to build up and operate flexible production systems (Warnecke, 1996).

Most factories focus on a specific production programme (Ueda 2001). They have to follow the market for the products in this programme. In a dynamic and often unpredictable market, this makes proactive adaptation all but impossible. Instead, they can only react – resulting in a continuous struggle for adaptation (figure 1) (Wiendahl, 2002). This struggle ties many resources and leads to permanently suboptimal operation of factories.

Therefore, methods are needed for effective consideration of dynamics during the planning phase in order to be able to design systems that are capable of handling the situation a factory operates in best and to make continuous adaptation unnecessary (if possible) or easy and cheap; section two of this paper illustrates the need for lifecycle oriented planning. Basic requirements to methods for lifecycle oriented

Fig. 1. Risk of volatile markets (Schuh, *et al.*, 2000).

planning are presented in section three of this paper. Section four contains the description of the methods, while section five describes key figures developed for use with this method. Section six concludes this paper.

## 2. THE REQUIREMENT FOR LIFECYCLE ORIENTED PLANNING

Production systems have to offer a certain degree of flexibility in order to achieve optimal performance over their complete lifecycle. Enterprises have identified the need for flexibility in principal; however, when it comes to making investments, they are often unwilling to spend money for flexibility, leading to short term decisions that are not optimal in the long run (Reinhart 1999). The right degree and the right kind of flexibility are often not identified; companies instead maintain large amounts of overcapacity, referring to it as "flexibility". This practice leads to lost money, either for overcapacities, or for lost opportunities or reactive adjustments that might become necessary. All three reduce the competitiveness of the company.

In current production system planning, often only initial investments, i. e. the costs for building the production system and for operating it to produce the first product(s) it has been designed to produce, are taken into account for decisions. Further products which might be produced after this first one are normally not or insufficiently conidered for the calculation of economic performance. Therefore, the factory concepts chosen are often rather cheap initially, but expensive if their complete lifecycle is compared to other concepts; the current practice of considering only a part of the system's lifecycle prevents investments in flexibility when even a slight increase in initial costs might lead to a substantial cost reduction over the system's lifecycle. Therefore, in many cases it is not the system that is optimal in the long term that is chosen. The fact that currently investments for adaptations throughout the lifecycle reach about 50 % of the initial investment costs shows that this is a relevant consideration.

This problem could be avoided by taking uncertain developments such as market uncertainties into account for production system planning. Why is it not done? There are many reasons.

One of them are strict cost targets set by the management for early stages of the planning process: current business practice is to calculate with an amortisation time of one or two years. If short amortisation times for investments are a necessity, there is no money for flexibility.

Besides the lack of determination to invest in flexibility that these targets show, there is another, maybe even more important reason that prevents investments in flexibility: it is very hard to quantify its effects and therefore its benefits. Therefore, no explicit and clear trade-off can be made between more flexible and less flexible production systems. The benefit of flexibility cannot be "proven". Therefore, companies need measures and methods for lifecycle oriented production system planning; in particular, a method to determine the optimal degree of flexibility is needed.

## 3. REQUIREMENTS TO PRODUCTION SYSTEM EVALUATION METHODOLOGY

For long term optimal decisions, a methodology is needed that can give an overview over the long term economic performance of a production system. It must evaluate costs and benefits of flexibility. More specifically, it must fulfil the following requirements:

1. It must give an overview over the cost occurring during the complete lifecycle of a production system (not only the lifecycle of the first product being produced by the system).

2. A quantified evaluation of the risk of different production system alternatives is necessary. Both the probability and the effects of risks have to be taken into account for this evaluation.

3. Market dynamics must be modelled to the extent possible. Also uncertain market developments have to be modelled and their probability quantified; they must not be neglected but explicitly taken into account for evaluation.

4. The calculation and analysis of the economic performance of different production systems in the case of various market developments must be possible.

5. Adjustments of the production system must be taken into account for evaluation. This applies both to adjustments definitely planned at the time of planning of the production system and to adjustments that are made during its lifecycle as a reaction to market requirements.

In order to ensure the applicability of the methodology in everyday use, the following additional requirements must be fulfilled:

6. The methodology must be easy to use. In many enterprises, people think about flexibility. However, their thoughts are not used because they are too complex or otherwise impractical.

7. All data needed for application of the methodology must be easily available.

8. The methodology shall motivate enterprises to invest money into flexibility. In particular, it shall support the view of flexibility as an "insurance against risk".

9. The methodology shall help to spread explicit handling of flexibility to as many enterprises as

possible. This will make it possible to plan and use flexibility adequately also in production networks.

10. It shall help to standardise lifecycle oriented evaluation, planning and controlling procedures.

## 4. METHOD FOR LIFECYCLE ORIENTED PLANNING OF PRODUCTION SYSTEMS

On the basis of the presented requirements, a method for lifecycle oriented planning for production systems was developed in the research project LicoPro ("Lifecycle design for global collaborative production")(Eversheim 2001).

Figure 2 gives an overview of the methodology. It consists of three modules: strategy module, configuration module and evaluation module. In the strategy module, information about market developments and about the enterprise and its available production resources are collected. One of the outputs are scenarios of demand figures over time for all products considered, with associated probabilities. The configuration module is then used to assist in the configuration of production system concepts; the configuration is based on the information collected in the strategy module. The evaluation module computes the economic performance of each of the production system alternatives conceived in the configuration module over its complete lifecycle for each demand scenario from the strategy module. In the next step, key figures are computed from the results of this evaluation that characterise a) the expected economic performance of each production system considered and b) the associated risk. Based on these figures, the production system to be implemented can be chosen by decision makers in the company.

The method also allows optimisation of production system alternatives: in case no sufficiently good system has been found in the first run, targeted modifications can be made based on key figures developed specifically for the analysis of system alternatives. The modified production system alternatives will then be evaluated in a further simulation run.

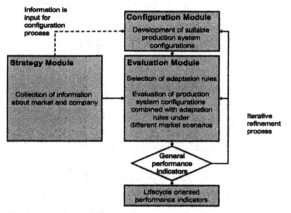

Fig. 2. Method for lifecycle oriented planning of production systems.

In the following subsections, the modules are described in more detail.

### 4.1 Strategy module

There are three categories of information about market developments and about the enterprise that are collected in the strategy module.

*Information about the market.* Sales figures over time for all products produced in the production system considered are needed to describe the market development. However, in practice they are normally not available; today, with highly dynamic markets, even less than in the past. Therefore, enterprises often only consider an "average case", along with best and worst cases. These three cases build the basis for their planning activities. The latter two are normally only used for sensitivity analyses, however; they are not considered in the economic performance calculations. In an increasingly dynamic environment, the probability that these average figures actually become reality is low. Nevertheless, all other available information about markets is neglected. This leads to decisions that are balanced towards certain production systems; however, the tendency is not due to reality, but only to the fact that the available data is filtered – unintentionally, rather than on the basis of well-funded deliberation.

In order to increase the quality of the decision, the authors therefore suggest to take into account many scenarios when making investment evaluations (Gausemeier 1996). Since information about possible future scenarios is not commonly available in enterprises, the authors suggest creating scenarios from individual events that may occur in the future and from their probabilities; this information is normally available. The probability of a scenario can be computed from the probabilities of the underlying events. These events are, of course, uncertain; therefore, the information created from it is also uncertain. However, considering multiple futures ensures that all data about the future that is available in the company is actually used for the decision process; this is the basis for a decision that is as good as possible.

When using the method, the first step is to choose time horizon and granularity. The time horizon generally depends on the planning task at hand; normally, it would be the complete lifecycle of the production system evaluated. The next step is to determine the number of scenarios considered. There is a trade-off between running time of the calculation and the degree of exactness to which the future is modelled. Subsequently, events are determined along with associated probabilities. Finally, the effect of events on the market – and, more specifically, on the sales figures of the products considered – has to be modelled.

The so called ScenarioWizard then computes $m$ market scenarios $S_1$ to $S_m$ from these data; these scenarios describe the production figures over time. Each scenario includes not only the production figures for one product, but for a complete production task, i. e. for all products considered (including their variants), and over the complete timeframe considered. For each period (for example, for each year), demand figures for each considered product are computed. This way, also product ramp-ups and introductions of new models can be modelled.

*Information about available production resources.* In this part of the module, information about production resources available inside and outside the company or the production network are collected and aggregated.

*Information about company preferences.* In this part of the strategy module, information about philosophies, strategies and concepts is collected. One example is the risk preference of the company. The data collected here will later (in the evaluation module) be used to achieve a weighted cost-benefit-analysis of the considered production systems.

### 4.2 Configuration module

Based on the information collected in the strategy module, production systems are configured in the configuration module. During the research project LicoPro, heuristics and solution catalogues have been developed to support this task. However, they are not in the focus of this article; therefore, they will not be described here in further detail.

### 4.3 Evaluation module

The goal of the evaluation module is to aid decision takers with their decision for one production system or another. In particular, it shall help to answer the question about the right degree and the right type of flexibility to be implemented. Note that even under lifecycle considerations, "right" does not necessarily mean "high"; important is that the flexibility fit the requirements of enterprise and production task. In some cases, the optimal production system under flexibility considerations can be inflexible and cheap.

The evaluation module evaluates the economic performance of each production system alternative combined with each market scenario, either via calculation or via simulation. The figure used to characterise economic performance is the net present value (NPV; see below for a description of the usage of the NPV in a production environment); the result of all calculations is integrated into an $(n,m)$-matrix $E$ containing the "fitness" of all production system alternatives.

Taking into account the probabilities associated with each market scenario, aggregated economic performance indicators for each production system alternative can be computed from this matrix. One

key figure is the averaged NPV (aNPV, see below) which characterises the expected performance of a production system alternative. For computing this figure, hence, all market scenarios are considered, not only in the sense of a sensitivity analysis, but as input affecting the economic performance of a production system alternative. In addition to the aNPV, figures characterising worst and best case performance, standard deviations etc. can be computed.

Based on these figures, risk indicators for each alternative can be calculated. Risk in this case means the loss probability of a production system. Losses can occur either if too much is invested into flexibility and if therefore flexibility remains unused; or if not enough flexibility is available and market chances cannot be exploited, resulting in lost opportunities.

It can be shown that in certain cases the investment into flexibility results in decreased loss probabilities, reflected either in better average performance or in better worst case performance. (Of course, it is also possible that best case performance is positively affected by installing flexibility.) Depending on the risk preference of the enterprise taking the investment decision (this preference has been identified in the strategy module), one production system or another might be preferable in this case.

In general, the developed method supports the view of flexibility as an "insurance against risk": investments in flexibility are, in this view, premiums enterprises have to pay to insure themselves against risks (Baumgarten 2003). As described above, the level of this premium has to be determined by the company in accordance with company philosophy and especially risk preference. The evaluation module support companies in this decision.

## 5. KEY FIGURES

In this section, the authors present three measures for the flexibility of production systems that are used in the evaluation module described above: net present value, averaged net present value, and value at risk. The usage of these figures for the evaluation of production systems has been prepared by the DaimlerChrysler research centre in Ulm (Bürkner, et. al., 2004).

### 5.1 Net present value (NPV)

Needed is a key figure that gives "an overview over the cost occurring during the complete lifecycle of a production system" (see requirement one).

As key figure for lifecycle oriented measuring of economic profitability, the net present value (NPV) can be used. The NPV is a financial key figure that puts in relation future benefits from and future efforts for a certain project, in this case for the implementation of a production system. More concretely, for the calculation of the NPV, all revenues and expenditures are discounted to the current moment and expenditures are subtracted from revenues. Essential for a realistic evaluation is the correct choice of the interest rate $r$. If the NPV is zero, the implementation of the evaluated production system is just as profitable as a fixed-interest capital investment with interest rate $r$; if the NPV is positive, the return on investment for the production system is better than that for the fixed-interest capital investment.

The calculation of this hypothetical value makes a comparison of the economic performance of different production systems possible: the performance of the production system with the higher NPV is better.

The NPV is suitable for lifecycle evaluation since it takes into account expenditures and revenues over any arbitrary period of time and discounts is appropriately. The NPV is computed as follows:

$$NPV = \sum_{t=0}^{T} \frac{Z_t}{(1+r)^t} \qquad (1)$$

with $t$ = current period, $r$ = interest rate, $Z_t$ = cash flow in period t, and $T$ = considered time horizon.

In addition to the NPV, an averaged capital value (averaged net present value, aNPV) can be calculated on the basis of the production figures planned (Bürkner, et. al., 2004). The aNPV is the result of a calculation in which the capital values associated with different scenarios of the planned production figures are weighted be their probability and thus represents the average value: it describes the expected NPV of a production system under consideration of market uncertainties.

*5.2 Value at risk (VaR)*

The second requirement to our methodology states that "a quantified evaluation of the risk of different production system alternatives is necessary. Both the probability and the effects of risks have to be taken into account for this evaluation."

The suitable key figure for this purpose is the value at risk (VaR) (Holton, 2003; Jorion, 1997). It describes a possible financial loss. The description of a VaR must always include a probability level and a time span; a typical value at risk statement is, for example: the VaR is € 2 million with a probability of 99 % and a timeframe considered of three years. This statement means that during the upcoming three years, the loss will, with a probability of 99 %, not exceed an amount of € 2 million.

There are some essential remarks concerning the VaR to be made:
- The value at risk is a one-sided, loss-oriented risk measure. In contrast to the two-sided fluctuation risk that can be measured, for example, by figures such as standard deviation, possible gains are not considered. Only the risk of losses is quantified.
- The value at risk is based on a probability distribution for future losses; it is a future-oriented, monetary risk evaluation figure.
- The value at risk does not, as its name suggests, indicate the complete value that is at risk, i. e. the maximum loss. It merely represents a loss limit that is exceeded with a small probability only (in the case of the 99 % value at risk, with a probability of 1 %). The maximum losses that can occur are hence ignored. However, these losses can nevertheless occur; this is the so-called remaining risk.
- The value at risk is no consistent method for the measurement of risks, meaning that different methods for computing the value at risk deliver different results. It is nevertheless suited for the comparative analysis of different production systems, provided that the same method of evaluation is used for all systems.

So far, the value at risk is mainly used in the financial sector (by banks and insurance companies) for measuring quantifiable risks (for example, stock risks). The authors suggest to also use the value at risk for the evaluation of the risk associated with the implementation of different production system alternatives (Bürkner, *et al.*, 2004). It can help to answer the question: what will happen in the worst case (for example in the case of no flexibility in the production system to react to market changes or other risks)? The use of the value at risk for the evaluation of different production system alternatives is, however, a challenge. It is particular difficult to get suitable data about risk probabilities in the future. Its application for production planning tasks is being further investigated by partners of the research project LicoPro.

## 6. CONCLUSION

The requirement for lifecycle oriented planning of production systems has been argued, and requirements to a method for lifecycle oriented design of production systems have been identified. A method for lifecycle oriented design of factories has been presented; it makes it possible to explicitly take into account uncertainty when designing production systems. It thus helps to achieve optimal performance over the complete lifecycle of a production system.

The described method has been implemented in an IT tool and successfully tested with industry partners in the LicoPro research project. It has proven useful and usable for practice.

## 7. ACKNOWLEDGEMENTS

We extend our sincere thanks to all partners in the research project LicoPro („Lifecycle design for global collaborative production"), who have contributed significantly to the development of the results described in this paper.

LicoPro is funded by the EU under the IST programme (contract number IST-2001-37603).

## REFERENCES

Baumgarten, H.; T. Sommer-Dittrich; M. Friese (2003): Einsatz von Realoptionen zur effizienten Simulation wandlungsfähiger industrieller Strukturen. In: *Simulation und Visualisierung,* **2003**, p. 9.

Bürkner, S.; J. Roscher; M. Friese (2004): Modellbasierte Gestaltung und Bewertung von Wandlungsfähigkeit in der Automobilindustrie. Submitted for publication to *Industrie-Management*

Eversheim, W.; M. Sesterhenn; J. Harre (2001): LicoPro – Lifecycle-oriented design of flexible and agile production systems. *Proceedings of the international IMS project forum.* Ascona

Gausemeier, J.; A. Fink; O. Schlake (1996): *Szenario-Management – Planen und Führen mit Szenarien. 2. bearb. Auflage.* Hanser, München.

Holton, G. A. (2003): *Value at Risk: Theory and Practice.* Academic Press, San Diego.

Jorion, P. (1997): *Value at Risk: The New Benchmark for Controlling Market Risk.* Irwin, Chicago.

Reinhart, G.; S. Dürrschmidt; A. Krüger (1999): Stückzahlflexible Montage- und Logistiksysteme: Integrierte Planung kapazitätsflexibler Systeme. *wt Werkstattstechnik online,* **1999**, pp. 413-418

Schuh, G., Th. Friedli, P. Kunz (2000): Diskontinuitäten auf dem Weg zur Produktion der Zukunft. *Industriemanagement 16,* **2000**, pp. 23-28.

Ueda, K.; A. Markus; L. Monostori; H. J. J. Kals; T. Arai (2001): Emergent Synthesis Methodologies for Manufacturing. *Annals of CIRP,* **2001**, pp. 535-551.

Warnecke, J.: Fabrikplanung. (1996) In: Betriebshütte – Produktion und Management, Teil 2. (Eversheim, W., G. Schuh (Eds)) Springer Verlag, Berlin.

Wiendahl, H.-P.; S. Lutz (2002): Production in Networks. *Annals of CIRP, 2002,* pp. 1-14

Wittmann, E.: Integriertes Risikocontrolling im Konzern. (2000) In: *Strategische Steuerung – Erfolgreiche Konzepte und Tools in der Controllingpraxis.* (Horváth, P. (Ed)) Schäffer-Poeschel, Stuttgart.

ELSEVIER

IFAC

PUBLICATIONS
www.elsevier.com/locate/ifac

# MANUFACTURING PROCESSES FOR INDUSTRIAL DISASSEMBLY: JETTING TECHNOLOGIES

**Eckart Uhlmann, Jan Dittberner, Adil El Mernissi**

*Institute for Machine Tools and Factory Management. Technical University Berlin, Germany*

In times of increasing shortage of natural resources and more environmentally influenced legislative restrictions industry will change its face. More emphasis will lie on disassembly technologies that ease remanufacturing to prolong a products life. Within the Collaborative Research Centre 281 research is done concerning the use of manufacturing processes in disassembly. Separating processes are divided into non-destructive and destructive. Non-destructive technologies are mandatory for the disassembly of re-usable parts. Destructive processes help to get fast access to re-usable parts, and often mean the fastest method to isolate material. Within these, two of the most efficient are water jet cutting and plasma arc cutting.
*Copyright © 2004 IFAC*

Keywords: Manufacturing processes, environment, ecology, maintenance

## 1. INTRODUCTION

In a world of increasing industrialisation and developing economies the amount of natural resources will thin out sooner or later (v. Weizsäcker *et al.*, 1997). Thus it is necessary to find new ways of producing goods. One solution is to recover parts, components, and material from used products by implementing recycling. An important step within the recycling run is the disassembly of the product. Depending on the value of the various parts of a product the component is either suited for re-use or for re-utilization. Though there are already some companies already working in the field of recycling there is no automated or partly automated disassembly industry yet. The focus lies on re-utilising material. Therefore, first a rough separation takes place with mostly simple, handheld tools. Then the rest is put into a shredder and subsequently sorted by a centrifuge. The achieved material purity is very poor.

To install a highly effective industrial disassembly, it is possible to use the existing separating manufacturing technologies for the disassembly operations. Depending on the recycling purpose of the component that has to be dismantled two groups of processes are available. To extract re-usable parts non-destructive technologies such as unscrewing, dismantling, and evacuating are required. Destructive processes and tools are used to separate material or to provide an easy access to re-usable components. They can be divided into processes that cause secondary damages and those without. The advantage of using secondary damage free processes such as sharing, splitting, or ripping apart is the lack of subsequent post treatment such as cleaning or replenishing. Disadvantages are high emerging process forces, slow progress, and resulting demolitions of the processed part as well as back stroke impacts on tools (Spur *et al.*, 1997).

Quality criteria for destructive disassembly processes with secondary influences are short processing times, economical operation and minimised secondary influences of re-usable parts. Secondary influences range from minor contamination on surfaces to the destruction of whole parts or from harmless gaseous particles to hazardous aerosols. They can consist of sparks, scrap material, slag, chippings, splinters, dust, steam, and vapour.

In the scope of the sub-project A1 "Separation of Insoluble Connections" of the Collaborative Research Centre 281 (CRC 281) a number of destructive processes and tools was examined. First the manufacturing technologies were theoretically grouped and their usability for disassembly estimated. The processes considered applicable were adapted and optimised according to the disassembly specific quality criteria. Among the examined processes that cause secondary influences were the jetting technologies water jet cutting, laser beam cutting and plasma arc cutting. The laser technology proved to be economically not competitive (Feldmann, et al., 1996, Uhlmann et al., 2002a, Uhlmann etal., 1999b).

Since there is no connection between tool and work piece jetting processes have the advantages of a nearly force free mode of action, of an omni directional machining, and of the free choice of starting and end point. In the following the accomplished research on both processes, within the mentioned sub-project, is described.

## 2. WATER JET CUTTING

At the end of the 19th century low pressure water jets were used to wash stone and other hard materials out of the ground. Today, with the development of ever more effective and powerful compressors and pumps ultra high pressure water jets are available. The range goes up to almost 4,500 bar. Combined with an abrasive medium it is possible to process almost every material. This so called abrasive water jet cutting was invented in the 1960ies and had its industrial breakthrough in the late 1980ies. Since then more and more applications were found.

### 2.1 Process

Water jets are produced by transforming the energy of pressurised and compressed water into the kinetic energy of the jet. This transformation takes place at a nozzle. The diameter of the nozzle determines the size of the jet and, together with the pressure, the application of it. Broad jets at low, middle, or high pressure are used for cleaning or deburring operations. With narrow jets at ultra high pressure is it possible to cut any given material. In the following, the paper will focus on ultra high pressure water jets and call them water jets (Uhlmann et al., 1999a).

There are three variants of water jet cutting: pure water jet cutting, abrasive water injection jet cutting, and abrasive water suspension jet cutting. For all three the generation of the pressure base on the principle for pure water jets. A hydraulic pump generates a primary pressure of approximately 100 bar. The oil is led to a bidirectional working pressure transformer. There the pressure of the hydraulic unit is transferred to de-ionised water from

the tab. The used transformation factor has of range of 1:30 to 1:45. To high pressurised water runs trough a pulsation damper to create a steady flow and leaves the system at the nozzle. Then meets with the work piece and cuts it. Due to the fact that not all kinetic energy is absorbed in the cutting process the residual jet is stopped by a catcher. It is filled by water or by balls of metal or glass. Pure water jets are capable of cutting soft materials such as food, wood, textile, rubber, synthetic foam and insulating compounds.

Abrasive water jets are created by adding an abrasive aid to the pressurised water. With abrasive injection water jets this done directly after the nozzle in a mixing chamber. The mix has then to go through a focussing tube to assure that particles and water are unidirectional. The injection system is the technological most relevant process variation. It is used for cutting hard materials such as steal, non-iron metal, or titanium in nearly all fields of industry. The examinations on adjusting water jet cutting to the requirements of disassembly were carried out on a system working with the injection principle (Uhlmann et al., 1999a).

For abrasive suspension water jet cutting an abrasive is added to a pressurised reservoir of water. This reservoir is used as a sub-stream from the main stream. The suspension is then added to the main stream prior to the nozzle. Compared to the injection system the suspension system provides a better mixing of water and abrasive, wastes less of the kinetic energy, and is much more subject to wear.

### 2.2 Experimental studies

A robotic abrasive water jet cutting device from Ingersoll Rand, Bad Nauheim/Germany was used for the examination of water jet cutting in disassembly. Again metallic materials were processed. The maximum working pressure was 3,200 bar, the average abrasive mass flow of 350 g/min and an abrasive garnet mesh 80 was used.

First, the focus is on the optimization of the maximum traverse rate where a work piece is cut through. Depending on the material and the work piece thickness, a model was build to predict the maximum traverse rate. Beside the material properties, the working distance, abrasive mass flow and the working pressure quantifies the traverse rate. It is highest at maximum pressure and an abrasive mass flow of 350 g/min.

Another task of the investigations on water jet cutting was to determine and explain the jet properties, particularly the influence of the secondary jet on inner components and assemblies. If water jet cutting is used for three-dimensional machining, damages occur caused by the primary and the secondary jet. The primary jet is the jet prior to machining and the secondary jet exists behind the work piece that is machined and dies in infinity or when striking inner components. Within disassembly the chances are high that the secondary jet strikes an inner component or other assemblies (Fig. 1). When the so-called secondary jet distance, the distance between the primary work piece and inner components, is low, re-usable parts are in risk of being damaged. Thus the objective is to minimize the energy of the secondary jet and to predict its influence on other components. The secondary distance is measured on a straight line which runs vertically to the primary work piece through the jet exit point.

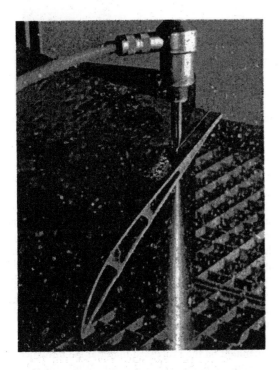

Fig. 1. Disassembly of turbine blades with an abrasive water jet

In analogy to the calculation of the kerf geometry, it is possible to calculate the depth of the damage for the secondary jet. Essentially, the damage of the inner components depends on the traverse rate and the secondary jet distance from the primary work piece. When neglecting the variations in cutting rate, the damage inflicted by the secondary jet during cutting with maximum traverse rate is zero. Thus the traverse rate must be seen in relation to the maximum traverse of the primary work piece. The closer the traverse rate gets to the maximum traverse rate the lower is the secondary jet energy and the possible damage. Yet, not only has the power of the secondary jet changed, but also the direction of the jet given by the exit angle.

The damage by the secondary jet was carried out in analogy to the tests on the primary abrasive water jet. Thus, the secondary jet spreads according to the same rules as the primary jet. As a result, the damage of inner components can be minimized and predicted. With this knowledge, it is possible to set cutting strategies for disassembly which avoid damages on re-usable components (Uhlmann et al., 1999a).

The abrasive water jet was integrated into the disassembly process of standard washing machines to open the plastic lye container and to separate it from the washing cylinder made of stainless steel. Fig. 2 shows the disassembly stages of a washing machine. It starts with the opening of the case, followed by the separation of the swinging system which is then stripped from weights, pump and motor. Afterwards the water jet opens up the lye container and the washing cylinder is removed.

### 2.3 Summary

The examinations on water jet cutting showed both chances and limitations. It is possible to predict the secondary influences on re-usable part by adjusting the parameters or the cutting direction and thus to avoid them or at least to minimise them. Water jet cutting proofed to be economically adequate for various disassembly operations, but only as a stand alone solution. The integration of this process into an automated disassembly line is very complicated. The used water can harm the transporting system and every subsequent process or tool.

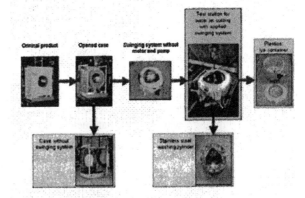

Fig. 2. Disassembly stages of a washing machine

### 3. PLASMA ARC CUTTING

### 3.1 Process

Plasma arc cutting is a thermal jetting technology. It uses the heat of plasma to melt, burn, or sublimate material. Plasma is the so called fourth state of matter. It is generated when adding even more energy to certain gases. The gases de-ionise and are thus electrically conductive. If the plasma nozzle and the work piece are used as anode and cathode a electric arc of high energy can exist (Bach, 1983).

Plasma arc cutting is used for some decades. It finds application in the ship building industry and for cutting sheet material. The main advantages are the fast and economical functioning as well as the high flexibility concerning the place of application. Disadvantages are the appearance of hazardous gases, the low cutting quality, and high thermal influence on the processed material (Prismeyer, 1997). Except for the emitted gases those points do not count for disadvantages in disassembly.

## 3.2 Experimental studies

Within disassembly planning the opening of washing machine housings was identified as the first possible automated disassembly step. For this, the application of plasma arc cutting was examined. With it, the panels of washing machines are separated in such a way that a frame remains which holds the inner components. For different washing machine cases type-specific disassembly sequences were elaborated. Inflammable materials like insulators or cable loops had to be avoided.

The aim of the first disassembly step is to reach inner components and assemblies as fast and cost-effective as possible. A robotic plasma arc cutting device PA-S 45 CNC by Kjellberg, Finsterwalde/Germany, was used. It works with a transmitted plasma arc at cutting currents of 40 A to 130 A, and a cutting voltage of 160 V. The resulting cutting power ranges between 6 KW and 20 kW. The technical examinations contained a qualitative gas analysis as well as a process optimization with regard to maximum cutting speed, minimum mechanical and thermal component damage, and minimum removed particle mass (Uhlmann et al., 1999b).

The maximum cutting speed was obtained in experiments with the standard experimental set-up by Tagucci during which the parameters cutting amperage, cutting gas, work piece material, and material thickness were varied. The cutting gases tested were air, Ar/H2, and Ar/H2/N2. The maximum cutting speeds could be obtained with air. How ever there is a high emission of removal particles. With increasing cutting amperage the maximum cutting speed increases, too. At a cutting amperage of 125 A a maximum cutting speed of 12 m/min can be achieved. The cutting amperage and the cutting gas with a minimum removal particle mass were determined.

Fig. 3 displays the removal particle mass as a function of the cutting amperage in the case of plasma arc cutting of a work piece similar to the washer case of the type Bosch V454. After an optimization the particle emission during plasma arc cutting of a washer case could be reduced by 95 % compared to the initial setting which was a cutting amperage of 125 A and the cutting gas air. The conclusion is that the removal particle mass is almost

irrespective of the cutting speed in the case of a defined separating cut and that it is only minimal when cutting gas Ar/H2 and a cutting amperage of 75 A are applied. A disassembly time of three minutes is calculated for the industrial disassembly of a complete washing machine case if optimized parameters are applied during the separation of the four lateral walls of the washer case. After that, the assemblies and components can be disassembled in a destruction-free manner (Spur et al., 2000, Uhlmann et al., 2002b).

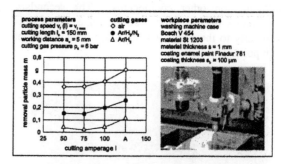

Fig. 3. Removal of particle mass as a function of cutting amperage and cutting gas during plasma arc cutting of a washing machine case

A special test stand was built to determine secondary influence on close components. During the process of cutting a part of the removed material evaporates or melts. The melted parts consist of cinder and sparks. They are accelerated by the cutting gas and thus get in contact with other components. To reuse these additional cleaning and rework processes have to be applied. Therefore another examination point was to minimize the secondary influence. The main influences on the amount of damaging particles have cutting gas and processed material.

Dependent on the processed material the influence of the cutting gas was very different. While for AlMg3 no influence was observed, the other examined materials showed contrary results. X6Cr17, stainless steel, is best cut with air, where the damages caused by secondary particles are minimal. Opposed to this St1203 is cut best with either Ar/H2 or Ar/H2/N2. This simple working steel is the most common material of washing machine housings.

Taking all examinations into account and considering that air is the cutting gas that allows highest feed speeds it proofed difficult to find the best parameter adjustment. The fact that for the most common material Ar/H2 causes fewest emissions and damages turned the balance to this cutting gas.

Including the results of the above described examinations, the design of the plasma arc device and the periphery were altered and integrated into a hybrid pilot disassembly system. Due to the high cutting speed, good automation and low cutting costs, plasma arc cutting was used in the pilot disassembly system to open washing machine cases from various suppliers (Fig. 4). Further examinations take place on cutting poles and wires by plasma arc cutting.

Fig. 4. Integration of plasma arc cutting in a pilot-disassembly system

### 3.3 Summary

Plasma arc cutting is the most effective disassembly process with secondary influences. Due to the optimisation of the process combined with an installed housing with a suction system it was possible to minimise the occurring emissions. Therefore the potential danger for workers within a hybrid disassembly system was reduced.

## 4. CONCLUSION

The described examinations show the independence of water jet cutting and plasma arc cutting from material and working direction. The processes are compared with other processes such as drilling, abrasive cutting and a self developed strike-cutting tool. Plasma arc cutting is the most flexible and fastest process for the disassembly of metallic consumer goods. Only when it comes to disassemble non-metallic components water jet cutting is even more effective, otherwise it is second choice. The flexibility of these processes is their main advantage. For specific tasks, other tools may have higher cutting rates or a lower risk of damaging inner components.

All results from the sub-project A1 "Separation of Insoluble Connections" are added to a so called "Process Selection System" (PSS) for Disassembly (Fig. 5). The PSS is designed to help with the planning of disassembly operations. For each disassembly operation it displays the possible tools and processes along with the necessary process parameters and the coasts per unit or cut meter.

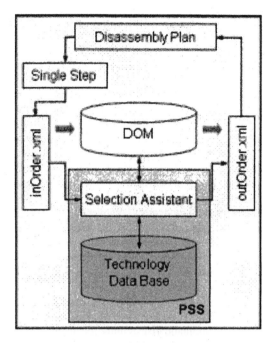

Fig. 5. Process Selection System (PSS) for destructive disassembly processes

## 5. OUTLOOK

The challenges for industrial disassembly are new life-cycle management laws, a growing recycling and maintenance industry, the handling of contaminations and harmful chemicals in recyclable products, and the re-use of components from recyclable products. An approach to this field is made by the Collaborative Research Centre "Disassembly Factories". Focuses lie on the disassembly of consumer goods and on the cleaning of components for re-use. Object is to develop processes and tools for a fast and economical disassembly. The sub-project "Separation of Insoluble Connections" does a basic approach to this.

Therefore a combination of these processes with a cryogenic pre-processing is in development. By cooling material to a temperature of nearly minus 200 °C by liquid nitrogen it is possible to embrittle it. Thus lower forces occur while cutting material and it is possible to increase the cutting speed.

## 6. ACKNOWLEDGEMENTS

Investigations and analyses described in this article were realised at the Institute for Machine Tools and Factory Management (IWF) of the Technical University Berlin within the scope of the sub-projects A1 "Separation of Insoluble Connections" of the Collaborative Research Centre 281 "Disassembly Factories for the Recovery of Resources in Product and Material Cycles" patronized by the Deutsche Forschungsgemeinschaft (DFG).

# REFERENCES

Bach, W. (1983) *Beitrag zum thermischen Schneiden dickwandiger Werkstücke.* Habilitation, University Hannover.

Feldmann, K. and Meedt, O. (1996) Recycling and Disassembly of electronic Devices. Life Cycle Modelling for Innovative Products and Processe (Krause, F.-L.; Jansen, H.), London, Chapman & Hall, p. 233-245.

Prismeyer, U. (1997) *Thermische Schneidverfahren und Werkstoffreaktionen in Hinblick auf die Entstehung von Staub und Aerosolen.* Dissertation, University Hannover.

Spur, G., Axmann, B., Elbing, F. and Seibt, M. (1997) *Fertigungsverfahren der Demontage. Normungsvorschlag zur Begriffsklaerung in der Demontage.* Institute for Machine Tools and Factory Management, TU Berlin.

Spur, G.; Uhlmann, E.; Elbing, F.; Dittberner, J.; Sundaresan, S. and Thantry P, B. (2000) Flexible automatic disassembly for the recycling of consumer goods. In: *Advances in manufacturing Technology XIV, Proceedings of the 16th National Conference on Manufacturing Research,* London/England, p. 407-411.

Uhlmann, E.; Axmann, B. and Elbing, F. (1999) Model of kerf and simulation of damage in abrasive water jet cutting, In: *Production Engineering.*

Uhlmann, E., Axmann, B. and Elbing, F. (1999) Fast Disassembly of Point Forming and Welded Connections. In: *Proceedings of the 4th World Congress R`99. Recovery, Recycling, Re-integration.* Geneva, Switzerland,P. III.328-333.

Uhlmann, E., Spur, G., Elbing, F. and Dittberner, J. (2002a) Innovative Machining Technologies and Tools for the Disassembly of Consumer Goods. In: *Proceedings of the 1st International Conference on Design and Manufacture for Sustainable Development,* Liverpool, England, P. 211-218.

Uhlmann, E., Spur, G., Elbing, F. and Dittberner, J. (2002b) Optimization of Plasma Arc Cutting for the Disassembly of Washing Machines. In: *Proceedings of the International Conference on Cutting Technology 2002,* Hannover, p. 61-66.

Von Weizsäcker, E. U.; Lovins, A.; Lovins, H. (1997): *Faktor Vier. Doppelter Wohlstand - halbierter Naturverbrauch.* Droemersche Verlagsanstalt Knaur, Munich.

ELSEVIER
IFAC
PUBLICATIONS
www.elsevier.com/locate/ifac

# PRODUCTION OF MODULARISED PRODUCT SYSTEMS

**Peter Jacobsen, Associate Prof.**
**Dept. of Manufacturing Engineering and Management, Building 423**
**Technical University of Denmark, Denmark**

Abstract: To day, more and more products are customized. Trends are not only to sell a product to the customer, but to sell a product system. The system can either be a combination of physical products or physical products together with some kind of service. Customers get in this way not a product but a solution. Modularisation is one tool used in designing the products.

Designing and controlling a production system making customized products in an economical way is not an easy task. In order to fulfil the Lean and Agile manufacturing philosophies the production is often carried out in networks. Here the decoupling point has a central role.

The scope for this article is therefore to analyse the possibilities for using modularisation in designing and controlling a production system. How will the development of modularised product systems influence the production system?

In the paper, a case will be used to support the ideas. *Copyright © 2004 IFAC*

Keywords: Modulation, Modelling, Product, Production systems, Systems concepts.

## 1. INTRODUCTION

The objective is to focus on the changes in our perception of the terms product and production and identify trends. The article will therefore identify the different elements in a product, followed by an analysis of the life cycle. The life cycle is a value chain for the product. One of the loops in the life cycle is production. The term production means a network of production nodes that have to be kept together by logistic.

In order to survive the competition, companies need to be innovative as the product life cycle is shortened. The design task is not only to design the product but to design a product system together with the production system model as well.

Simulation is obvious as a tool to verify the idea.

## 2. PRODUCT SYSTEM

Traditionally a product is considered as a physical item fulfilling the market requirements. Item features could be functionality, price, environmental issues, etc. As the product life cycle is getting shorter and market requirements are customized, modularization of the product is used as a tool (Pine, 1993). In order to be more innovative, services are becoming a marketing issue as well.

Rapid development in the electronic area has resulted in design of products with an inbuilt knowledge. At the same time, the products are getting smaller and smaller. The manufacturing technology is able to produce and assemble products down to two mm. (Fleischer, *et al.* 2003)

At the same time, it is also recognized that parts with originally different functions are merged together. Examples are televisions, mobile phones and computers that merge into one product.

Previously a customer bought a product from a dealer fulfilling the functional requirements. The user was responsible for the product and for its use during its life cycle. To day, companies are trying not only to provide a customer with a product alone, but also to sell a solution or system. The solution can be system of different products from the same company.

The requirements to a production system have in this way increased considerably. As the concept for a production system has broadened by also including areas as service and healthcare, a new understanding for a product has to be developed.

Therefore, as definition a product is a system consisting of varying percentages of physical parts, knowledge and service activities all adding value to the customer (Jacobsen, *et al.*, 2004). In some cases, the physical part will be 100% of the system and in other cases, there will be no physical part included in the system.

### 2.1 Elements in a product system

The *physical product* (see figure 1) consists of a basic part together with add-on and interchangeable units. The basic part is the lowest collection of components in order to form the basis for building up a product around it with knowledge, services and possibilities of enhancing it with new functional units and services. The purpose of add-on units is to increase the value of the basic part by adding new units on it including a new range of services. The basic part could be a car, which gets a GPS, internet service, etc. as an add-on unit. The increasing number of add-on units is provided with knowledge and services, ex the cruise controller in a car. The value of the product increases also by substituting one interchangeable unit in the basic product with another. An example could be to change an existing GPS in a car with a better version.

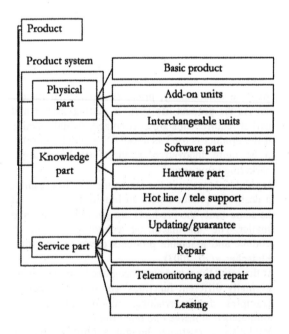

Fig. 1. Elements in a Product System

To increase the functionality and value of a product, it is often provided with *knowledge*. In some examples, the inbuilt knowledge automates some of the actives related to the use of the product, in other examples the knowledge is supporting the usage of the product. The percentage of importance that the knowledge content has in relation to the products itself is highest for industrial products, where the tendency of mechanical parts being more electromechanical is continuing rapidly due to the development of electronic technology. More and more electronic devices containing new service possibilities are built in the products and in this way increasing the value of the product. In the food sector, the products have information on it about last sales date, the content of different additives, and often which country, that has produced it. Many services inside the service sector have a build in knowledge, so the customers can do the service himself. In the health care area, the patient gets increased information about the treatment taken place; different micro sensors, medication pumps etc. handle remote control of medication etc.

*Service* is a value offered together with the product at the time of sale. It is not a service which the customer/user themselves does later. The listed types in figure 1 are not complementary but supplementary. Leasing is a special way of possessing a product. The idea is that the customer does not own the product, but leases the product from the producer or product provider. In this way, the producer will have a close relationship to the customer. The benefit for the customer is that he will get an improved service, as he automatically gets a substitute if his product breaks down. An additional agreement will also secure that he can update his product to newer versions. An environmental issue would be that products containing poisonous elements are returned to the producer, and not scattered in the nature.

Lately in the newspaper (Berlingske, 2004) there was an article about a refrigerator, that could automatically phone the owner, if there was no milk or meat left.

Hot-line support is well known from software systems or trade of televisions where the customer can phone and ask about the function of the product. Monitoring means activities where the functions of a product are recorded and messages given to the user or to the product provider. Monitoring is increasingly used in many sectors, and tele monitoring is the next development step where the product provider can follow, diagnose etc.

Possibilities for service related to a product could be delivery service, installation, operation service, monitoring, maintenance, and service operators.

Most of service elements are electronic add-on or interchangeable units assembled to the physical part during the production. Some of the elements makes a dialog between the user and the product system. Other makes a dialog between user and some service people.

All 3 elements in the product system can be divided into a basic unit, add-on units and interchangeable units according to the modularization principles.

## 2.2 Life Cycle Loops

Figure 2 illustrates a product life cycle divided into 4 loops. In the first loop, the concept loop, the customer and the sales people agree on the best valuable solution for the customer. Terms used in this loop are as example, "to which side do You want to open your freezer?", or "Which type of door do You want to have in your freezer?". The solution is based on a catalogue of possible solutions from the three other loops. The result is a product system. In the next loop, the configuration loop, the system is configured and all relevant parameters determined. The terms used here are different from the previous loop, as the customized solution now turns into technical terms such as "element A124" together with "element V56". The third loop represents the production where a lean production plan is chosen. Based on all possible partners, an optimal solution/model creates a virtual production network. As a result, all partners are informed. The last loop is the disposal bop. The four loops depend on each other and are decided concurrently.

Throughout the life cycle, the product should gain a value. According to the Lean philosophy, this value might be positive if the customer wants it, and is ready to pay for it. It might be negative, but avoidable, and it might negative and waste.

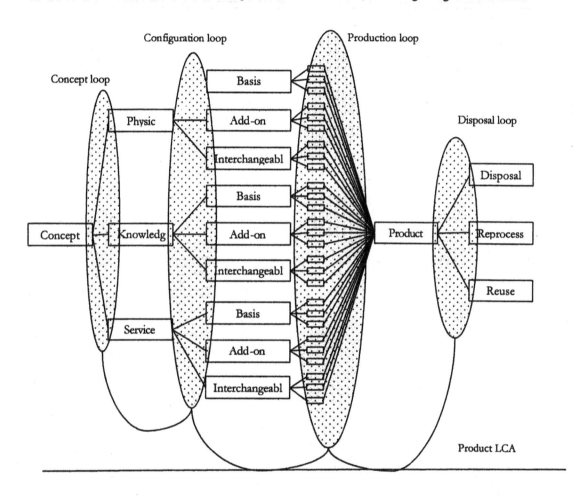

Fig. 2. Product Life Cycle

## 3. PRODUCTION LOOP

A production system can (Jacobsen, et al., 2002) be defined as "a viable and agile network system that transforms product specifications, and raw material into customized product systems. In the transformation are used workflow, suitable manufacturing and control processes based on environmental principles and with a combination of human intelligence wherever necessary". Production is changing from just producing parts fulfilling some functional customer requirements to be more service oriented. As seen in figure 2 nearly all production can be characterised as assembly production. The build-in of knowledge and services in a physical product makes it difficult to design a production system, as it is different design procedures that are going to be combined. The service industry traditional is more labour intensive and more unpredictably than the industrial production. In order to produce the product system, the production net work needs to have strong competences on both the traditionally mechanical production and the electronic production. This is often not the situation for just one company. The wishes for customized and modularized products require that the production systems need to be economical and have a high degree of flexibility. It will be concentrated around core competencies and production in small units. As for the product, the modularization concept can describe a production system and help creating the production units. Often production is separated from assembly at the decoupling point in order to optimize the Lean/Agile philosophies (Womack, *et al.*, 2003) and (Kidd, 1994). The trend is also that companies work in virtual networks (Vesterager, *et al.*, 2001). The disadvantage of creating a production system in a network is that it might be a problem to optimize the logistic of transportation between the units.

### 3.1 Design Procedure

The following will present a procedure for designing a production system. It is based on a project (Wiid, *et al.*,2004) creating a procedure for developing production systems. It has been adjusted to accept product systems instead of just physical products. It consists of 5 phases as indicated in figure 3.

The first two and half phases are analyzing phases where all the basic considerations are decided. The last part of phase 3 and the beginning of phase 4 covers design activities. The last one and a half phases are carrying out activities. Compared with the loops in figure 2, the analyzing phase can be compared with the conceptualization loop, the design phase with the configuration loop and the final implementation phase with the production loop

It should be emphasized that the procedure is iterative.

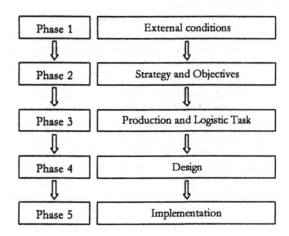

Fig. 3. Developing procedure

### Phase 1, External Conditions

It is an analyzing phase where the input is a product. By investigating the partners on the market, i.e. suppliers, customers, collaborators, potential invaders, the product is modularized and formed into a product system according to figure 1. The new is, that the analysis is broader as not only the physical production is treated but also companies that have a competence on knowledge and service. As shown in figure 4 it is question of setting up a network or framework of producers, service centre and suppliers supporting the product system. Some companies and suppliers have competences on special areas, which in the figure are indicated by different patterns.

By setting up this type of network, a virtual company between the producers can be created.

Examples of supporting tools for this phase is Porters Five Forces (Johnson, *et al.*, 2002), Swot Analysis, Business Excellence and PESTEL framework (Johnson, *et al.*, 2002).

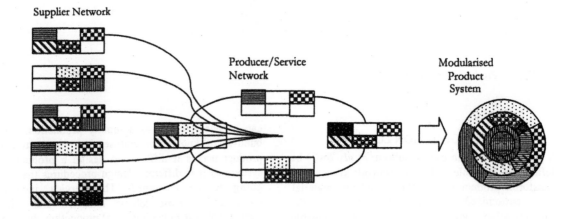

Fig. 4. Supply network of possible partners

*Phase 2, Strategy and Objectives*
The purpose with this phase is to set-up critical goal parameters that are common for all partners. It secure that all partners are moving in the same direction by having consensus for the future, that means on the strategy focus and the success criteria for obtaining these. After describing the future the focus is on the present situation where a swot analysis is made on four important criteria's, ex. economy, market, business and organization.

By comparing, the present with the future different initiatives are defined. The game map tool is shown in figure 5.

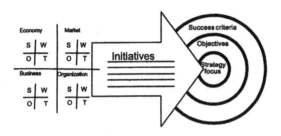

Fig. 5. Game Map

*Phase 3, Production and Logistic Task*
This phase is mainly used in situations where an existing system is modified. The existing product portfolio is analysed and possibilities for setting up a modularised product system is considered. A Current State Value Stream map is used to lean the production and optimise the logistic. As covered in phase 1 possibilities for defining virtual companies are analysed.

*Phase 4, Design*
The main activity in this phase is to define the design task. The decision about which net constellation to select depends on a logistic calculation. The different lean tools (VSM, 5S, etc) are used and the whole network verified, validated and optimised by a simulation tool. At the order time an individual bill-of-material is generated. This BOM profile limit the number of partners in the production network. A capacity and logistic calculation result in a final production plan At the same time as the product system is configured the production net is created as indicated in figure 6. The basic items can be produced according to the lean principles, whereas the add-on and interchangeable units will have a decoupling point. The production before the decoupling point can follow the lean principles, but based on a low demand and transport distances a buffer must be accepted. An optimal solution would be a principle of nearness like the "smart car – smartville" production. (Jakobsen, 1999).

Possibilities for using agents in the organization are analysed (Wilson, 1998). They are "loose" and can be used whenever necessary. They do not load the budget, as they are only paid, when used. On the other hand, they can only fit into special job functions

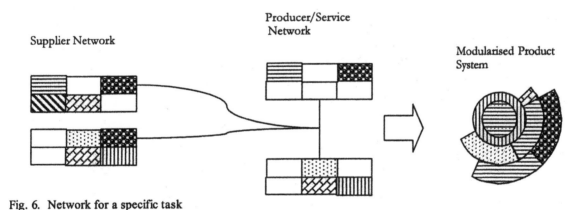

Fig. 6. Network for a specific task

*Phase 5, Implementation*
In the final phase, the design is implemented and tested. As indicated in figure 1, service activities can mainly be divided into 2 areas, one area that concern user-system interface and one area that deals with a user-producer interface. The first one includes activities where the user takes action based on the message from the system. It could be the refrigerator phoning the user with the information about low content of milk. Activities in this area normally do not involve the producer of the product system.

Activities in the second area, user-system interface, are mainly planned and therefore the producer of the product system can scheduled them. Examples are messages about service for a car, updating of software or periodical end of a leasing agreement.

They are items sold in a product system as a unit but after a time frame they will come back as planned activities. The result is that the producer net work either have to put hours in or deliver a new interchangeable unit.

The producer net work knows which core competences and resources to plan for.

## 4. CONCLUSION

The paper has described the change from a physical product to a product system. One of the reasons for developing a product system is the market competition, which is hard and companies need to be innovative. The innovative is to sell a product system that includes services and knowledge fulfilling the requirements from the customer. The product is customized and can consist of varying amount of physical, knowledge and service units. The basis for this development is the rapid development inside electronics.

Based on the product system concept the production system of the future will change. They will need to have competences on not only traditional mechatronics but on services as well. A production network will therefore often be a solution.

Different elements in a product system have been identified and the relation to a production system analyzed. The result is a procedure for designing a production system.

## 5. REFERENCE LIST

Berlingske Avis, 29.08.2004. www.berlingske.dk

Fleischer, J., T. Volkmann and H. Weule (2003) Factory Planning Methodology for the Production of Micro Mechanical Systems, *CIRP Seminar on Micro and Nano Technology 2003* Micro Engineering, Copenhagen 2003

Jacobsen, P., L. F. Pedersen, P.E. Jensen and C. Witfelt (2002) Philosophy Regarding the Design of Production Systems, *Journal of Manufacturing Systems*, **V20 N6** 2001/2002

Jacobsen, P. and L. Alting (2004), A New Perception of Product Definitions, *the 5th International Conference on Integrated Design and Manufacturing in Mechanical Engineering*, Bath 2004

Jakobsen, Allan (1999) Smart Car Production, *Effektivitet*, 1/99

Johnson, Gary and Kevan Scholes (2002) *Exploring Corporate Strategy* (6th Edition) Pearson Education Limited

Kidd, P.T. (1994) *Agile Manufacturing: Forging New Frontiers*, Addison-Wesley Pub Co, 1994

Pine II, J. (1993) *Mass Customization Products and Services Strategic Innovation* July/august 1993

Vesterager, J.,L. Larsen, J.D. Pedersen, M. Tolle, and P. Bernus (2001) *Use of GERAM as Basis for a Virtual Enterprise Framework Model*, Kluwer Academic Publishers

Wiid, H. and M.H. Olsen (2004) *Udvikling af Produktionssystemer*, Master thesis, DTU

Wilson, K J (1998), Companies Use Nontraditional Benefits to Attract and Retain Employees, *Iron and Steelmaker*, **Vol. 25, Issue 10**, swets

Womack, J.P. and D.T. Jones (2003) *Lean Thinking: Banish Waste and Create Wealth in Your Organization*, LEAN Enterprise Institute, Free Press

ELSEVIER
IFAC
PUBLICATIONS
www.elsevier.com/locate/ifac

# AGENT-BASED PRODUCT LIFE CYCLE DATA SUPPORT

Niemann, J.[1], Ilie Zudor[2], E., Monostori, L.[2], Westkämper, E.[1]

[1] IFF Institute of Industrial Manufacturing and Management, Universität Stuttgart, Fraunhofer IPA,
Nobelstrasse 12, 70569 Stuttgart, Germany
[2]MTA SZTAKI, Computer and Automation Research Institute, Hungarian Academy of Sciences, Kende u. 13-17,
H-1111 Budapest, Hungary

Abstract: The paper presents a framework based on autonomous, co-operative agents for
life cycle oriented data support. The framework identifies internal and external data
sources for product optimisation. Depending on their types, the data are stored and
permanently updated in different decentralised locations (web, company and machine).
Initiated by the product manufacturer the databases will be permanently updated with
changes or experiences from current operations. The data can be acquired from machine
control, Auto-ID or other sources. Copyright © 2004 IFAC

Keywords: Life cycles, data logging, production systems, agents

## 1. INTRODUCTION

Industrial manufacturing companies concentrate
their businesses more and more on engineering,
assembly and services. They follow new paradigms
to add value by customer orientation, systems
management and services in the life of products
(Anderl et al., 2000; Anderl et al., 1997; Niemann
and Westkämper, 2004).

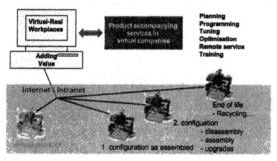

Fig. 1: The modern paradigm of life cycle
management

Machine manufacturers and other branches like the
automobile industry reduce their own capacities to

the dominant or core technologies. Manufacturing of
parts and components are done by suppliers or
specialised companies. More and more the profit
becomes a result of business operations in design,
engineering, final assembly and service. These
phases of production are the core competencies of
companies, which produce strong market or
customer oriented products and add value in the
products life cycle.

Traditionally, product and process designers have
been concerned primarily with product (process) life
cycles up to and including the manufacturing step.
Nowadays, the focus is shifted from the production
to the products themselves covering all life cycle
phases, e.g. material acquisition, production,
distribution, use, and disposal. In other words,
product stewardship is emphasised; the
responsibility of companies goes beyond its
operations to include the responsibility for their
products' performance throughout the product life
cycle (Figure 1).

In the processes of design and engineering, the
functionality of products is defined. By assembly,

maintenance and disassembly the real configuration, functionality and specific or characteristic properties for usage of products are finished (as build) or changed. In the usage phase special know-how on the design and characteristic properties, like specific process knowledge to optimise utilisation and performance, is required. The increasing technical complexity promotes product-near services and assistance of manufacturers. At the end of such developments, there are new business models for selling only the functionality of capital intensive products, rather than the products themselves.

Behind these tendencies there is a new paradigm: linking products in the Manufacturers Network from beginning to end for adding value and maximum utilisation. For this paradigm manufacturers need life cycle management systems, tools and technologies, which master the permanent product reconfiguration. Crucial for this is a holistic life-long product data support. The paper presents such a framework for a holistic life cycle information support system for manufacturing machines.

## 2. LIFE CYCLE PLATFORM

A future development has to take into account even the possibilities to implement all basic data of products into their internal information system. This would help to support all operations done with the product and surrounding activities with actual documentation (Feldmann, 2002; Gu, 1997; Kärkkäinen et al., 2003; Kimura, 2000).

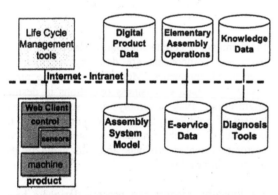

Fig. 2: Platform for the integrated management of products life cycle

Future management systems for the life cycle are open systems which operate on standards in communication and allow implementing product or customer specific IT applications. Figure 2 summarises such a platform with basic functions for communication and specific systems to support products and operations with data in all phases of their life. In order to perform this the product itself has to be monitored and different information about the actual product status have to be gathered, evaluated and disseminated on different levels. Therefore the implementation of a life-long product

monitoring system is essential. The different activities are performed by different actors as there are product manufacturer, user, service companies or the recycler. All these different actors become partner in a network established for holistic product optimisation over the entire life cycle.

## 3. LIFE CYCLE INFORMATION SUPPORT

A general model must cover the entire machine life cycle beginning with the machine design and ending with the machine 'death' or recycling ('end-of-life'). Similar to a patient's file at a doctor's practice, this digital machine file can be considered as a document where all machine data and events have been logged. The different types of data can be divided into data for life cycle product tracing (discrete events, static) and life cycle tracking (continuous process behaviour, dynamic) (Figure 3).

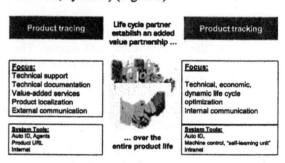

Fig. 3: Life cycle tracing and tracking

Product tracing mainly focuses on global parameters of the product concerning e.g. the physical location or activities performed over the life cycle. Mainly engineering and product data management (EDM/PDM) data are gathered to document conditions of 'discrete events'. Main objective is to create a record of activities along the product life cycle (Tichkiewitch and Brissaud, 2001).

The product tracking focuses on actual condition-orientated activities to mainly influence the current situation. The central objective is a technical and economic product optimisation in a dynamic environment.

The following sections describe these two dimensions more detailed. Merging these two dimensions means to get a detailed overview and controlling instrument (organisational, technical and economic view) to master product life cycles in turbulent and dynamic environments.

### 3.1 Life cycle product tracing

The activities in this dimension can be described as a life-long product tracing starting from design up to recycling.

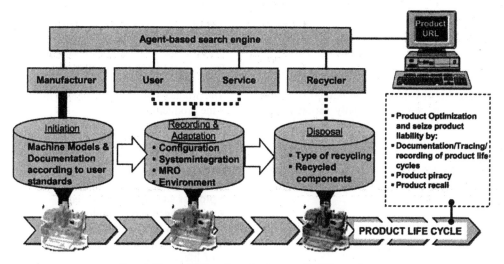

Fig. 4: Traceability of products over the life cycle

The main objective is to perform technical support and documentation for the product. The communication mainly aims at providing data concerning due dates, resources' identification codes and modification data for external life cycle partners. This allows recording a machine life cycle file, which provides information for subsequent (and previous) life cycle partners and helps to optimise future machine generations. Another aspect is to meet the requirements concerning laws of product liability and to prove/prevent product piracy. By electronic parts documentation falsified or irregular parts can be identified to avoid machine breakdowns or safety problems. In special cases the knowledge about the actual 'whereabouts of parts' can reduce cost in case of product recalls.

In order to prevent data overflow in one central data base all data gathered will be physically saved in decentralised systems. A software agent serves as a search engine, which provides information from other life cycle partners upon request. The agent operates with references to deliver a prompt reply to authenticated requests. The system will be operated via a web front-end with a product specific URL. The product specific URL avoids the current problems of different standards concerning data acquisition and supply (Figure 4).

The main objective is to create a network with external life cycle partners to access available additional product knowledge (Alting and Legarth, 1995; Niemann and Westkämper, 2004).

### 3.2. Life cycle product tracking

The vision of the digital factory of the future is to integrate all data of manufacturing resources into one (factory) planning environment. Due to a dynamic environment and operation processes the data are not static but have to be updated permanently. This applies also to manufacturing facilities and its configuration changes over the life span. The assigned data and documents have to follow the life cycle and have to be adapted accordingly on a permanent basis. Therefore the models and the knowledge about construction, design and configuration of the machine initially have to be set up by the manufacturer of the product (Westkämper et al., 2000; Gu, 1997; Kimura, 2000). The package also has to contain CAD product data, as well as the latest release of process models, parameter settings for optimal machine and tool operation. The models are integrated into/ attached to the machine according to the software standards of the customer and constitute an integral part of the delivery. After machine delivery to the customer the machine is integrated into the digital factory models at operator's side and all subsequent changes along the machine life cycle will be locally added to the given models (localised on the machine). From this stage on the data administration and updating is performed by the user.

The innovative approach here is to store all product-related data as well as models and knowledge about efficient machining operation decentralised directly on the machine. Therefore the machine knows about itself and how to fulfil its tasks best in close interaction with the machine control.

In terms of process control, it is however necessary to satisfy other requirements which make it possible to apply sensor-based and measurement techniques to the monitoring and guidance of processes. This also allows determining the current status of machines in an ongoing manner. Sensors record the relevant data from machines and processes. The process models feature the interactive links between input parameters and quality parameters relating to the components being manufactured and the current status settings of machines involved in the production process. This makes it possible to compensate for system-based deviations. An in-situ simulation can be used to determine status settings in advance for preventive purposes, and can also help to raise accuracy by adjusting parameter settings.

By this way the machine will be equipped with a 'self-learning unit' which is initially set-up by the manufacturer and is permanently fed with additional user knowledge and real experiences from current machining processes. This additional internally oriented knowledge chain leads directly to better process results and improved cost ratios over the product life cycle.

## 4. DATA MINING TECHNOLOGIES

In our approach we propose the implementation of product related information on two levels (in the products themselves and external databases), depending on the information type, using agent-based concepts and the Auto ID technique for external data retrieval.

### 4.1 Agent based data supply

Contemporary agent based techniques are widely used in quite a lot of application fields because of the advantages they can offer (Huhns and Stephens, 1999; Weis, 1999), provided by the characteristics of multiagent systems related to technological and application needs, natural view of intelligent systems, complexity management, speed-up and efficiency, robustness and reliability, scalability and flexibility, costs, development and privacy.

Some of the agent based architectures regard the elements of manufacturing systems, such as machines, operations, human operators, and even the manufactured product and its components themselves, as agents (Van Brussel et al., 1998; Monostori and Ilie-Zudor, 2000). Our concept complies with these approaches, but concentrates on the product agent.

To really cover the entire life cycle of a product is needed that the product agent (PA) that represents the physical product, does not end its life as in traditional approaches with accomplishing the manufacturing process, but further exists until the last phase of the product's life cycle, the disposal.

There is a large amount of information, which can facilitate the process of disposal. A part of this information is available at the production and might be incorporated in the product agent. An other part of the information relates to the use of the given product together with the modifications made on it during its life cycle, e.g. repair data (Ilie-Zudor and Monostori, 2001).

Taking the versatility of the products and the complexity of the disassembly/disposal process into account, the availability of as exact information as possible, is of key importance.

The product agent should include information on:
- the product life cycle,
- user requirements,
- design,
- process plans,
- bill of materials,
- quality assurance procedures,
- process and product knowledge
- maintenance and reliability
- the supply and demand of parts
- as well as information related to the product's end-of-life, such as:
  - the possible steps of disassembly,
  - the data that would help establishing the point of maximum financial profit (from this point onward the disassembly is not worth anymore; the data might change in time, and at the end of life of product, those should be actualised),
  - which parts can be sent to: recycle, reuse, incinerate or landfill,
  - type of waste that subparts from disassembly represent (solid, hazardous, liquid),
  - the data about the environmental impact of subparts as the disassembly proceeds.

A part of the necessary information can be common to a set of products centrally (e.g. the product type specific information), the other part can be intrinsic data of the given product (e.g. product specific knowledge related to the production and usage) (Ilie-Zudor, E. and Monostori, L., 2002).

Product take-back requires manufacturers to be responsible for collecting and dealing with products at end-of-life. The presented concept of product agent may facilitate this process.

### 4.2 Auto systems for product identification

For information collection during the PA's life cycle we suggest the introduction of Auto ID techniques. The introduction of Auto ID technologies is seen as a new way of controlling material flow. Auto-ID is the broad term given to a host of technologies that are used to help machines identify objects.

Auto ID encompasses various technologies such as: bar code technologies, Radio Frequency Identification (RFID) tags, smart cards, magnetic inks, biometrics, optical character reading, voice recognition, touch memory and many more (Auto-ID Center). Its main principle consists in the application of a tag containing information about and on products (parts and finished), which will be later read by a device called Reader (or interrogator).

The information contained on the tag, bar code etc. is not standardised at the moment, and different developers encompass different quantity of information, as this drastically influences the costs of the tag. Examples of information stored on a tag can be found in the approaches discussed in (Alting and Legarth, 1995) and in (Gu, 1997).

The location for maintaining product-related information should be treated decentralised, therefore we propose the separation of information on two levels according to the categories of product tracing or product tracking related data.

The concept proposed implies that components of the products have been applied Auto ID tags, which connects them to the local network and the Internet where the relevant data about them are stored. When the product reaches its end-of-life and will be brought back in the agent-based environment, the tag will make the automate connection with the database containing the data regarding its disassembly and disposal treatment, as well as with the software embedded in product agents for negotiating the tasks the product needs to be performed.

The PA drives its own disassembly process conform to the prescriptions for its end-of-life (e.g., the product will not be disassembled in all its components from the assembly process, but parts from the same material are kept together, as well as parts for which the disassembly is not worth it will be kept together for disposal).

Furthermore, tagged disassembled components are scanned and will be automatically synchronised with similar treatment components (e.g. parts that will be melted) and when reaching the minimum quantity necessary for processing start-up, will negotiate their own treatment process accomplishment. This may help considerably reducing storing costs of parts by their timely registration for processing.

A similar procedure is to be applied when/if the product needs maintenance or repair during its lifetime. When brought at the service centre, the tag will provide the data necessary for its identification, which will allow automatic connection to the database containing product service information. After service, the database will be updated with the new product-related data (what, where, when).

Beside the general advantages that Auto ID technology may bring to the entire supply chain to which products belong, there are particular advantages for PA management at products' end-of-life:
- eliminate human error from data collection when the product is returned to the manufacturer and needs liaison with it's end-of-life processing data,

- reduce inventories, as the components will spend the shortest time possible in stores, due to the automatically synchronisation with other parts of the same type
- improve safety and security, as the toxic or otherwise damaging materials will be kept in the inventories less time,
- eliminates inconsistencies or delays associated with lack of expert knowledge when a member of the team is unavailable at the arrival of the product, etc.

## 5. LIFE CYCLE SUPPORT NETWORKS

Today, the Internet offers a great variety of usable tools for life cycle management. The Internet uses tools, engines and robots (behind the interfaces) to search for information and knowledge. There are new standards for b2b (business to business) and b2c (business to consumer). Leading companies in the automotive and machine industries use the Internet as a platform for logistics and the administration of processes between OEM and suppliers. The technologies offer a broad spectrum of tools for managing the link between manufacturers and users wherever they are located. Examples of this include the management of the logistics of component supply for assembly and maintenance or technical support in the usage phase. The basic architecture of the Internet and of Intranets in companies' information technologies is illustrated in Figure 5. By this the machine is integrated into the local factory environment as well as linked to a world-wide net of supporting activities performed by other life cycle partners.

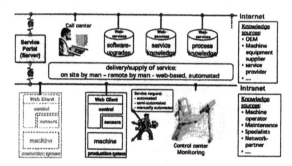

Fig. 5: Knowledge sources and sinks for life support networks

The figure shows a structure for the network of services based 'around the machine' to optimise the product life cycle. The network is characterised by connections which allow the transfer of knowledge and information automatically as well as manual. The nodes serve as a provider, server and distributor of knowledge. In this way, complex structures are generated which consist of knowledge sinks and sources whereby communication via the web is made possible using transparent interfaces.

Security techniques (firewall, en- or decryption) have to be adapted to the needs of manufacturers and be able to be operated in closed areas with service partners. Both the architecture of a product's control systems and Internet availability at each work place are important for assembly operation and maintenance. The diagnosis of machine functions and the monitoring of usage can be integrated into these systems and linked to the Internet. The same diagnosis systems are required for all changes in configuration.

## 6. SUMMARY AND OUTLOOK

The paper describes a general framework for the integration of different data sources and sinks along the product life cycle. The objective is to use these data for product optimisation in all different life stages. The relevant data will be stored in a decentralised way on the machine, the company's intranet and the internet. The data retrieval and supply can be realised by application of modern Auto-ID systems and agent-based search routines. Future machines will be equipped from the manufacturer with digital and holistic product documentation up to process models and simulation features to support product and parameter optimisation. By this the machine continuously gains knowledge about itself and is able to support and trigger manual or automated optimisation activities thanks to is 'self-learning unit'.

## REFERENCES

Alting, L. and Legarth, J.B. (1995). Life cycle engineering and design. CIRP keynote paper, Annals of the CIRP 1995, Vol. 44(2): 569-580.

Anderl, R., Daum, B. and John, H., 2000, Produktdatenmanagements zum Management des Produktlebenszyklus, in: ProduktDaten-Management 1, pp. 10-15.

Anderl, R., Daum, B., John and H., Pütter, C. (1997). Cooperative Product Data Modelling, in: Krause, F.-L., Seliger, G., Life Cycle Networks: Proceedings of the 4th CIRP International seminar on Life Cycle Engineering, 26-27 June 1997, Berlin, Germany. London u.a.: Chapman and Hall, pp. 435 – 446.

Auto-ID Center: http://www.autoidcenter.org.

Feldmann, K. (2002). Integrated Product Policy - Chance and Challenge: 9th CIRP International Seminar on the Life-Cycle Engineering, April 09-10, 2002, Erlangen, Germany, Bamberg: Meisenbach.

Gu, P., Hashemian, M. and Sosale, S. (1997). An integrated modular design methodology for life-cycle engineering, Annals of the CIRP, 46/1:71-74.

Huhns, M.N. and Stephens, L.M. (1999). Multiagent systems and societies of agents, in *Multiagent Systems*, ed. Weiss, G., ISBN 0-262-23203-0.

Ilie-Zudor, E. and Monostori, L. (2001) Agent-based support for handling environmental and life-cycle issues, Lecture Notes in Computer Science; 2070: Lecture Notes in Artificial Intelligence, Engineering of Intelligent Systems, Springer, pp. 812-820.

Ilie-Zudor, E. and Monostori, L. (2002). An agent-based approach for production control incorporating environmental and life-cycle issues, together with sensitivity analysis, Lecture Notes in Computer Science; 2358: Lecture Notes in Artificial Intelligence, Developments in Applied Artificial Intelligence, Springer, pp. 157-167.

Kärkkäinen M., Holmström J., Främling K. and Artto K. (2003). Intelligent products - a step towards a more effective project delivery chain, *Computers in Industry*, Vol. 50, No. 2.

Kimura, F. (2000). A Methodology for Design and Management of Product Life Cycle Adapted to Product Usage Modes, The 33rd CIRP International Seminar on Manufacturing Systems, 5-7 June 2000, Stockholm, Sweden.

Monostori, L. and Ilie-Zudor, E. (2000). Environmental and life cycle issues in holonic manufacturing, Proceedings of The 33rd CIRP International Seminar on Manufacturing Systems, June 5-7, Stockholm, Sweden, pp. 176-181.

Niemann, J. and Westkämper, E. (2004). Life cycle product support in the digital age. In: ElMaraghy, Waguih (Chair); CIRP u.a.: Design in the Global Village / CD-ROM : 14th International CIRP Design Seminar. May 16-18, 2004, Cairo, Egypt. Windsor, Ontario, CA, 2004, o.Z.

Tichkiewitch, S. and Brissaud, D. (2001). Product models for life-cycle. Cirp Annals Manufacturing Technology 2001, 50/1:105-108

Van Brussel, H.; Wyns, J.; Valckenaers, P.; Bongaerts, L. and Peeters, P. (1998). Reference Architecture For Holonic Manufacturing Systems, *Computers in Industry, Special Issue on Intelligent Manufacturing Systems*, Vol. 37 (3), pp. 255-274.

Weis, G. (1999). Multiagent Systems, A modern approach to distributed artificial intelligence, 1999

Westkämper, E., Alting, L. and Arndt, G. (2000). Life Cycle Management and Assessment: Approaches and Visions Towards Sustainable Manufacturing, Annals of the CIRP, 49/2:501-522.

ELSEVIER
IFAC
PUBLICATIONS
www.elsevier.com/locate/ifac

# INFORMATION TECHNOLOGY IN MANUFACTURING

**G. Chryssolouris, D. Mourtzis, N. Papakostas**

*Laboratory for Manufacturing Systems and Automation*
*University of Patras, Greece*

Abstract: This paper describes the advent of information technology in manufacturing during the last decades, attempting to outline the systems and applications of today as well as the challenges to be addressed in the future. Specifically, the technologies addressed include Computer Aided Design (CAD), Computer Aided Engineering (CAE), Process Planning (CAPP) and Manufacturing (CAM), Product Data / Lifecycle Management (PDM / PLM), Simulation, Automation, Process Control, Shop Floor Scheduling, Decision Support and Decision Making, Manufacturing Resource Planning (MRP II), Enterprise Resource Planning (ERP), Logistics, Supply Chain Management and e-Commerce systems. *Copyright © 2004 IFAC*

Keywords: Information technology, Information integration, Computer-aided design, Computer-aided engineering, Computer-aided manufacturing, Computer-integrated manufacturing, Manufacturing Systems.

## 1. INTRODUCTION

The benefits related to the introduction of IT into manufacturing systems have been thoroughly investigated during the last decades. Starting from numerically controlled (NC) machines to machining centres, flexible manufacturing cells and systems as well as CAD, CAM and CAPP applications, early interest in their use in manufacturing was mainly focused on cost advantages (Cagliano and Spina, 2000). On the other hand, systems focusing on the manufacturing inventory control in the 1960s, and Material Requirements Planning (MRP) systems in the 1970s, when companies could no longer afford maintaining large quantities of inventory, were complemented by new capacity planning, sales planning and forecasting tools in conjunction with scheduling techniques (Umble, *et al.*, 2003), leading to the materialisation of systems known as closed-loop MRP. In the 1980's, manufacturing companies were aiming at improving productivity and quality, while reducing costs and lead times (Cagliano and Spina, 2000), taking advantage of the increased and affordable Information Technology (IT) power.

When technology allowed for stand-alone systems to be integrated into unified systems (end of 1980's, mostly), the concept of Computer Integrated Manufacturing (CIM) was devised, promising to provide improvements in efficiency, operational flexibility, product quality and time to market, while responding in a faster way to changing customer needs (Cagliano and Spina, 2000). CIM projects, however, often associated with the concept of the un-manned factory, did not always yield successful results, mostly due to their poor implementation (often requiring tough and unexpected organisational changes) and the poor understanding of the strategic potential of information technology in general (Cagliano and Spina, 2000). Nevertheless, new computer architecture alternatives, the advances in microprocessor technology, the advent of the Internet era, the standardisation of software interfaces, the wide acceptance of formal techniques for software design and development (the Unified Modelling Language – UML) and the maturity of certain software products (Relational Database Management Systems – RDBMS and CAD systems, for instance) paved the way for facilitating the integration among diverse software applications, while allowing for transactions to take place in a fully electronic way via Internet. The evolution of information systems (Figure 1 – Birchfield, 2002) over the last decades has played a crucial role in the adoption of new information technologies in the environment of manufacturing systems.

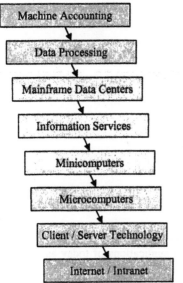

Fig. 1. Evolution of Information Systems.

## 2. COMPUTER AIDED DESIGN, ENGINEERING, PROCESS PLANNING AND MANUFACTURING

The 21st century modern photorealistic, Graphical User Interface (GUI) - based CAD successors, using parametric technology, integrating Finite Element Analysis (FEA) tools as well as full simulation capabilities, seem to have nothing to do with the primitive, slow, requiring costly computers to run, text command-based CAD systems of the 1980's.

New CAD-based technologies, known as Rapid Prototyping (RP), capable of producing parts directly from CAD models in a few hours, bring new perspectives in manufacturing. These technologies have been claimed of being able to cut down new product costs by up to 70% and time to market by 90% (Kruth, 1991). All RP processes require input from a 3D solid CAD model, usually in the form of layers, which are sent to the RP machine for the production of the physical part by adding one layer above the other.

CAE systems are used for reducing the level of hardware prototyping during product development and for improving understanding of the system (King, et al., 2003). These systems are based on simulation principles, including (King, et al., 2003): computational fluid dynamics using three-dimensional mesh and simplified Navier-Stokes equations to predict fluid flow, temperature and pressure; FEA for analysing materials for structural characteristics, thermal performance, and electro-magnetic fields and one-dimensional fluid analysis for predicting the flow of a fluid around circuits (e.g. pipes).

Process Planning activities determine the necessary manufacturing processes and their sequence in order to produce a given part economically and competitively (Cay and Chassapis, 1997). CAPP systems, an essential component of CIM environments, aim at automating process planning tasks so that the process plans are generated consistently (Cay and Chassapis, 1997). During the last decades (Eversheim, 1980; Alting and Zhang, 1989), mainly two approaches have emerged: the variant approach relies on standard plans coming from previously manufactured parts, while the generative approach includes the automatic generation of process plans without referring to

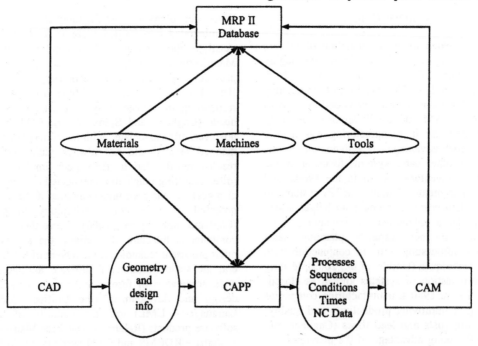

Fig. 2. CAPP integration with CAD/CAM systems

existing plans of previously manufactured components (Cay and Chassapis, 1997). Semi-generative CAPP systems combine both approaches. CAD/CAPP integration issues have been extensively investigated. Interfaces supporting automatic feature recognition – identifying manufacturing features of a workpiece from geometric descriptions automatically – have been demonstrated in some cases (Cay and Chassapis, 1997).

CNC machining was one of the most important developments for manufacturing technologies in the twentieth century, allowing for mass production of consumer products and flexibility in cases of specialized parts (Yeung, 2003). From the first NC machine tool in 1952, at MIT, USA, the computer numerical control (CNC) of the 1970's till the use of PC and Open CNC technology of the 1990's, new CAD / CAM software products have been developed, featuring sophisticated programming capabilities for NC code generation (Newman, et al., 2003). The programming language, however, has not changed and is based on the G/M code programming, developed in the 1950s (then became ISO 6983). Although most CAD/CAM systems are capable of generating CNC tool paths, the planning of the machining process, including decisions regarding roughing and finishing, number of passes and sequence of paths, relies on the programmer's knowledge (Yeung, 2003). Recent approaches have started to deal with this problem, in most cases, however, addressing specific environments and conditions (Yeung, 2003). Other systems, such as Computer Aided Quality (Mbang and Haasis, 2004) systems, have also started to emerge and to become part of the engineering workflow.

Bridging CAD, CAE, CAPP, CAM systems (Figure 2; Ssemakula, 1990) and integrating them into CIM systems constitute a major issue in today's manufacturing world. Standardisation activities, however, such as the Standard for the Exchange of Product Model Data (STEP), the ISO 14649 (representing data models for Intelligent Computer Numerical Controllers; Newman, et al., 2003), as well as advanced approaches, including AI techniques and distributed agent-based methods, have proven that fully automated software environments for engineering tasks are not far from reality (Cay and Chassapis, 1997).

PDM/PLM systems allow for performing a variety of data management tasks, including vaulting, workflow, lifecycle, product structure, view and change management. PDM systems are claimed as capable of speeding up the process of distributing engineering information, while centralizing control of the overall engineering design work. PLM systems constitute a collaborative product development supply ecosystem that enables manufacturers to manage a product from its early concepts to its retirement (Connolly, 2002). These systems tie everything together, allowing engineering, manufacturing, marketing, and outside suppliers and channel partners to coordinate activities. In the past,

the whole procedure of shipping new designs to a set of locations and confirming their receipt could take as long as weeks. Technically speaking, today's PDM/PLM systems mainly focus on the administration of computer files, without, however, having much access to the actual content of these files, since CAD systems are used instead, for developing product models, since geometry data make up for the major part of the product-defining characteristics (Weber, et al., 2003). PDM/PLM systems are not usually capable of capturing the knowledge behind characteristics and properties of a product and therefore they cannot support on a continuous basis the entire product development process (Weber, et al., 2003). Only recently have PDM/PLM systems gained some new role in the product development process. Today's state-of-the-art PDM/PLM solutions in one of the most complex industrial domains, the automotive industry, employ concepts, such as the Generative Template: a solution aiming to reduce design cycle time in several development processes by employing computer models to incorporate component and knowledge rules that reflect design practice and past experience. The future trend is to expand the use of PDM/PLM systems on an enterprise-wide basis. In such a case, integration issues with CAx and MRP II systems will have to be addressed and resolved.

## 3. MANUFACTURING CONTROL

Automation in production systems is considered as a way to improve flexibility (Desforges, et al., 2004). In an excellent review, Birchfield summarizes the IT developments during the past 50 years in the hydrocarbon process industry, one of the leading industrial sectors in process automation (Birchfield, 2002). In the mid-1950's, fluid flow sensors, pneumatic transmission of process data, pneumatic controllers and valve actuators were the most prominent forms of automated control. Data were presented to the user through full graphic panel boards displaying the process in the control room with process indicators and controllers mounted on the display. Electronic instrumentation started prevailing by the early 1960's. The first on-line process computer control project in this industry was implemented at the Texaco Port Arthur catalytic polymerization unit using a TRW computer (Birchfield, 2002). Low-cost mini-computers were used in pilot demonstrations of set point supervisory control and direct digital control in the mid-1960's (Birchfield, 2002). In the late 1960's Programmable Logic Controllers (PLCs) were introduced, as an alternative to the large cost associated with the replacement of the complicated relay based control systems. By the early 1970's, the computing power of general-purpose computers expanded significantly and low-cost microprocessors were incorporated into electronic instruments, thus increasing functionality and flexibility. At that time PLCs started offering data communication capabilities. By 1978, chip technology, providing integrated circuit chips with over than 64000 transistors, contributed to the architecture of the modern control system, consisting of several major building locks, organized hierarchically (Birchfield, 2002). In the 1970's control panel boards in the central control rooms were replaced by CRT monitors in the operators' workstations (Birchfield, 2002). It is the time, when instrumentation and control computers started getting integrated in a cohesive control system (Birchfield, 2002). The control systems in the process industry of the 60's and 70's, based on PID (Proportional, Integral, Derivative) control of single loops, were replaced in the late 80's by multivariable model-based predictive controllers implemented in single programs, which could prevent constraint violations before they occur, rather than reacting afterwards (as PID controllers do) (Birchfield, 2002). PID controllers, however, are still heavily used in microelectronics manufacturing (Edgar, et al., 2000), executing fixed process recipes without feedback of important process outputs. By mid 90's, on-line measurements were effectively integrated into control systems and model-based multivariate controllers were installed, thus managing the dynamic process complexities and interactions more efficiently. The recent trend in advanced process control is towards larger controllers: while in the early 1990's a typical multivariable predictive controller could handle 12 controlled or constraint variables using 6 variables or control valves, applications in early 2000's report over 400 variables and 200 valves (Birchfield, 2002). At the same time PLC functions have more or less been standardized (the latest standard – IEC 1131-3 – has tried to merge PLC programming languages under one international standard). However, PLCs are being replaced by personal computers in certain applications, lately.

New developments in the use of wireless technologies in the shop floor, such as Radio Frequency Identification (RFID) as a part of automated identification systems, involve the retrieval of the identity of objects monitoring items moving through the manufacturing supply chain, enabling accurate and timely identification information (McFarlane, et al., 2003). Smart sensors and actuators able to process information related to calibration, fault detection, diagnosis and others, appeared during the 1990's and allowed the control of complex functions or processes (Desforges, et al., 2004).

Integration of control systems with CAD / CAM and scheduling systems as well as real-time control based on the distributed networking between sensors and control devices (Ranky, 2004) currently constitute key research topics.

## 4. SIMULATION OF MANUFACTURING SYSTEMS

Problems related to the design and operation of manufacturing systems are often so complex that their representation with mathematical modeling is not possible, without having to make a significant number of assumptions (Baldwin, et al., 2000). In such cases, simulation environments and software products has been successfully used instead. Computer simulation has become one of the most widely used techniques in manufacturing systems design, enabling decision makers and engineers to investigate the complexity of their systems and how changes in the system configuration or in the operational policies may affect the performance of the system / organization (Baldwin, et al., 2000). Simulation models are categorized into static, dynamic, continuous, discrete, deterministic and stochastic. Since late 1980's, simulation software packages have been providing visualization capabilities, including animation and graphical user interaction features. Interestingly, most simulation applications have been used for training education purposes beyond modeling of real systems. Computer simulation offers the great advantage of studying and statistically analyzing what-if scenarios, thus reducing overall time and cost required for taking decisions based on the system behavior. A major difficulty, simulation modelers often encounter, is in transforming the real world's multi-dimensional, visual and dynamic characteristics into the one-dimensional, textual and static representation required by traditional simulation languages (Ülgen and Thomasma, 1990; Adiga and Glassey, 1991; Xu, et al., 2000). Simulation systems are often integrated with other IT systems, such as CAx, FEA, Production Planning and Optimization systems.

An extension to Simulation technology – Virtual Reality (VR) technology – enabled engineers to immerse in virtual worlds and interact with them like if they were real. Instead of keyboards and screens, the interface offered in VR environments includes displays for the eyes, gloves for the hands and headphones for the ears. Virtual-reality based systems have been used for supporting manufacturing activities, such as factory layout, planning, operation training, testing and process control and validation (Xu, et al., 2000; Chryssolouris, et al., 2002).

Other applications include the verification of human related factors in assembly processes by employing desktop 3D simulation techniques, replacing the human operator with an anthropometrical articulated representation of a human being, called "mannequin" (Chryssolouris, et al., 2000a).

## 5. ENTERPRISE RESOURCE PLANNING AND MANUFACTURING OPTIMIZATION

Enterprise Resource Planning (ERP) has been associated with a quite broad spectrum of definitions and applications over the last decades (Jacobs and Bendoly, 2003). The MRP systems in the 1970's, were complemented with additional capabilities, leading to the materialization of systems known as closed-loop MRP. The manufacturing resources planning (MRP II) systems of the 1980's incorporated the financial accounting and management systems. The MRP II concept was further expanded to incorporate all resource planning and business processes of the entire enterprise, including areas such as human resources, project management, product design, materials and capacity planning (Umble, et al., 2003). That was the time when the ERP concept was devised to integrate smaller, otherwise isolated, systems so that real-time resource accountability across all business units and facilities of a corporation could be maintained (Umble, et al., 2003). The elimination of incorrect information and data redundancy, the standardization of business unit interfaces, the confrontation of global access and security issues (Umble, et al., 2003), the exact modeling of business processes, have all become part of the objectives list to be fulfilled by an ERP system. ERP systems are highly complicated information technology systems. Large implementation costs, high failure risks, tremendous demands on corporate time and resources (Umble, et al., 2003), complex and often painful business process adjustments are included among the main concerns pertaining to an ERP implementation. It is very important for an organization to take into consideration a set of critical factors for increasing the chances for successful ERP implementations, such as (Umble, et al., 2003): clear understanding of strategic goals, commitment by top management, excellent project management, organizational change management, an experienced and capable implementation team, data accuracy and data migration from legacy systems, extensive education and training of end users, focused performance measures, multi-site issues. ERP implementations usually prove to be huge and complex projects, often resulting in cost and schedule overruns. Statistics show that (Standish Group, 1998): only 10% of ERP implementations are considered fully successful in terms of functionality, estimated costs and time frames, the average cost overruns reach a 178%, the average schedule overruns reach a 230%, the average implemented functionality reaches a 41% of what originally desired.

ERP systems often provide Supply Chain Management (SCM) solutions or provide interfaces for interacting in an integrated way with other external information technology systems. SCM solutions deal with the current trend of manufacturing companies to maximize their communication and collaboration capacity by integrating their operations with those of their business partners (Chryssolouris, et al., 2004a). On top of these operations, e-commerce applications allow for commercial and business transactions to be carried out in a fully electronic way. The Internet growth and the associated software technologies provide the means for the realization of these trends.

ERP systems often incorporate optimization capabilities. Optimization has become increasingly important during the last years, since fierce competition has actually put a lot of pressure on

companies to save costs and time from virtually every manufacturing process. From the optimization problems related to refinery planning (addressed with mathematical programming models) and shop-floor scheduling and production planning (Chryssolouris, 1992) to the today's complex decision making problems (Chryssolouris, et al., 2000b; Chryssolouris, et al., 2004b), requiring embedded simulation engines and fast optimization algorithms, such as simulated annealing, tabu search and evolution programs, the vast improvement on computing power has played a significant role in real industrial applications. In all these years, however, the lag between academia and real industrial practice is still substantial (Chryssolouris, 1992).

A new generation of decentralized, factory control algorithms has recently appeared in literature, known as 'agent-based'. A software agent a) is a self-directed object, b) has its own value systems and a means to communicate with other such objects, c) continuously acts on its own initiative (Baker, 1998). A system of such agents, called a multi-agent system, consists of group of identical or complementary agents that act together. Agent-based systems encompassing real-time and decentralized manufacturing decision-making capabilities, have been reported (Papakostas, et al., 1999). In such a system, each agent, as a software application instance, is responsible for monitoring a specific set of resources, namely, machines, buffers, or labor that belong to a production system and for generating local alternatives upon the occurrence of an event, such as a machine breakdown.

## 6. CONCLUSIONS

During the last decades the advances in IT have affected greatly the operation of the manufacturing systems. This paper attempts to describe the main application areas of IT in manufacturing. The standardization and integration efforts of the last 20 years give a short idea of what to be expected in the near future: true, full scale, seamless integration of all IT systems in the manufacturing environment, including e-commerce and SCM applications and real time decision making and control. The Internet growth, the new wireless communication technologies, the GRID technology, Database Systems able to manage enormous quantities of information, are some factors expected to influence IT in the manufacturing of the future.

## REFERENCES

Adiga, S. and C.R. Glassey (1991). Object-oriented simulation to support research in manufacturing systems. *International Journal of Computer Integrated Manufacturing*, 29, 2529-2542.

Baker, A. (1998). A Survey of Factory Control Algorithms That Can Be Implemented in a Multi-Agent Hetararchy: Dispatching,

Scheduling, and Pull, *Journal of Manufacturing Systems*, 17/4, 297-320.

Baldwin, L.P., T. Eldabi, V. Hlupic and Z. Irani (2000). Enhancing simulation software for use in manufacturing, *Logistics Information Management*, 13/5, 263-270.

Birchfield, G. (2002). Advanced Process Control, Optimization and Information Technology in the Hydrocarbon Processing Industries – The Past, Present and Future. *Aspen Technology Publications*, http://www.aspentech.com/publication_files/Advanced_Process_Control.pdf

Cagliano, R., and G. Spina (2003). Advanced manufacturing technologies and strategically flexible production, *Journal of Operations Management*, 18, 169-190.

Cay, F. and C. Chassapis (1997). An IT view on perspectives of computer aided process planning research, *Computers in Industry*, 34, 307-337.

Chryssolouris, G. (1992). *Manufacturing Systems - Theory and Practice*, Springer – Verlag, NY.

Chryssolouris, G., D. Mavrikios, D. Fragos, V. Karabatsou (2000a). A virtual reality-based experimentation environment for the verification of human-related factors in assembly processes, *Journal of Robotics and Computer-Integrated Manufacturing*, 16/4, 267-276.

Chryssolouris, G., N. Papakostas and D. Mourtzis (2000b). A Decision-Making Approach for Nesting Scheduling: A Textile Case, *International Journal of Production Research*, 38/17, 4555-4564

Chryssolouris, G., *et al.* (2002). A Novel Virtual Experimentation Approach to Planning and Training for Manufacturing Processes, *International Journal of Computer Integrated Manufacturing*, 15/3, 214-221.

Chryssolouris, G., S. Makris, V. Xanthakis and D. Mourtzis (2004a). Towards the Internet-based supply chain management for the ship repair industry, *International Journal of Computer Integrated Manufacturing*, 17/1, 45-57

Chryssolouris, G., N. Papakostas and D. Mourtzis (2004b). Refinery short-term scheduling with tank farm, inventory and distillation management: an integrated simulation-based approach, *European Journal of Operational Research*, (accepted for publication).

Connolly, J. (2002). A look at PLM and RPD, *Time Compression Technologies*, 7/5, 16-17.

Desforges, X., A. Habbadi and L. Geneste (2004). Distributed machining control and monitoring using smart sensors/actuators, *Journal of Intelligent Manufacturing*, 15, 39-53.

Edgar, T.F., *et al.* (2000). Automatic control in microelectronics manufacturing: Practices, challenges, and possibilities, *Automatica*, 36, 1567-1603.

Eversheim, W., *et al.* (1980). Application of automatic process planning and NC-programming. In: *The CASA / SME Autofact West Conference*.

Jacobs, F.R. and E. Bendoly (2003). Enterprise resource planning: Developments and directions for operations management research, *European Journal of Operational Research*, 146, 233-240.

King, G.S., R.P. Jones and D. Simner (2003). A good practice model for implementation of computer-aided engineering analysis in product development, *Journal of Engineering Design*, 14/3, 315-331.

Kruth, J.P. (1991). Material Incress Manufacturing by Rapid Prototyping Techniques. *Annals of the CIRP*, 40/2, 603-614

Mbang, S. and S. Haasis (2004). Automation of the computer-aided design – computer-aided quality assurance process chain in car body engineering, *International Journal of Production Research*, 42/17, 3675-3689.

McFarlane, D., *et al.* (2003). Auto ID systems and intelligent manufacturing control, *Engineering Applications of Artificial Intelligence*, 16, 365-376.

Newman, S.T., R.D. Allen and R.S.U. Rosso Jr (2003). CAD/CAM solutions for STEP-compliant CNC manufacture, *International Journal of Computer Integrated Manufacturing*, 16/7-8, 590-597.

Papakostas, N., D. Mourtzis, K. Bechrakis, G. Chryssolouris *et al.* (1999). A Flexible Agent Based Framework for Manufacturing Decision Making. In: *Proceedings of the FAIM99 Conference* (23-25 June), Tilburg, Netherlands, Begell House Inc., 789-800.

Ranky, P.G. (2004). A real-time manufacturing / assembly system performance evaluation and control model with integrated sensory feedback processing and visualization, *Assembly Automation*, 24/2, 162-167.

Ssemakula, M.E. (1990). Process planning system in the CIM environment, *Computers and Industrial Engineering*, 19/1-4, 452-456.

Standish Group International (1998).

Ülgen, O.M. and T. Thomasma (1990). SmartSim: An Object Oriented simulation program generator for manufacturing systems, *International Journal of Computer Integrated Manufacturing*, 28, 1713-1730.

Umble, E.J., R.R. Haft and M.M. Umble (2003). Enterprise resource planning: Implementation procedures and critical success factors. *European Journal of Operational Research*, 146, 241-257.

Weber, C., H. Werner and T. Deubel (2003). A different view on Product Data Management / Product Life-Cycle Management and its future potentials, *Journal of Engineering Design*, 14/4, 447-464.

Xu, Z., Z. Zhao and R.W. Baines (2000). Constructing virtual environments for manufacturing simulation, *International Journal of Production Research*, 38/17, 4171-4191.

Yeung, M.K. (2003). Intelligent process-planning system or optimal CNC programming – a step towards complete automation of CNC programming, *Integrated Manufacturing Systems*, 14/7, 593-598.

ELSEVIER
IFAC
PUBLICATIONS
www.elsevier.com/locate/ifac

# Flexible Process Models in Manufacturing based on Skeleton Product Models

**Bley, H.; Bossmann M.; Zenner, C.**
Saarland University
Institute for Production Engineering/CAM
P.O. Box 151150, D-66041 Saarbrücken, Germany
Phone: +49 681 / 302-3787, Fax: +49 681 / 302-4372, E-mail: bossmann@cam.uni-saarland.de

Abstract:

In this article a new feature concept to assist the development of reusable structures of product and process variants is described. The aim is a closer connection between product and process models. Therefore intelligent elements that can be interpreted in processes are needed in design systems. This can be achieved by an associative connection of models using a new feature concept leading to a roughly structured process tree that presents a skeleton process model.

Keywords: product strategy, processing technique, processes, variability, parallelism, manufacturing processes, interaction mechanism

## 1. INTRODUCTION

As a result of growing international competition, all industries have to increase the range of products and to reduce the time between new models. One issue to manage these challenges is to develop standards to support the development of reusable structures of product variants and the Simultaneous Engineering process. In product design exist many new methods advancing the virtual product realisation. So new concepts have to be developed that can use these design methods to save production information that are also defined by a design expert.

If there is a possibility to integrate production data in the product model following processes will be supported better and can be made more parallel. Parallel development processes lead to a reduction of costs and a better flexibility for following divisions in case of product modifications. Further results are a bigger time quota and a better product transparency caused on an improved teamwork of the divisions. In this context, the Digital Factory builds the basic condition to this aim.

## 2. DIGITAL FACTORY

Talking about IT-support in production planning, the buzzword Digital Factory is mentioned. But, what can be understood by a Digital Factory? It is not a computer game where virtual products are produced in a virtual factory.

The Digital Factory represents an approach that tries to improve the IT-assistance in the area of production planning, where real production systems that are used to manufacture real products are developed. As it is shown in figure 1, there is a relatively high degree of IT-assistance in product design and production. In product design, 2D and 3D CAD systems as well as EDM/PDM systems that are helpful for a structured administration of product data are used to fulfill requirements like short development times with a high quality of the product design. In production, ERP and PPC systems like SAP R/3 support the production process and allow e.g. for an automated materials procurement or purchase order disposition. But, in production planning that bridges the gap between product design and production there is only a very low degree of IT-assistance. If planning tools are installed, they will be used as isolated software applications, mostly without interfaces to other tools. In this context, most companies just try to combine their existing planning tools in order to realize a better IT-support in production planning (Bley and Franke, 2003).

This combination of isolated tools represents a first step towards a Digital Factory. But, the approach of the Digital Factory represents a combination of applications. Its aim is to integrate the different tools into an environment to use the information of preliminary planning phases directly and to avoid redundancies. Further more, the Digital Factory represents more than the visualization of future resources and production systems. It allows to test and to optimize the production system before its realization (Schiller and Seuffert, 2002). In this context, besides a reduction of planning time, the early detection of planning mistakes also leads to a better planning quality and a reduction of costs (Bley and Zenner, 2004).

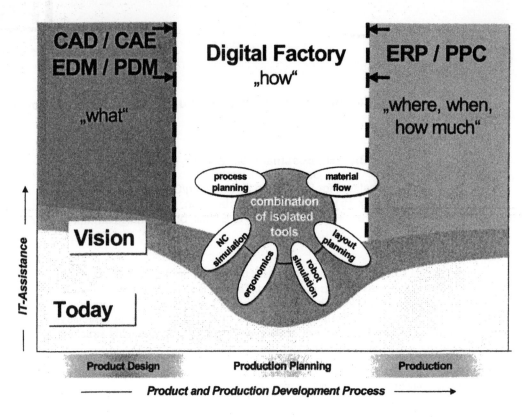

Figure 1: IT-assistance within the product and production development process

But, the Digital Factory does not only provide IT assistance in the production development process. It also provides a planning methodology that supports the daily work of the planning experts (BOSCH 2002). According to (VDI 4499, 2004), the Digital Factory is defined as a generic term for an extensive network of digital models, methods and visualization that are integrated by a continuous data management. In combination with the product, it supports the integrated planning and realization and the continuous optimization of all essential processes and resources within the real factory. At the moment, the main focus of the implementation of a Digital Factory can be seen in the combination and integration of the isolated software applications and the improvement of the functionalities of those tools. But, the integration of product design and production planning by bridging the gap between CAD systems and process planning tools also represents a high potential and a further research activity for several companies and industries, especially for the automotive industry and its suppliers. Therefore, a solution approach towards the integration of product design and process planning based on skeleton models in product design and the use of the feature technology is presented in the following.

## 3. FLEXIBLE DESIGN METHOD

Design methods support the design work with many new concepts that are developed so fast that they hurry ahead the operative insertion. These methods gain more and more tasks due to the fact

that they generate a benefit for the company by supporting routine work. One of these design methods is called skeleton modelling and is presented in the following.

### 3.1 Skeleton Modelling

Skeleton modelling is a continuous concept from the requirement specification to the 3D model. Skeleton modelling could be an important corner support in digital product development after standardization as a shorter product development phase can be reached. The skeleton modelling method bases on reference geometry (point, centre line, layer and flat) whose positions are modified by using different parameter configurations. These model references represent the know-how that is always usable without new consideration about the design and another model intelligence. Thus, skeleton models could be also the basis for the placement of design and production features that are described more detailed in chapter 3.2.

Position modifications of the features are made by the reference objects in the skeleton model, in which type and shape of the features can be modified by other parameter configuration. Skeleton models represent geometrical references and parameters to generate rough part, machining part and the finished part to support the production process by digital planning. There are different skeleton model hierarchy levels existing (figure 2). The highest skeleton model level is the assembly skeleton that consists of global product parameters and references. The second level of skeleton modelling is the sub-assembly skeleton that

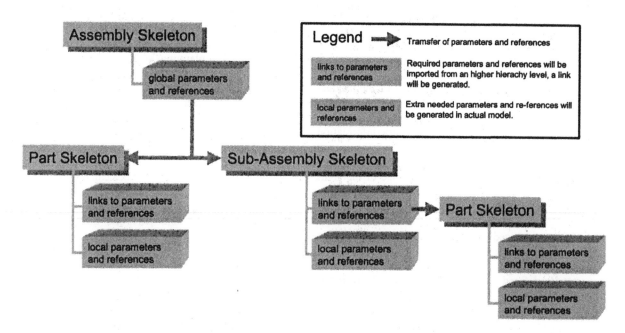

Figure 2: Skeleton model hierarchy with parameter and reference disposal

gets global parameters and references from the assembly skeleton, and makes parameters and references available to the basic skeleton level. The part skeleton is the basic skeleton model that only carries its own parameters and references. Sub-assembly skeletons and part skeletons can be at the same hierarchy level, so global parameters and references are imported directly to the part skeleton.

### 3.2 Features Technology

The Guideline VDI 2218 describes the "Feature Technology" within the product development process and exemplifies its implementation on the basis of selected practical examples. Shortly the use of knowledge in form of "Features" enables a high degree of ripeness of a serial product as well as an optimised product quality. At the same time Simultaneous and Concurrent Engineering will be effectively supported. According to the Feature Modelling Experts group "FEMEX", a feature is defined as "an information unit (element) representing a region of interest within a product" (Weber,1996). Thus, a feature is described by an aggregation of properties of a product.

The content of information of features is dependent on the support of the design system by new feature concepts.

Mostly feature concepts start with a high degree of ripeness or with the finished model as shown in figure 3. Therefore new enhanced feature concepts are needed in early product development phases to support the following processes as production planning and quality management by measuring the finished products.

Implementation of new feature concepts in early product development phases could generate the desired profit of a simultaneous engineering process. Such a feature concept could be the basis of an automatic use of other feature concepts in later development phases that are already in operative insert by using manual initiation.

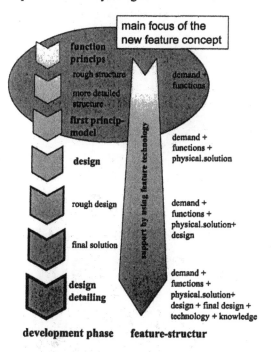

Figure 3: Feature technology in the product development phase (Vajna and Podehl, 1998)

Advantages of skeleton modelling with feature concepts are

- parametric structures,
- reusability of the models,
- better integration of process information,
- automatic insertion of measurement operations,
- product modifications as a result of automated synchronisation.

Feature concepts can build the basis for a horizontal connection between product development and production planning (Franke 2003).The described features for the skeleton modelling are called skeleton elements. These elements represent functions or simple geometry areas of the product model. Geometry areas can be e.g. the skeleton element shaft that carries the design information of the rotating geometry and the semantic information of the element type shaft.

So process planning gets the manufacturing process "drilling" and the countered reference of the shaft that have to be drilled. Another example is the skeleton element "seat of the bearing" that is interpreted as a part of the skeleton element shaft with enhanced information about following processes like grinding caused by a higher surface quality. The aim of the new feature concept is a totally defined product model that can be read out completely. All components can be interpreted in processes. The process geometry is referenced to the geometry of the skeleton elements.

Figure 4 shows an example of a skeleton model with skeleton elements.

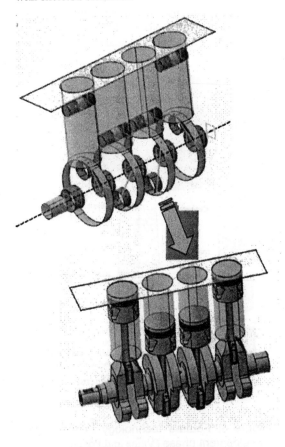

Figure 4: New feature skeleton model and 3D-model of a four cylinder engine

## 3. FLEXIBLE MANUFACTURING MODEL AS A PART OF DIGITAL FACTORY

In process modelling, there are a lot of methods too that are mainly used in conglomerates because of enormous software costs that make Digital Factory for small and medium sized enterprises prohibitive. The connection of digital product models and digital processes takes place at a later time and supports new design methods like specified skeleton modelling. The parametric models should be used for planning processes to get a flexible process model, without generating additional work and expense in the design work. Realising an effective connection between design and process models the feature information has to be read out (semi-) automatically and has to be interpreted in process elements. Mainly this is an interpretation of a feature element type into a process element type. In a next step slave applications can be read out and provided to process planning. The feature request is realised by using macros that allow to make a search in the whole part tree and to generate processes after skeleton element location. Figure 5 gives an overview of the workflow. The reference elements for the processes also have to be an element of the feature to make an automated referencing possible. Therefore the feature needs geometry information of the design element to reference the geometry to the process. The result of this concept is an associative connection between the product model and the associated process model. After activation of the feature macro the process planning gets a roughly structured process skeleton model that includes the process interpretations of all feature elements in the product model.

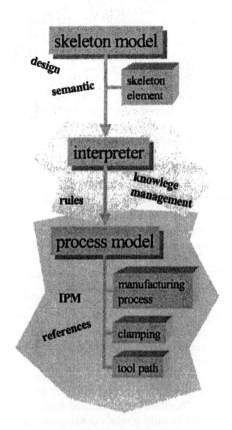

Figure 5: Knowledge transfer of the feature concept

Basically each production process gets a clamping. The interpretation is used for generating a clamping for each manufacturing technology and each manufacturing direction inside of one technology class like drilling and milling.

120

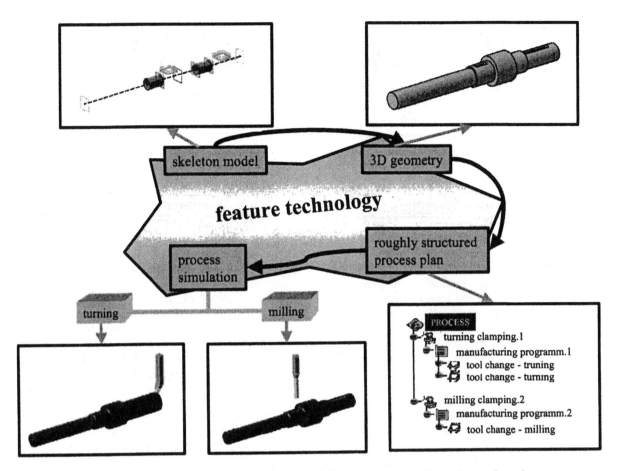

Figure 6: From the skeleton model to process plan and the process simulation of manufacturing processes

Each clamping is coupled with a process, so the result of a first feature element interpretation is a roughly structured process plan (Hassis, *et al.*, 2004). To intensify the new feature concept the product model generates automatically "in process models" that can reference to each clamping supporting process simulations by assigning rough part and finish part. Another benefit is an automatic creation of manufacturing drawing for each manufacturing segment. A first rough structure is built up by the orientation of feature elements that generate a clamping for all orientations and all manufacturing processes. A feature macro recognizes automatically every skeleton element on the imported solid, from holes to bosses and pockets, and creates associative features that are used to generate tool paths automatically by using the customer's own CNC programming preferences. This roughly structured process tree has to be overworked by the planning division that creates tool movements between all manufacturing processes in one manufacturing class to connecting them to one sub process. Further more the process planning division also has the challenge to minimize the cycle time by reorganization of the given process tree. After optimization and expansion of the process tree a process skeleton model for product variants of the product skeleton model is built up. If the solid model is changed, the associative features and the tool paths will be updated automatically. So the main result is a

flexible CAD/CAM connection with a process simulation.

## 4. IMPLEMENTATION

At this time the skeleton elements exist to build up 3D design models and to couple them to manufacturing systems. Today, a benefit already exists because the referencing to given structures in the part tree is more comfortable than a referencing to unreachable geometry information in the part model. The macro for an automated referencing is in work.

In a first step the 3D-geometry is built up by using the design information of the skeleton elements. Therefore the skeleton elements serve as references for each geometry class (shaft, hole, pocket, ...). The skeleton element carries the design information that is needed in 3D design models to built up the volume model and in the process model to realize the interaction by referencing the geometry to the manufacturing process.

Therefore the feature implementation needs all information about the downstream production processes that should be supported to build up the feature with all needed reference information. The result is not a universal feature, but a feature combination concept that supports connections to other existing feature concepts. It is not the aim to build up an universal features concept, because their initialization effects a lot of settings that are not all needed. Therefore feasible feature concepts have to support the work effectively by a minimal assignment. The feature concepts should be targeted to the special task. The task that is described in this paper is an automated CAD/CAM connection by the use of a roughly structured process plan.

## 5. SUMMARY AND OUTLOOK

The standardization of skeleton modelling with skeleton elements is an enhanced feature concept without generating additional work for the design division by using parametric models. The connection of product and process models supports the Simultaneous Engineering process. The planning division gets automated process information from the product model. Flexible associative models in production planning support routine work due to use the process model for all variants of the product by online connection of the product model and the process model. A geometrical change automatically induces a change in the process model. In this paper a connection of product model in design systems with metal cutting manufacturing is shown. The aim is the implementation of further elements that are connected to other manufacturing processes too.

The result of connecting the new feature concept with processes and knowledge management is a rough structured process plan by using feature information to generate processes and geometry references for the manufacturing processes. The connection of product, process and resource in early development phases creates the benefit of a rough cost estimation as well that supports the political decisions to further proceeding. In the future, the combination of material flow simulation between the manufacturing systems will be another important goal. (Bley, *et al.*, 2004).

## REFERENCES

Bär, T., Haasis, S. (2001)
Verkürzung der Entwicklungszeiten durch den Einsatz von Skelett-Modellen und Feature-Technologie, *VDI-Bericht Nr.1614*, Düsseldorf, Germany

Bley, H., C. Franke. (2003).
Continuity of Assembly Process Planning in the Digital Factory, *ProSTEP iViP Science Days, pp. 124-131*, Dresden, Germany.

Bley, H., Franke, C., Zenner, C. (2004).
The User as the Main Focus for the Actual Realization of the Digital Factory, *International Conference on Competitive Manufacturing, pp. 337-342*, Stellenbosch, South Africa, 2004

Bley, H., C. Zenner. (2004).
Handling of Process and Resource Variants in the Digital Factory, *Proceedings of the 37th CIRP International Seminar on Manufacturing Systems*, pp. 61-66, Budapest

Bosch (2002).
40 Prozent weniger Planungszeit, *Sonderdruck aus Automobil Produktion Ausgabe 5/2002* (in German).

Franke (2003).
Feature-basierte Prozesskettenplanung in der Montage als Basis für die Integration von Simulationswerkzeugen in der Digitalen Fabrik, Schriftenreihe Produktionstechnik, Universität des Saarlandes, Germany, ISBN 3-930429-57-8

Hassis, S., Frank, D., Rommel, B., Weyrich, W. (2000). Feature-basierte Integration von Produktentwicklung, Prozessgestaltung und Ressourcenplanung, *Konstruktion (2004) Nr.4*, Springer-Verlag, Berlin.

VDI, (2004), VDI-Richtlinie 4499: Digitale Fabrik, preliminary definition of Digital Factory (in German).

Vajna, S., Podehl, G. (1998)
Durchgängige Produktmodellierung mit Features, *CAD/CAM-Report 17 (1998) 3*, S.48-53.

Schiller, E., W.-P. Seuffert. (2002).
Digitale Fabrik bei DaimlerChrysler, Sonderdruck Automobil Produktion 2/2002, pp. 4-10 (in German).

Weber, C. (1996)
What is a Feature and What is its Use? – Result of FEMEX Working Group I; *Proceeding of the 29th International Symposium on Automotive Technology and Automation 1996, pp. 109-116*; Florence, ISBN 0-94771-978-4

# DRY ICE BLASTING AND LASER FOR CLEANING, PROCESS OPTIMIZATION AND APPLICATION

**Eckart Uhlmann, Adil El Mernissi, Jan Dittberner**

*Institute for Machine Tools and Factory Management, Technical University Berlin, Germany*

In recent years cleaning has become more and more important for production. Manufacturing, maintenance, and recycling are not possible without cleaning processes. Today almost all branches of industry use cleaning technologies. However, some are a potential hazard for the human health and the environment. The Institute for Machine Tools and Factory Management at the Technical University Berlin develops future-orientated, eco-efficient cleaning technologies for hard surfaces and cooperates with companies to integrate these technologies into the process chains of production, maintenance and recycling. This paper presents system developments, customized process optimizations and fundamental research of dry ice blasting and laser cleaning. *Copyright © 2004 IFAC*

Keywords: Manufacturing processes, ecology

## 1. INTRODUCTION

Cleaning is an important production step in manufacturing, maintenance, repair, and recycling processes. The European industry is forced to change the current cleaning processes not only due to technical aspects, but especially because of new ecological and economical circumstances. Main reasons are an increased environmental awareness, a new environmental legislation, rising waste disposal costs, and increased sewage taxes. At the moment mainly chemical, mechanical and aqueous surface treatment processes are used. However, these processes are characterized by a lack of flexibility regarding contamination and basic material, an inadequate quality due to the changing demands, and by abrasive and corrosive influences on the component to be treated. Energy and time intensive follow-up treatment of washing and drying of components are necessary and adds to the complexity and size of surface treatment equipment. Moreover, chemical, solvent, and sound emissions cause health risks for workers and environment. Often contaminated sewage from surface treatment processes is not treated and collected sufficiently. This leads to water and soil pollution. Additionally there is an extensive loss of production capacity caused by long down-times for disassembling, cleaning, and assembling.

Industry has recognized the urgent need for action. Accordingly the cleaning technology is currently in a state of change. It can be assumed that the substitution of conventional cleaning technologies will continue within the next years and that the application of new technologies will increase steadily. Dry ice blasting and laser cleaning are two of these new cleaning technologies. This Paper is a fundamental research. It presents a description of the active mechanisms and some applications of the cleaning process by dry ice blasting and laser.

## 2. DRY ICE BLASTING

### 2.1 Process

Dry ice blasting is a compressed air blasting process. It uses solid carbon dioxide ($CO_2$) –the so called dry ice- in form of pellets as a one-way blasting medium. Under atmospheric pressure solid carbon dioxide has a temperature of -78.5 °C. The production is based on fluid carbon dioxide, which is a by-product from hydrogen, ammonia, and ethanol production as well as from oil and gas refineries. For the production of dry ice pellets liquid carbon dioxide with approximately -20 °C and 20 bar is expanded to atmospheric pressure. Due to the Joul-Thomson-effect during the expansion it cools down and $CO_2$-snow is generated. With help of a hydraulic stamp the $CO_2$-snow particles are pressed through a mold and cylindrical dry ice pellets are formed. Pellet parameters that effect the cleaning process are density, hardness and shape of the $CO_2$-pellets (Axmann and Elbing, 1997; Axmann and Elbing, 1998). Fig. 1 shows the components of a dry ice blasting system.

After the cleaning process, no residues of the solid blasting medium are left as the dry ice immediately sublimes when striking the surface, it attains the gaseous phase. This is a great advantage to other cleaning processes where the medium either requires complex processing or involves costly disposal as the impurities are dissolved in the medium. Also the fact that dry ice pellets turn to gas has the advantage that no media remains in boreholes or cavities of the machined parts. Subsequent cleaning or drying is not necessary. Dry ice pellets do not produce any residues and are only slightly abrasive and corrosive, and flexible concerning different material and contamination (Uhlmann and Elbing, 1998). Dry ice blasting work time is saved because application can take place on site without having to remove sensitive parts. Machining times are distinctly shorter than those when sand blasting is used. Time savings can be as high as 50 % to 75 % when blasting moulds. This less aggressive method also serves to prolong the service life of the moulds. One disadvantage inherent in jet processes that are operating in fully automatic mode is that they expose operators to extreme working conditions. Sound pressure levels of up to 125 dB(A) were measured for dry ice blasting during machining with equipment which was not hermetically sealed. Therefore, the process should be conducted with hermetically sealed equipment and staff should be equipped with hearing protection. Moreover, the $CO_2$-concentration in the surrounding air may be elevated.

For dry ice blasting, the pellets are collected in a storage element and injected into an air stream by means of an breaker plate wheel used as a dosing device. The pellets are accelerated by compressed air through a laval nozzle and depart from the nozzle at almost the velocity of sound

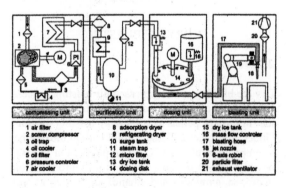

| | |
|---|---|
| 1 air filter | 8 adsorption dryer |
| 2 screw compressor | 9 refrigerating dryer |
| 3 oil trap | 10 surge tank |
| 4 oil cooler | 11 steam trap |
| 5 oil filter | 12 micro filter |
| 6 pressure controler | 13 dry ice tank |
| 7 air cooler | 14 dosing disk |

| | |
|---|---|
| 15 dry ice tank | |
| 16 mass flow controler | |
| 17 blasting hose | |
| 18 jet nozzle | |
| 19 6-axis robot | |
| 20 particle filter | |
| 21 exhaust ventilator | |

Fig. 1. Dry ice blasting parameters

## 2.3 Active mechanisms

The active mechanisms in dry ice blasting are based on three effects (Fig 2) (Spur, *et al.*, 1998a; Uhlmann, *et al.*, 1999):

- The thermal effect supplied leads a regional cooling of the part where the pellets strike the surface. As a result elasticity is lost, and the adhering coating embrittles and shrinks. Cracks are being formed. Due to the different thermal expansion coefficients of coating and substrate, the bond dissolves when the adhesive energy is exceeded. The coating partially chips off.
- The mechanical effect duo to kinetic energy of pellets and air stream contributes to the removal of the coating (Spur, et al., 1998b).
- The sublimation effect duo to the sudden increase in volume by 800 fold resulting from the sublimation when the pellets strike the surface of the part supports the process. Gas flows underneath the adhering coating. The removal of material is hence based on a combined thermo-mechanical effect.

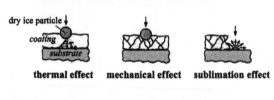

**thermal effect**   **mechanical effect**   **sublimation effect**

Fig. 2. Active mechanisms of dry ice blasting

## 3. LASER CLEANING

In recent years, an increasing interest in using laser technology for surface cleaning was observed. Compared to the conventional methods of surface cleaning laser offers significant advantages such as (Wissenbach and Johnigk, 2002):

- contact-free and hence force-free cleaning,
- cleaning of sensitive surfaces,
- low thermal and mechanical influence of substrate,
- high precision,
- no blasting medium and no chemicals used,
- high grade of automation and control, and
- selective cleaning.

### 3.1 Process

The process of laser cleaning is based on local and contact-free influence of pulsed or continuous laser beam. The process relies on the fact that damage threshold of the contamination is lower than of the underlying substrate material. The energy density of the laser pulse is adjusted so that it exceeds the damage threshold of the contamination layer, but is well below the damage threshold of the substrate material. In this way the surface structure of the underlying material is not damaged. Once the surface is clean, further use of the laser will have no harmful effect. The process is therefore self-limiting making laser cleaning very suitable for the cleaning of fragile substrates.

*3.3 Active mechanisms*

Different active mechanisms appear depending on composition and thickness of the contamination, and basic material properties. For the cleaning process the following mechanisms are important (Fig 3):

- ablation by evaporation or decomposition of the contamination layer, and
- ablation by thermally induced stresses or by a laser beam-induced shock wave.

The laser light is rapidly absorbed by the top few microns of the contamination instantaneously vaporizing this thin layer and forming highly compressed plasma (unstable high-pressure ionized gas). The creation of this plasma generates a shock wave which propagates into the bulk contamination breaking it up and ejecting the contamination as fine particles. The ejected particles are then captured by an exhaust extraction system (Wissenbach and Johnigk, 2002).

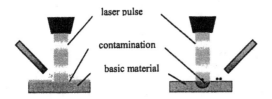

Fig. 3. Active mechanisms of laser cleaning

Often several mechanisms operate simultaneously. The respective portion depends on the material properties and the setting parameters, in particular of the wavelength, power density, and time of interaction. The absorption of the contamination and the basic material is of special importance for the cleaning process. If the contamination absorbs the laser beam well and the underlying substrate material exhibits a small absorption factor, the process of removal is stopped automatically with the impact of the laser beam on the reflecting substrate.

## 4. EXPERIMENTAL INVESTIGATIONS

*4.1 Experimental set-up*

Experimental investigations on dry ice blasting were conducted with a compressed air system by Kaeser Kompressoren, Coburg/Germany, which comprise a stationary single-stage screw compressor DSB 170, an equalising vessel with a volume of 0.4 $m^3$, a steam trap Eco-Drain 13, and a micro filter element FX to collect oil and water. The screw compressor with a nominal motor performance of 90 kW provides for a maximum blast pressure of 14 bar and a maximum volume flow of 11 $m^3$/min. The mobile blasting system used, MicroJet by Green Tech & Co. KG,

Hofolding/Germany, accelerate the blasting medium according to the pressure principle. To handle the nozzle a 6-axis articulated arm robot of the type mantec r3 was used. The machining area and the articulated arm robot are surrounded by a sound protection of the size 3 m x 3 m x 3 m and connected to an exhaust removal with a filter for particles. To accelerate the blasting medium, depending on the application, a round and flat nozzles of Linde, Munich/Germany, were used. The dry ice blasting test stand is displayed in (Fig 1). As blast medium cylindrical dry ice pellets from Messer Griesheim GmbH, Krefeld/Germany, were used. The pellets were 3 mm in diameter with a length between 2 mm and 18,5 mm. The medium length was 6,7 mm.

For the realization of the experimental investigations on laser cleaning the test stand displayed in Fig. 4 were used. The laser is an electro-optically pulsed Q-switched Nd:YAG solid body laser with an infrared laser radiation at a wave length of 1,064 nm.

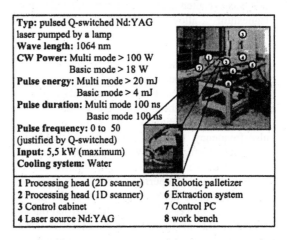

| Typ: pulsed Q-switched Nd:YAG laser pumped by a lamp |
|---|
| **Wave length:** 1064 nm |
| **CW Power:** Multi mode > 100 W |
| Basic mode > 18 W |
| **Pulse energy:** Multi mode > 20 mJ |
| Basic mode > 4 mJ |
| **Pulse duration:** Multi mode 100 ns |
| Basic mode 100 ns |
| **Pulse frequency:** 0 to 50 (justified by Q-switched) |
| **Input:** 5,5 kW (maximum) |
| **Cooling system:** Water |

| 1 Processing head (2D scanner) | 5 Robotic palletizer |
|---|---|
| 2 Processing head (1D scanner) | 6 Extraction system |
| 3 Control cabinet | 7 Control PC |
| 4 Laser source Nd:YAG | 8 work bench |

Fig. 4. Technical proprieties of laser cleaning system

The most important components of laser cleaning systems are the laser source, the beam guidance system, the handling system, the control system, and the extraction system. For the handling a 3-axis-robotic palletizer with a working area of 500 mm x 500 mm x 250 mm, a maximum speed of more than 20 mm/s and a positioning accuracy of more than 100 µm was constructed and realized in cooperation with Bauer&Mück GmbH, Berlin/Germany. It is possible to direct the process by optics guidance or transfer by glass fiber. Two different processing heads were used, on that were attached to different optics. In the processing heads the laser beam is moved over the surface by beam deflection systems (scanners). The scanners are a linear scanner with a woblle galvanometer and a two-dimensional scanner with XY galvanometer. They are connected to the control system. The processing head with the linear scanner can be used by the robotic palletizer or as a hand-held unit, which increases the flexibility of the system. For the laser cleaning the complete extraction of the by-products is necessary for quality and operational safety. An extraction system with particle filter and activated-carbon filter for the outflow of all by-products is used.

With dry ice blasting the removal of the paint from metallic parts, the removal of silicone seals as well as the removal and cleaning of mounted printed-circuit boards were examined. The experimental investigations of laser cleaning were focused on the removal of the paint from metallic parts and the cleaning of exchange engine components. The coated metallic parts processed by dry-ice blasting were sheet steel made of DC01 A with a workpiece thickness of 1 mm and coated with stove enamelling paint Finadur 781 of Mankiewicz, Hamburg/Germany. The metallic parts AlMg3 and X5CrNi 10-18 were de-coated by laser beam. The layer had thicknesses between 15 μm and 70 μm. The mounted printed-circuit boards were coated with conformal coatings on the basis of epoxy, polyurethane, and silicone resin. The epoxy coating of the type SL 1305-25 AQ had a thickness of 10 μm and the polyurethane coating SL 1308 FLZ of 50 μm. Both were produced by Lackwerke Peters, Kempen/Germany. The removing from silicone seals was executed at components from an aluminum magnesium alloy of the A-Klasse of DaimlerChrysler.

During testing phase the setting parameters of the dry-ice blasting process pressure, mass flow of dry ice, working distance, blasting angle, blasting time, and feed speed of the nozzle were varied. For the laser cleaning investigations the setting parameters pulse frequency, the feed speed, distance of points and laser performance were varied. For the evaluation of the tests, the removal width b and the therefore the removal performance P were measured. The measuring of the roughness and of the surface profile of the processed surfaces was carried out at a Talysurf 120 L of Taylor-Hobson, Leicester/United Kingdom. To evaluate the surface topography and the structure of the material, a scanning electron microscope DSM 950 and an optical microscope PJN 322 of the company Zeiss, Oberkochen /Germany were used.

### 4.2 Removal of paint from metal components by dry ice blasting

An important application for dry ice blasting is the removal of paint from metal components. For this purpose, the parameters for processing sheet steel material were optimized (Uhlmann and Elbing, 1998). The aim was to find the optimum settings for stove enamelling paint with different coat thicknesses between $s_L = 15$ μm to 70 μm. Fig. 5 shows maximum feed speed $v_{fmax}$ versus working distance a and blasting angle ß as well as the removal performance P versus feed speed $v_f$ for different coat thicknesses $s_L$ of stove enamelling paint for paint removal with dry ice blasting working with a round and a flat nozzle.

Using a round nozzle, the maximum feed speed for stove enamelling paint was in the range of $v_f = 1.0$ m/min and 4.0 m/min, for the flat nozzle between $v_f = 0.2$ m/min and 0.7 m/min. The highest

feed speed was obtained with a blasting angle of ß = 90 ° and a working distance of a = 150 mm for the round nozzle and of a = 50 mm for the flat nozzle. The reason for the increase of the maximum feed speed with increasing blasting angles is the increase in kinetic energy applied to the paint. Furthermore, the resultant thermal stress leads to cracking. This means increased pellet penetration into the cracks so that the paint is removed by the sublimation effect.

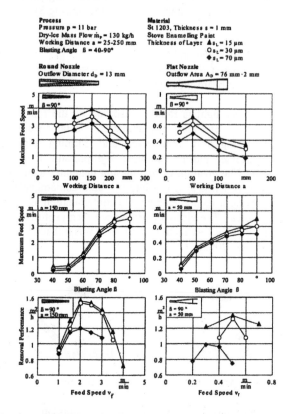

Fig. 5. Maximum feed speed versus working distance and blasting angle as well as removal performance versus feed speed for different thicknesses of layer of stove enamelling paint for paint removal from sheet steel material with dry ice blasting working with a round and a flat nozzle

With an increasing feed speed the removal performance increases up to maximum. By increasing the feed speed even more occur a strong decrease of the removal performance. The maximum removal performance is not in the very low or very high feed speed range. Instead, it can be found where the ratio between removal performance and feed speed is lowest. In the case of stove enamelling paint the removal performance decreases with an increase in coat thickness at constant feed speed.

### 4.3 Removal of paint from metal components by laser

The experimental investigations on the removal of paint from metal components with laser covered the determination of the relevant setting parameters as well as their influence on the cleaning result, on the thermal and on the mechanical surface. The distance of points, pulse frequency, feed speed, beam forming and the beam performance were defined as relevant setting parameters. The distance of points determines the space between machined areas by the pulsed laser system. For the evaluation of the mechanical and thermal influence the mean total height Rtm was measured and the processed surfaces were observed with an optical microscope. As a result, the modulation of feed speed and the beam performance shows a significant influence on the mean total height $R_{tm}$ (Fig 6).

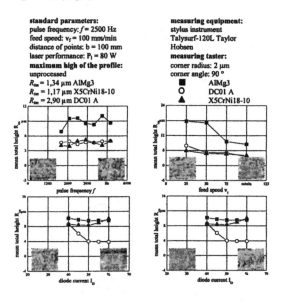

Fig. 6. Mean total height $R_{tm}$ versus pulse frequency, feed speed, point distance and diode current and photos of the processed surfaces as well as sub-surfaces

With increasing feed speed and beam performance the mean total height $R_{tm}$ decreases. The maximum cleaning performance was obtained with a distance of points of 100 μm, a pulse frequency of 3500 Hz, a beam performance of 91 W and a feed speed of 150 mm/min. During processing with glass fiber beam guidance, the maximum cleaning performance was determined with a distance of points of 50 μm, a pulse frequency of 4000 Hz, a beam performance of 91 W, and a feed speed of 40 mm/min. In summary, the cleaning performance with direct guided laser beam is four times higher than the cleaning performance with glass fiber guided laser beam.

### 4.3 De-coating and cleaning of mounted printed-circuit boards

A other application of dry ice blasting is the de-coating and cleaning of mounted printed-circuit boards (Uhlmann, et al., 1999). Objective is the reuse of components and printed-circuit boards. The process is applied for the removal of conformal coatings within repair and recycling and before the de-soldering process of damaged or expensive components. With regard to dry ice blasting, the blasting pressure was reduced to 5 bar and the dry ice

mass flow to 80 kg/h. Fig. 7 shows the paint removal performance P versus the feed speed $v_f$ for the removal of epoxy, polyurethane, and silicone coatings from mounted printed-circuit boards by dry ice blasting working with a round and a flat nozzle.

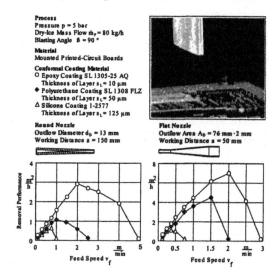

Fig. 7. Removal rate versus feed speed for the removal of epoxy, polyurethane, and silicone coatings from a mounted printed-circuit board by dry ice blasting working with a round and a flat nozzle

With an increasing feed speed, the removal performance increases up to maximum. By increasing the feed speed even more occur a strong decrease of the removal performance. Using a round nozzle, the maximum removal performance for the epoxy coating with a typical layer thickness of 10 μm was P = 3 m²/h, for the flat nozzle P = 7 m²/h. The highest removal performance was obtained with a feed speed of $v_f$ = 2 m/min for both round and the flat nozzle. For the round nozzle the maximum removal performances for the polyurethane coating with a thickness of 50 μm and of the silicone coating with a thickness of 125 μm were P = 1 m²/h and P = 0.7 m²/h. The optimum feed speeds were v = 1 m/min for the polyurethane coating and v = 0.7 m/min for the silicone coating. For the flat nozzle a removal performance of more than P = 4 m²/h with an optimum feed Speed of $v_f$ = 1.5 m/min for the polyurethane coating was obtained. The removal performance for the silicone coating was P = 1 m²/h with a feed speed of $v_f$ = 0.4 m/min. The experimental results show that it is possible to remove different types of conformal coatings at different layer thicknesses with dry ice blasting. The obtained feed speeds were very high compared to the removal of paint from sheet steel material.

### 4.4 Removal of silicone seals by dry ice blasting

The seal between the components in automotive engines is provided with silicone. Within repair and maintenance two typical problems of silicone occur: Decomposition of the components by the bonding effect of silicone and for removal of the seal remainders is difficult. There exists no satisfying technology yet. At present, the silicone seals are removed mechanically by manual scraping-off. However, this is a very time-consuming and expensive process. There is no known chemical solvent or cleaning agent that silicone can be solved in. Due to its elastic characteristics it is not possible to remove silicone completely by blasting with sand, glass or steel. Moreover, these procedures lead to damages of the components. In addition, the geometry of the components is very complex.

The experimental investigations on the removal of silicone seals were executed at components from an aluminum magnesium alloy of the A-Klasse of DaimlerChrysler (Fig. 8). For the process optimization mass flow of dry ice, pressure, working distance, angle, and feed speed was varied. A high performance nuzzle as well as a beam hose with a length of 8 m for transporting of the dry ice pellets were used. The experimental optimization yielded a mass flow of dry ice from 130 kg/h, a pressure of 11 bar and a working distance of 100 mm at a free angle of 90 °. Depending of thickness and adhesion of the silicone residues as well as the breadth, depth and accessibility of seal joint the processing with a feed speed up to 1 m/min by complete cleaning was possible. To assess possible material impairment through dry ice blasting the surfaces and pre-surfaces of the materials after the processing were analyzed. This was performed with the help of a surface measuring laser and a scanning electron microscope. No significant modifications of the surface roughness and the material structure occurred the feed speeds, which are relevant for removing silicone seals.

Fig. 8: Removal of silicone seals with dry ice blasting.

### 4.5 Cleaning of exchange engine components by laser

In the context of a feasibility study exchange engine components were cleaned with a laser. The cleaned components were cylinder heads, pistons and outlets with strongly adherent contaminations (Fig. 9). These components were contaminated with agreed, carbon-containing impurities (Uhlmann and El Mernissi, 2003). The cleaning of the examined exchange engine components by laser was successful. The contamination were completely removed. A damage of the basic material was prevented by an exact adjustment of the relevant parameters.

Fig. 9. Cleaning of piston with laser

## 5. CONCLUSION

The executed feasibility studies have shown that dry ice blasting and laser cleaning are flexible usable and environmentally beneficial manufacturing processes. It is possible to use both to clean different contaminations from different basic materials ant to de-coat different coating systems. Particularly in cleaning a series of companies apply these processes successfully.

## REFERENCES

Axmann, B. and F. Elbing (1997). Strahlverfahren. In: Zeitschrift für wirtschaftlichen Fabrikbetrieb, Vol. 92, No. 3, pp. 76-77.

Axmann, B. and F. Elbing (1998). Kompetenzzentrum Strahlverfahren. In: Zeitschrift für wirtschaftlichen Fabrikbetrieb ZWF, Vol. 93, No. 3, pp. 57.

Spur, G., E. Uhlmann and F. Elbing, (1998a). Experimental Research on Cleaning with Dry ice Blasting. In: Proceedings of the International Conference on Water Jet Machining WJM '98, Krakow/Poland, pp. 37-45.

Spur, G., B. Axmann, and F. Elbing (1998b). Model for Kerf Profile and Jet Velocity in Abrasive Water Jets. In: Proceedings of the 5th Pacific Rim International Conference on Water Jet Technology, New Delhi/India, pp. 407-420.

Uhlmann, E. and F. Elbing (1998). Reinigen und Entschichten mit $CO_2$-Trockeneisstrahlen. In: Schwerpunktseminar Reinigen/Abtragen, Institut für Werkstoffkunde, Universität Hannover, Hannover/ Germany.

Uhlmann, E., B. Axmann and F. Elbing, (1999). Cleaning and Decoating in Disassembly. In: Proceedings of the 4th World Congress R '99 - Recovery, Recycling, Re-integration. Geneva, Schwitzerland.

Uhlmann, E. and A. El mernissi (2003). Arbeitsschutz und Sicherheitstechnik in der Demontage. In : Arbeits- und Ergebnisbericht, Teilprojekt A4, Sonderforschungsbereich 281 "Demontagefabriken zur Rückgewinnung von Ressourcen in Produkt- und Material-kreisläufen", TU Berlin, S. 129-160.

Wissenbach, K. and C. Johnigk (2002). Laserstrahlreinigung bei der Metallverarbeitung. In: Fraunhofer IPA Technologieforum F 83, Stuttgart, S. 21-57.

ELSEVIER
IFAC
PUBLICATIONS
www.elsevier.com/locate/ifac

# DYNAMIC LIFE CYCLE CONTROL OF INTEGRATED MANUFACTURING SYSTEMS USING PLANNING PROCESSES BASED ON EXPERIENCE CURVES

**Niemann, J., Westkämper, E.**

*IFF Institute of Industrial Manufacturing and Management, Universität Stuttgart,
Fraunhofer IPA, Nobelstrasse 12, 70569 Stuttgart, Germany*

Abstract: The paper presents a framework for controlling the life cycle of so called "integrated manufacturing systems". In the process, the economic efficiency of continuous improvement measures taken during its further life are evaluated in a simulation and monitored using real production data. To do so the paper identifies relevant data sources and describes the integration into a simulation model. To improve the performance of the manufacturing system the model describes general starting points. Before implementation the changes will be simulated and evaluated. After implementation the data from reality will be compared with the forecast to improve the model. *Copyright © 2004 IFAC*

Keywords: Production planning, life cycle, simulation

## 1. INTRODUCTION

Today's manufacturing companies have the option of evaluating and optimizing the usage of investments from a life cycle point of view. In order to avoid outage losses, investment goods need to be continuously adapted during their life cycle to compensate for the technical, economical and organizational drivers of change which occur. To achieve this, influencing factors critical to success need to be identified and reproduced in order to benefit from the behavior of the overall system in a simulation for dynamically evaluating and optimizing a series production. To do this, segments need to be structured hierarchically in model cells and their effects in the network depicted. In this way, using an effect model, segment activities can be optimized pro-actively over their life cycle. In order to meet the requirements of modern manufacturing systems, the model is based on a system-related, structured and integrated production segment. By integrating a planning concept based on experience curves, it is possible to achieve continuous technical/economical controlling to maximize segment usage over the entire life cycle.

## 2. INTEGRATED MANUFACTURING SYSTEMS

In the 90's, production-orientated topics such as the modularization or segmentation of manufacturing execution and planning processes were met with increasing resonance. The concepts of the "fractal company" (Warnecke and Hüser 1993) or of the "modular factory" (Wildemann 1998) with "customer-orientated manufacturing by segmenting production" are management concepts with systematic integrated thinking in business processes. Technologically-orientated production segments are manufacturing systems or product- and market-orientated organizational units possessing extensive autonomy and integrated responsibility as far as costs, deadlines and quality are concerned. The term "integrated manufacturing system" describes the comprehensive integration and optimization of all company activities with the central aim of producing goods or services in a profitable and customer-orientated manner. This is attained by focusing on methodical principles and strictly applying them in order to achieve the objective set and also by optimizing employee potentials when initiating and taking improvement measures. Through this form of understanding, the term not only addresses the actual production field but also includes and integrates the peripheral areas of technical and organizational support (Korge and Scholtz 2004, Hinrichsen 2002, Spath 2003). As a result, this concept does not only take into consideration the actual production and the areas providing direct support (maintenance, control, etc.) in optimization processes but also includes logistics, work preparation and quality assurance (Figure 1).

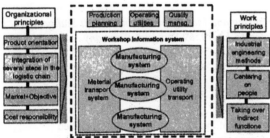

Figure 1: Operating principles of an integrated manufacturing system.

Therefore, system limits are no longer fixed for integrated manufacturing systems according to space- or time-related points of view. The main criterion when setting up the system is much more the necessity or function of a process in creating and supplying a product with regard to the criteria of costs, quality and time. In this way, an integrated manufacturing system can be viewed as being a networked unit of all the steps and elements of the production process involved in creating and supplying the product in a more far-reaching sense. The optimization of this extended overall system promises greater potentials in comparison with those which could be attained by improving individual elements singularly. The optimization of the overall system can be seen as being a continuously-running process which is essentially controlled by man and which functions according to well-known process-oriented and industrial engineering principles and methods. The system is controlled and (continuously) improved using robust processes, standardized work sequences and methodically-supported learning processes (Korge and Scholtz 2004, Hinrichsen 2002, Spath 2003).

### 3. MAXIMUM EXPLOITATION OF USAGE OF MANUFACTURING SYSTEMS OVER THEIR LIFE CYCLE

Such types of manufacturing system invariably require a substantial capital investment. However, these machines and systems usually have a considerably longer life span than the products or generations of product they produce. Therefore, the life cycle of integrated manufacturing systems is not so much influenced by technical wear but rather by

the danger of the system ageing technically as a result of rapid market developments. For this reason, permanent planning activities over the life span of the manufacturing system are required in order to guarantee investment security. Optimum economic decisions regarding system configurations need to be made constantly in order to achieve maximum system usage.

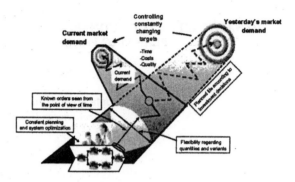

Figure 2: Controlling volatile products and production targets (altered in accordance with (Schuh et al. 2003)).

Figure 2 shows the planned life of an investment. The investment is calculated and made based on an assumed production program and an assumed lifespan. However, when the system is actually in use, numerous unexpected changes take place regarding order numbers and specifications. The true characteristic values of the system also deviate considerably from those calculated, with the result that the profitability of the investment needs to be permanently monitored and improvement measures constantly taken because of the volatile production targets. Due to limited forecasting, these factors cannot or can only marginally be calculated. It is for this reason that continuous adjustments and alterations which are life cycle-orientated need to be made. Their benefit (effect) must then always be evaluated with regard to the remaining lifespan of that particular manufacturing segment. Any measures or investments carried out beforehand represent "sunk costs" for the periods of time taken into consideration and are not relevant to later planning measures.

### 4. HIERARCHICAL STRUCTURING OF INTEGRATED MANUFACTURING SYSTEMS

Because system limits have been fixed very broadly, a deep multi-layered interlocking of effect relationships now exists between the individual system elements. Due to the high level of complexity, it is wise to evaluate the integrated effect of individual optimization measures with the aid of simulation studies. However, for the simulation, the individual objects (system resources) need to be structured horizontally and vertically according to system-related criteria in order to describe the effect relationships according to type, intensity and effect. In this way, subsystems are formed which reduce the degree of complexity. As the target function is to optimize the overall life cycle usage of the integrated manufacturing system, the individual objects (resources) are described according to the functions of costs, quality and time.

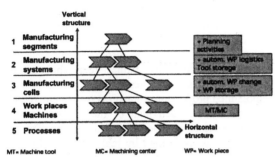

Figure 3: Structural layout of an integrated manufacturing system.

Figure 3 describes the horizontal and vertical structure of an integrated manufacturing system. Processes requiring resources with corresponding process times are placed at the bottom level. Planning activities are placed at the top hierarchical level as they are usually integrative, i.e. are active for one, several or all of the subsystems. (Westkämper 2004)
By using such a formal structure, it is possible to configure integrated manufacturing systems and to formally describe their composition.

## 5. INTEGRATION OF EXPERIENCE CURVES INTO LIFE-CYCLE PLANNING

Experience curves have the effect of reducing the costs, amount of time, etc., required to manufacture a product unit. Costs are reduced by a factor in percent each time the quantity manufactured is doubled. This effect was first discovered in aircraft manufacturing and has since been proven empirically. However, nowadays the knowledge obtained from the experience curve theory has been widely applied in practice and is used to forecast piece costs over time. It is essential that these effects do not occur automatically, but rather are always based on operational measures which result in cost reduction. (Hieber 1991) These measures are induced and implemented by production employees or planning departments to achieve an economizing effect in later production. In order to implement experience curves in planning the life cycle of manufacturing systems, effects which up till now have always been product-orientated need to be transferred to a plan for an integrated manufacturing system which is life cycle-orientated. Figure 4 lists the general points associated with the concept of such improvement measures. In order to be able to describe and model these effects, it must also be clarified which sub-systems (segment elements)

within the system-related structuring of the integrated manufacturing system will be affected by the measures planned. Therefore, for each effect point, the consequence of the measure must be described according to the dimensions of costs, time and quality.

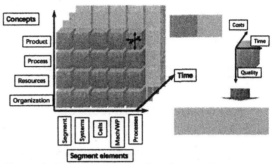

Figure 4: Sources and reductions in experience curve effects over time.

(Cost) Savings may be the result of rationalization effects, scale effects, learn or practice effects or of technical advancements. (Hieber 1991) However, the description of the effects on the individual objects (partial models) only represents a statistical description of the situation; the main aim is to evaluate the consequences of the measures on the overall system and over its entire life-cycle.
Each total process time is thus made up of the execution time and the preparation time. (REFA 1978) If the sum of all total process times from all the orders is calculated and then multiplied by all of the resource costs involved, it is possible to depict the segment-related costs. Experience effects are then finally expressed as a reduction in production costs (cost-evaluated total process times). If this is documented continuously over time and also forecast for future orders using the work plans, life-cycle related segment activities can thus be optimized. Figure 5 illustrates this behavior in the form of a graph.

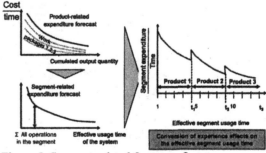

Figure 5: Segment-related forecast of costs.

In order to create effect models, the elements (resources) of the manufacturing system first need to be defined. The set-up can be made as detailed as desired. In order to derive core statements, it is sufficient to classify resources into the categories of personnel, operating utilities and auxiliary means (tools, large transportation aids, etc.). (Lorenzen

1997) These are then assigned to being time-related resource cost functions with the result that operating utilities and auxiliary means are made up of hourly machine rates and hourly wage rates for personnel. Quality defects, maintenance activities, etc., are integrated by norming activities to the "effective system usage time". Such disturbances represent unproductive periods of time and their elimination serves to increase productive usage time. An increase in productive time periods represents a reduction effect in relation to fixed costs, thus inducing a decrease in the experience curve.

## 6. INTEGRATION INTO A SIMULATION ENVIRONMENT

By setting up a simulation for integrated manufacturing systems, it is possible to create a controlling model which enables usage to be planned, controlled and evaluated over the entire life-cycle. Planning is carried out based on the simulation of operating sequences and is constantly verified. The elementary model cells form the basis for modelling segment activities. The aim is to optimize the maximum usage of the overall system over the planned life cycle. Maximization of usage is defined as being the maximum economization in costs of the aggregated life cycle costs compared with the ex-ante life cycle cost calculation at $T_0$ (Figure 6). The reference curve created as part of the investment planning is based on the assumption of a future production program right up to the planned end of the manufacturing system's life.

All current and future manufacturing orders are then continuously planned into the usage phase of the system, aggregated at segment level and then compared with the reference curve. Time- and cost-orientated assessments of rationalization measures to avoid opportunity costs make up simulation input values.

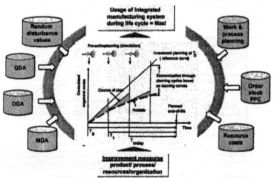

Figure 6: Controlling the life cycle of integrated manufacturing systems.

In this way, cost effects due to order fluctuations, product variants, program changes and alterations in segment configuration (change drivers) can be evaluated over its life cycle and an increase in usage continuously induced by implementing appropriate measures. Thus by making events dynamic, it is possible to make a cost-benefit estimation of the measures over the planned remaining lifespan of the manufacturing system. The method to be developed starts in the phase where an order quote is made. As part of the pre-calculation, it provides a cost forecast based on experience curves for expected average piece costs per batch. Planning is carried out using previous production data filed in the PPS system from identical or similar orders ("starting point of the experience curve"). Planning data from work and process planning and also resource-related planning values or disturbance values (e.g. shift models, maintenance, set-up times, assumptions concerning unplanned production stoppages) is then entered into the simulation-aided plan of the experience-based curve expected for the production orders. The data basis for this is formed by real data recorded from production data acquisition. In addition, data from the MDA(machine data acquisition)/ODA (operating data acquisition) and QDA (quality data acquisition) is compressed in order to be able to evaluate the actual consumption of resources from a monetary point of view. Alterations are then entered as new forecasts or planned values into the next planning cycle whilst taking the new order and preparation times into consideration in the altered work plans. The controlling model thus supplies data concerned with events, planning and evaluation by integrating the experience curves into a time scale over the entire lifecycle of a segment. In this way it is possible to create a closed control loop which constantly supplies updated references for planning cycles. The continuous recording of time-related developments and the addition of all batches/orders produced by the manufacturing system over its entire life cycle thus permits the life cycle of the system to be evaluated in detail and to be dynamically controlled.

## 7. SUMMARY

The development of an experience curve-based method for controlling the life cycle of integrated manufacturing systems permits the consequences of technical, organizational and product-related drivers of change to be continuously simulated and evaluated over a system's complete life cycle. In order to achieve maximum usage of a production segment adapted to the situation at that moment in time, the effects and overall benefit of improvement measures taken can be assessed in advance. By coupling this with real production data, the planning model can be verified based on a closed control loop. The industrial engineering improvements made are reflected in work plans with reduced order and preparation times and also in lower resource costs.

# REFERENCES

Warnecke, Hans-Jürgen; Hüser, Manfred: Revolution der Unternehmenskultur : Das fraktale Unternehmen. 2. edition Berlin et al.: Springer, 1993

Wildemann, H.: Die modulare Fabrik: kundennahe Produktion durch Ferti-gungssegmentierung, 5.edition, München, 1998

Korge. A, Scholtz, O. Ganzheitliche Manufacturing systems, Produzierende Unternehmen innovativ organisieren und führen, in: wt Werkstattstechnik online vol 94 (2004) H.1/2, p. 2-6.

Hinrichsen, S.: Ganzheitliche Manufacturing systems - Begriff, Funktionen, Stand der Umsetzung und Erfahrungen. In: FB/IE - Zeitschrift für Unternehmensentwicklung und Industrial Engineering, Darmstadt, 51 (2002) 6, p. 251-255.

Spath, D. (Hrsg.): Ganzheitlich produzieren. Innovative Organisation und Führung. Stuttgart: LOG_X Verlag GmbH 2003

Schuh, G., Sesterhenn, M., König, R., Lebenszyklusgestaltung kollaborativer Produktionssysteme, in: zwf Zeitschrift für wirtschaftlichen Fabrikbetrieb, 98 (2003) 1-2, p. 17-21

Westkämper, Engelbert: Das Stuttgarter Unternehmensmodell: Ansatzpunkte für eine Neuorientierung des Industrial Engineering. In: REFA Landesverband Baden-Württemberg: Ratiodesign: Wertschöpfung - gestalten, planen und steuern. Bodensee-Forum. 17. und 18. Juni 2004, Friedrichshafen. Mannheim, 2004, p. 6-18

Hieber, Wolfgang Lothar: Lern- und Erfahrungseffeke und ihre Bestimmung in der flexibel automatisierten Produktion. München: Vahlen (Controlling Praxis) 1991

REFA Verband für Arbeitsstudien und Betriebsorganisation: Methodenlehre des Arbeitsstudiums - Teil 2: Datenermittlung. 6. volume München : Hanser, 1978

Lorenzen, Jochen: Simulationsgestützte Kostenanalyse in produktorientierten Fertigungsstrukturen. Berlin u.a.: Springer, 1997 (iwb Forschungsberichte 107). München, Techn. Univ., Fak. für Maschinenwesen, Diss. 1996

ELSEVIER

IFAC

PUBLICATIONS
www.elsevier.com/locate/ifac

# ADAPTATION OF CONTROL PROGRAMS TO ASYNCHRONOUS I/O UPDATES

Richard Šusta*

* Department of Control Engineering,
Faculty of Electrical Engineering, Prague
http://dce.felk.cvut.cz/susta/

Abstract: The paper discusses the problems with adaptation of industrial control programs from synchronous to asynchronous I/O update when peripherals are redeveloped with the aid of control area networks (CANs). CANs decrease necessary wires between controllers and technology, but they have lower data update rates in comparison with I/O buses. To accelerate and optimize I/O polling, I/O data can be updated asynchronously to evaluations of control algorithms.
If a control program was written with the assumption of synchronous I/O updates, its adaptation to asynchronous I/O updates requires its partial modifications to exclude possibility of dataraces. The simplest and robust way represents simulating of synchronous I/O updates by copying all input values into an array before each evaluation of the program, but such spare buffering generally increases responses to events. Therefore, the paper presents conditions for running the program (or its parts) without spare buffering of I/O data. *Copyright © 2004 IFAC*

Keywords: PLC, scan dataraces, static verification, transfer sets, APLCTRANS

## 1. INTRODUCTION

Controlling or supervising a manufacture system usually needs cyclic transfers of great amounts of data between I/O hardware and computers. For that reason, industrial programs frequently operate directly in a cyclic manner, which requires that the programs have limited execution times and always terminate normally under any circumstances, otherwise a fatal error occurs. We will call such cyclic manner programs as *I/O handler programs*, in short *IOH-programs*, according to event handlers utilized in operating systems that have very similar properties.

The execution of an IOH-program can be scheduled by many ways, most frequently as:

**continuous task** — its new execution begins after finishing a previous one;

**periodic task** — the program is started at regular time intervals; or

**event-driven task** — it just waits for events to occur.

Event-driven tasks usually deal with extraordinary situations or with fast I/O data. The implementations of numeric control algorithm depending on sampling period need periodic tasks and continuous tasks are suitable for the rest of operations.

IOH-programs rarely access I/O hardware directly. Such manipulation are usually too slow, require special processing and permissions, so they are also either not recommended or reserved only for special situations, for instance the sampling of analog data for a discrete control algorithm.

---

[1] This research is partially supported by the Rockwell Automation Services, Kolín, Czech Republic

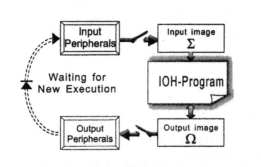

Fig. 1. IOH-Program in synchr. environment

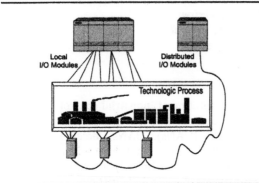

Fig. 2. Local and distributed I/O modules

Thus, IOH-programs ordinary run embedded in some proper hardware or software environment, as depicted in Figure 1. Such environments are also offered by robust PLCs (programmable logical controllers), which firmware synchronizes the evaluations of IOH-programs with I/O data by repeating three consequent steps:

**Input scan:** Hardware inputs are polled, or sampled respectively, and their values are stored into inner memory $\Sigma$, called *Input image*;

**Program scan:** user's IOH-program is executed once. It calculates new outputs and writes them into inner memory $\Omega$, called *Output image*; and

**Output scan:** after termination of the program, the values in output image $\Omega$ are copied into corresponding peripherals.

These steps correspond to a regulator with cyclic sampling of I/O data. On classical computer, if an IOH-program is scheduled as a continuous task then it can be programmed by endless loop, for instance in Pascal:

```
repeat
  Read_Inputs(Σ);
  Execute_Program(Σ,Ω);
  Write_Outputs(Ω);
until false;
```

IOH-programs scheduled as periodic and event-driven tasks can be programmed as event handlers of timers or I/O peripherals.

However, the synchronous updating of I/O has many advantages, its major drawback is polling I/O data in one-stroke, which increases I/O scan times inadequately, especially, when I/O peripherals are connected by control area networks (CANs), as depicted in Figure 2.

Distributed I/O modules offer many indisputable advantages, of which we mention only significant reduction of the length of all necessary wires, but there are also some drawbacks. In the contrast to local I/O buses, which speed is practically determined only by their electrical parameters, the transfer rates of networks are limited by much more factors. Polling all I/O data in one stoke, as it is perfomed by synchronous updating, concentrates all data transfers into short intervals and easily causes overloading of networks.

An asynchronous updating of inputs and outputs allows much better optimization of network traffic. In this case, input $\Sigma$ and output $\Omega$ images are usually converted into sets of *tags*.

Tags correspond to typed variables of classical programming languages, but with the addition of possible bounding their values to external sources, either local or remote, so their values can come from input modules (*input tags*) or from other computers, (*consumed tags*). [2]

If an input source is connected through local I/O bus, then its corresponding destination tag can be updated as fast as the communication devices can process the information, otherwise its update time is specified by an entered requested packet interval (RPI) which defines the required maximum amount of time between the updates, if the update is periodic, or a maximal delay between a change and sending new value, if its updating is bound to changes of a data source.

The opposite roles are played by *output* tags, which values are written into output modules, or *produced* tags, which are send to other connected computers.

One output or produced tag can be transmitted into more destinations, but any input or consumed tag is always updated at most from one source. Thus the value of any tag

(1) is always transmitted only unidirectionally, i.e. from its source to all possible destinations, if any; and

(2) can be also updated:
  - in shorter time than requested RPI and
  - during the execution of some instructions.

---

[2] Typical representatives of tag-based systems are PLCs of ControlLogix family manufactured by Allen Bradley, Rockwell Automation.

An IOH-program, which is scheduled as a continuous task with asynchronous I/O updating, can be programmed on a classical computer with the aid of two or more threads:

| *The main thread:* | *Other threads:* |
|---|---|
| **repeat** | **repeat** |
|   IOH_Program$(\Sigma, \Omega)$; |   IOManager$(\Sigma, \Omega)$; |
| **until false;** | **until Terminated;** |

where *Terminated* stands for the request to terminate the IOManager thread. Periodic and event-driven tasks can be again programmed as event handlers of timers or I/O peripherals.

The I/O update manager, which performs asynchronous updating, runs as a parallel program, which arises a possibility of dataraces. In parallel programming, dataraces are usually excluded by synchronization tools, as locks or mutex objects for protecting critical sections, but this method does not generally assure regular updates of all I/O data.

For that reason, utilizing synchronization tools need not be allowed necessarily in all tag-based environments. [3] But excluding mutual synchronization between I/O update manager and a IOH-program also means excluding many methods known from parallel programs. Therefore we must search another solution.

## 2. MODEL OF UPDATING ENVIRONMENT

In this paper, we aim to adaptation of a given IOH-program, which was written for an environment with synchronous I/O updating, to a new environment with asynchronous I/O updating.

First, we create model of the both environments to analyze their properties. We will distinguish between them by utilizing *TAG-IOH* for an environment with asynchronous I/O updating into tags and *S-IOH* for a synchronous environment depicted in Figure 3.

To isolate S-IOH and TAG-IOH differences from other difficult questions of parallel programming, we will assume:

(1) an IOH-program will be scheduled as a continuous task,
(2) its execution time is always limited by $t_{ep}$ constant, and

[3] For example, the instruction set of tag based PLC family ControlLogix allows only few instructions for predefined interlocked operations, i.e., only during their execution, it is granted that their operands will not be updated. These instructions do not involve other operations, with the exception of adding possible random time delays to them, so they do not allow synchronizing between program and I/O update manager.

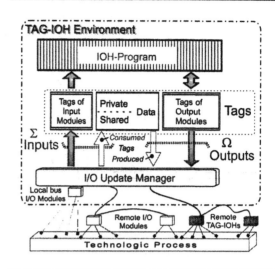

Fig. 3. $\Sigma$ and $\Omega$ in TAG-IOH environment

(3) the members of $\Sigma$ and $\Omega$ sets, which are accessed by the program, do not share mutually their memory locations, i.e., each member is mapped to its own storage.

Sets $\Sigma$ and $\Omega$ are specified for S-IOH environment by Figure 1. For TAG-IOH environment, we create $\Sigma$ as the set of all input and consumed tags, as depicted in Figure 3.

Similarly, $\Omega$ will contain all output tags and all produced tags. Finally, all tags or variables accessed by the program are included into $V$ set of variables.

The third assumption in the list above assures that $\Sigma, V$, and $\Omega$ are three disjoint sets, which we utilize in the following definition.

*Definition 1.* (IOH-program). Let be $\Sigma$, $\Omega$, and $V$ three disjoint sets of variables. We denote by $\mathcal{P}$ any program that satisfies the following:

- it always terminates and its execution time is less than a given constant $T$,
- it utilizes $\Sigma$ as its inputs, $\Omega$ as its outputs, $V$ as its internal variables, and
- it accesses nothing outside *storage* $S = V \cup \Sigma \cup \Omega$.

We denote by $\mathbb{P}\langle \Sigma, V, \Omega, T \rangle$ the set of all such IOH-programs, i.e., $\mathcal{P} \in \mathbb{P}\langle \Sigma, V, \Omega, T \rangle$.

Our the model of S-IOH environment consists of one thread with endless loop, as shown on Figure 4. Maximum execute times of the input updates, the program runs, and output updates are given by constants $t_{ei}$, $t_{ep}$, and $t_{eo}$.

Notice that the program has also write access into $\Sigma$ input image, when it is embedded in S-IOH environment. It is sometimes suitable for quick

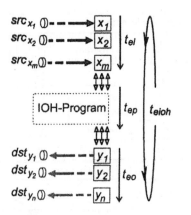

Fig. 4. Model of S-IOH environment

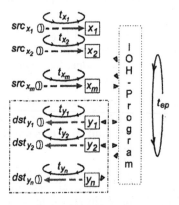

Fig. 5. Model of TAG-IOH environment

readdressing of inputs without many changes in the program, if required.

TAG-IOH model requires specifying RPI times in addition to storage $S$. In the following paragraphs, we denote a common index set by $I$, i.e.,

$$I \overset{df}{=} \{1, 2, \ldots |I|\} \qquad (1)$$

Let all $x_i \in \Sigma$, $i \in I$, $|\Sigma| = |I|$, be periodically updated from some $src_{x_i}$ external sources in randoms intervals, whose lengths $t$ of time have unknown probability distributions, so $\mathcal{P}$ program may make only one assumption that $t$ satisfies $t_{x_i} \geq t > 0$ where $t_{x_i}$ stands for some given RPI time of $x_i$ tag.

Similarly, we assume that the values of all $y_j \in \Omega$, $j \in I$, $|I| = |\Omega|$ are periodically copied into $dst_{y_j}$ external destinations in some random intervals $t$ satisfying $t_{y_j} \geq t > 0$, where $t_{y_j}$ are given RPI times.

In this case, $\mathcal{P}$ program is executed on a TAG-IOH by the way that corresponds to $m + n + 1$ threads (see Figure 5 ) where $m = |\Sigma|$ and $n = |\Omega|$. The

thread of user's IOH program performs endless loop with maximum length of program scan time equals to $t_{ep}$.

*Remark 2.* We utilized so $m + n$ thread only for simplification of the model. In reality, too many active threads decrease performance of operating systems. Therefore, the practical solution of I/O update manager could, for example, consists of one thread program that dispatches events, as incoming network packets, to sub-handlers. But properties of this solution will be similar to our thread model.

To optimize I/O polling, TAG-IOH environment can allow disabling automatic output updates, which is emphasized by the dashed rectangle in Figure 2. In such case, the program must request the updates for each group of output tags after finishing their evaluations.

In contrast, all $\Sigma$ input tags are always updated independently to the execution of IOH program and, therefore, $\Sigma$ *are read only data in TAG-IOH* because the program cannot reliably store any temporary values in them.

We may consider all possible write accesses into $\Sigma$ tags as program errors, which we employ in the adaptation.

## 3. I/O LATENCIES

There are two important differences between S-IOH and TAG-IOH environments caused by distinct I/O updates - I/O latencies and dataraces. In this section, we consider the latencies.

*Definition 3.* Let $x \in \Sigma$ be an input. Suppose the existence of two functions that return last times before a given $t$ time, when

- tch$(x, t)$ - data source of $x$ has changed its value, and
- tup$(x, t)$ - stored value of $x$ was updated.

*Input latency of $x$ in time $t$* is defined by

$$\text{til}(x, t) \overset{df}{=} t - \text{tch}(x, \text{tup}(x, t)) \qquad (2)$$

*Definition 4.* Let $x, y \in \Sigma$ be any two inputs. their mutual input latency in a given time $t$ by $|\text{til}(x, t) - \text{til}(y, t)|$.

In other words, an input latency specifies a time delay at time $t$ between actual value of an input and its value read from $\Sigma$ in time $t$, as depicted in Figure 6. Mutual input latencies specify time delay between samples stored in $\Sigma$.

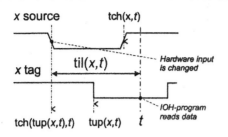

Fig. 6. Input latency

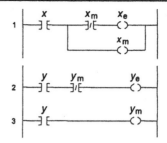

Fig. 7. Edge detection

| | Input latencies | |
| | One Input | Mutual |
|---|---|---|
| S-IOH | $\leq t_{ei} + t_{ep}$ | $\leq t_{ei}$ |
| TAG-IOH$_{(-B)}$ | $< RPI$ | $\leq RPI + t_{ep}$ |
| TAG-IOH$_{(+B)}$ | $\leq RPI + t_{ep}$ | $< RPI$ |

(-B) normal, (+B) with I/O bufering

Table 1. Input latencies of models

Input latencies are summarized in Table 1. For S-IOH model, maximal input latency equals to $t_{ei} + t_{ep}$, i.e., the duration of input sampling plus the execution time of IOH-program. Its mutual input latency depends only on $t_{ei}$ because all inputs are read during one input scan.

When discussing TAG-IOM model we need to distinguish if IOH-program emulates synchronous environment, which is a natural solution for preventing problems with asynchronous sampling — the input values are read into an array before running IOH-program and the array values are utilized instead of actual input tags. Similarly, outputs are written in another array and copied to output tags at the end of the execution. We will call this approach as *I/O buffering* .

Input latencies of TAG-IOH model without an I/O buffering are determined only by actual RPI times of $x_i$ inputs in the question ($t_{x_i}$ in the model). Mutual I/O latencies are given as the maximum of RPI times of inputs involved, to which we must add the execution time of an IOH-program, at most $t_{ep}$, if the first variable is read at beginning and the second before the end of the program. The similar conclusion can also be derived for $\Omega$ outputs and their latencies.

If an I/O buffering is employed in TAG-IOH model then input latencies are increased by the execution time because values stored in a buffer are not updated. On the other hand, mutual input latencies are frozen to the moment of buffering and do not depend on the execution time.

However, mutual latencies are non-zero, neither for S-IOH nor for TAG-IOH. They are not a characteristic behavior of a TAG-IOH — it may only emphasize them.

## 4. DATARACES IN TAG-IOH

*Dataraces*, known from parallel programs, are discussed in many papers, for instance in Choi et al. (2002), and usually defined as two memory accesses which satisfy four *datarace conditions*:

(1) the two accesses are to the same memory locations and at least one of the accesses is a write operation;
(2) the two accesses are executed by different threads;
(3) the two accesses are not guarded by a common synchronization object (lock); and
(4) there is no execution ordering enforced between two accesses, for example by thread start or join operations.

All inputs $\Sigma$ and outputs $\Omega$ of TAG-IOH satisfy datarace conditions. We pick up only such accesses to them, which could result in an erroneous behavior on TAG-IOH, but not on S-IOH, which means inputs $\Sigma$. They only are changed by updating processes of TAG-IOH model, and at any moment, unlike S-IOH. Thus, some codes will not work correctly with TAG-IOHs.

Possible dataraces in outputs $\Omega$ export problems to the external units, about which we made no assumption. Therefore, we will not study $\Omega$ outputs here.

We demonstrate this fact on two examples of well known rising edge detection in inputs $x$ and $y$, see Figure 7. The result $x_e$ (rung 1) of the edge detection will be 1 for one program scan, in which $x$ input has just changed to 1, otherwise $x_e = 0$. The result $y_e$ behaves similarly.

The ladder diagram can be converted into the following codes:

$$x_s := x \quad x_e := x_s \wedge \neg x_m; \quad x_m := x_s;$$
$$y_e := y \wedge \neg y_m; \quad y_m := y;$$

The first line copies $x$ variable into temporary $x_s$, stored into an evaluation stack, then $x_e$ rising edge is evaluated with the aid of $x_s$. The second line describes mathematically identical operation, but performed without temporary storage for $y$ input.

To analyze the program, we define its all traces.

*Definition 5.* Let $\mathcal{P} \in \mathbb{P}\langle \Sigma, V, \Omega, T \rangle$ be a IOH-program. We define *set of $\mathcal{P}$ traces* as a subset $\mathrm{trace}(\mathcal{P}) \subset \mathbb{P}\langle \Sigma, V, \Omega, T \rangle$ that includes all possible programs, which code consists of instructions executed during one execution of $\mathcal{P}$ program. *Set of traces* $\mathrm{trace}(\mathcal{P})$ *is deterministic* if it holds for any $\mathcal{P}_{\mathrm{tr}} \in \mathrm{trace}(\mathcal{P})$ that $\{\mathcal{P}_{\mathrm{tr}}\} = \mathrm{trace}(\mathcal{P}_{\mathrm{tr}})$.

Any deterministic set of traces contains only IOH-programs with single unchangeable streams of instructions, but it does not still assure its usability, because some special IOH-programs can have sets of traces with huge cardinalities. On the other hand, these programs will be probably excluded from analyses for their complexity in any case.

To express possible change of the values of input tags $x$ and $y$, we mark the accesses to them by superscripts to express the information about instant of time, in which the value of an input tag was sampled.

*Definition 6.* Given $x \in \Sigma$ input tag, $\mathcal{P} \in \mathbb{P}\langle \Sigma, V, \Omega, T \rangle$ program and that have deterministic set of traces. If $x$ value is read in some discrete time $t$ in a $\mathcal{P}_{\mathrm{tr}} \in \mathrm{trace}(\mathcal{P})$, then we denote such read *instant of $x$* tag by $x^{\langle t \rangle}$.

The set of traces of the rising edge detection contains one IOH-program, the original program itself. Utilizing integer times, we mark the access to the value of $x$ tag by $x^{\langle 1 \rangle}$ and two accesses to the value of $y$ tag on the rung 2 and 3 by $y^{\langle 2 \rangle}$ and $y^{\langle 3 \rangle}$. We obtain the program:

$$x_s := x^{\langle 1 \rangle} \quad x_e := x_s \wedge \neg x_m; \quad x_m := x_s;$$
$$y_e := y^{\langle 2 \rangle} \wedge \neg y_m; \quad y_m := y^{\langle 3 \rangle};$$

The tag $x_e$ still depends only on single time instant of $x_s$, but $y_e$ depends on $y^{\langle 2 \rangle}$ and $y_m$ where is stored $y^{\langle 3 \rangle}$. If $y^{\langle 2 \rangle} = 0$ and $y^{\langle 3 \rangle} = 1$ in some program execution accidentally due to updating the value of $y$ tag, then the rising edge of $y$ will not be detected.

*Lemma 7.* Let $\mathcal{P} \in \mathbb{P}\langle \Sigma, V, \Omega, T \rangle$ be arbitrary IOH-program. Given $x \in \Sigma$ input. If it holds for all $\mathcal{P}_{\mathrm{tr}} \in \mathrm{trace}(\mathcal{P})$ that $\mathcal{P}_{\mathrm{tr}}$ code contains only one instant of $x$, that $\mathcal{P}$ has no dataraces in $x$.

If $x$ variable is read only once then no datarace can exist. Lemma gives simple method, but their application is limited to trivial cases. For example, it does not exclude the presence of dataraces in $\mathcal{P} \in \mathbb{P}\langle \{x\}, \{i_1, i_2\}, \emptyset, T \rangle$ program, which is evidently datarace free: $i_1 := x^{\langle 1 \rangle}; \quad i_2 := x^{\langle 2 \rangle};$

*Proposition 8.* Let $\mathcal{P} \in \mathbb{P}\langle \Sigma, V, \Omega, T \rangle$ be arbitrary IOH-program. Given $x \in \Sigma$ input. If it holds for all $\mathcal{P}_{\mathrm{tr}} \in \mathrm{trace}(\mathcal{P})$ that all values of all $V \cup \Sigma$ variables were derived only from one instant of $x$, that $\mathcal{P}$ has no dataraces in $x$.

The structure of the proof is straightforward. If each value variable $y \in V \cup \Sigma$ depends only on one instant, a possible change of $x$ value will have no influence.

The proposition does not contains 'iff' clause because it can announce false alarms, for example, in $y = 1 \wedge x \wedge x$, but it concerns mainly to cases when one or more instants of variable are redundant and have no influence on a result, which are rare situations.

Proposition 8 offers possibility applying dataflow analysis methods that were developed for optimizing compilers. Many publications studies this problem, for instance Sathyanathan (2001) or Zheng (2000).

In the following subsection, we present a method applicable to some subset of IOH-programs. Even if it does not process sufficiently special operation, as pointer of arrays, it has simple implementation. We use it as an overview of the problem.

*4.1 Testing dataraces*

We will analyze the dependencies of variables on input $\Sigma$. First we define sets for storing this information.

*Definition 9.* ($\psi$-pair). Let $S$ be arbitrary non-empty set of variables. We define *$\psi$-pair on $S$* as $\psi\langle x, X \rangle$, where $x \in S$ and $X \subseteq S$, and two operators:

$$\mathrm{dom}(\psi\langle x, X \rangle) \overset{df}{=} X$$
$$\mathrm{co}(\psi\langle x, X \rangle) \overset{df}{=} x$$

where $\mathrm{dom}(\psi\langle x, X \rangle)$ and $\mathrm{co}(\psi\langle x, X \rangle)$ stand for *domain* and *codomain* of $\psi\langle x, X \rangle$. We denote a set of all $\psi$-pairs on given $S$ by $(S)^{\psi*}$.

*Definition 10.* Any subset $\tilde{X} \subseteq (S)^{\psi*}$ satisfying that $\mathrm{co}(\psi\langle x, V \rangle) = \mathrm{co}(\psi\langle x, V \rangle)$ implies $i = j$ for all $\psi\langle x_i, V \rangle, \psi\langle x_j, V \rangle \in \tilde{X}$, is called a *$\Psi$-set on $S$*. We denote the set of $\Psi$-sets for $S$ variables by $\Psi(S)$, i.e., $\tilde{X} \in \Psi(S)$.

In any $\Psi$-set $\tilde{X} \in \Psi(S)$, one $\psi$-pair exists at most for each variable $x_i \in S$ with the codomain equal to $x_i$. We also define codomains of $\Psi$-sets as the sets of codomains of all its $\psi$-pairs.

**Definition 11.** Given a $\Psi$-set $\tilde{X} \in \Psi(S)$. We define its codomain as:

$$\mathrm{co}(\tilde{X}) \stackrel{df}{=} \left\{ x_i \mid \psi \langle x_i, V \rangle \in \tilde{X} \right\} \quad (3)$$

**Definition 12.** A composition $\tilde{Z} = \tilde{X} \circledcirc \tilde{Y}$ of two given $\tilde{X}, \tilde{Y} \in \Psi(S)$ is $\tilde{Z} \in \Psi(S)$, $|\tilde{Z}| = |\tilde{X}|$, containing $\psi$-pairs $\psi \langle z_i, Z_i \rangle \in \tilde{Z}$. These $\psi$-pairs are constructed by the following algorithm:

**First step:**

for all $\psi \langle x_i, X_i \rangle \in \tilde{X}$, $i \in I$, $|I| = |\tilde{X}| = |\tilde{Z}|$
do begin:

$\quad \psi \langle z_i, Z_i \rangle := \psi \langle x_i, X_i \rangle$

$\quad$ for all $\psi \langle y_j, Y_j \rangle \in \tilde{Y}$ do begin:

$\quad\quad$ if $y_j \in X_i$ then $Z_i := (Z_i - \{y_j\}) \cup Y_j$
$\quad$ end

end

**Second step:**

for all $\psi \langle y_i, Y_i \rangle \in \tilde{Y}$, $i \in I$, $|I| = |\tilde{Y}|$
do begin:

$\quad$ if $y_i \notin \mathrm{co}(\tilde{Z})$ then $\tilde{Z} := \tilde{Z} \cup \{\psi \langle y_i, Y_i \rangle\}$

In words, the first step tests if the codomain of a $\psi$-pair from $\tilde{Y}$ is in domain of any $\psi$-pair from $\tilde{X}$. If it is satisfied, $\psi$-pair from $\tilde{Y}$ replaces by its domain the variable in the domain of $\psi$-pair in $\tilde{X}$.

The second step adds to the result all $\psi$-pair in $\tilde{Y}$, which codomains are not in the codomain of the result.

*Proposition 13.* The composition $\circledcirc$ is associative on $\Psi(S)$.

Proof: $\Psi$-sets are derived by simplifying transfer set theory (see Šusta (2003)). If we assume the existence of some abstract $\odot$ binary operation that satisfies idempotency ($x \odot x = x$) and commutativity ($x \odot y = y \odot x$) laws for any two arbitrary tags $x, y$, then we can create mapping of $\psi$-pairs into transfer sets. For instance, a $\psi$-pair $\langle x, \{x, y, z\} \rangle$ is mapped into $x := x \odot y \odot z$ assignment, which has direct conversion to $\{\hat{x}[\![x \odot y \odot z]\!]\}$ transfer set. The associativity was already proved for transfer sets.

The analogy between $\psi$-pair and expression allows utilizing them for testing variable dependency. We create *instantized input set* $\Sigma^{\langle t \rangle}$ of all instants of input tags in a given program trace $\mathcal{P}_{tr} \in \mathrm{trace}(\mathcal{P})$.

We describe dependencies in the instructions of some trace by *psi*-pairs, which we consecutively compose by $\circledcirc$ operation to one $\Psi$-set $D$. Finally, we test dataraces by the following algorithm that

composes mutually $\psi$-pairs to find out all dependencies.

*Algorithm 1.* Testing dataraces:

**Input:** Given a non-empty instantized storage $S^{\langle t \rangle} = \Sigma^{\langle t \rangle} \cup V \cup \Omega$ and $\Psi$ set $\tilde{D} \in \Psi(S^{\langle t \rangle})$.

**Initialization:** Loop index $i = 0$.

**Step 1:** Utilizing $\langle x_i, X_i \rangle \in \tilde{D}$ we perform for all $j \in I$, $|I| = |D|$, $j \neq i$: "If $\langle x_j, X_j \rangle \in \tilde{D}$ satisfies that $x_j \in X_i$, then $X_i := (X_i - \{x_j\}) \cup X_j$."

**Step 2:** If $i < |D|$ then we increment $i$ and repeat Step 1, otherwise we proceed to the final test.

**Final test:** If any $\mathrm{dom}(\langle x_i, X_i \rangle) \in \tilde{D}$ contains two different instants of one input, then $x$ has possible datarace.

The algorithm requires the maximum amount of memory $|V \cup \Omega| * |S^{\langle t \rangle}|$, contains finite loop and always terminates at most after $|D| * (|D| - 1)$ steps. [4] Each loop cycle adds dependencies of one variable into all domains of such $\psi$-pairs, in which domains it is presents, therefore, if any tag depends on more instants of some input, then they must appear at least in one domain set of the result.

## 5. CONCLUSION — ADAPTATION OF IOH-PROGRAM

The adaptation of a program can be suitable mostly in two cases to short the expensive commissioning phase:

- Peripherals have been redeveloped with the aid of control area networks (CANs) and we want to adapt major parts of our old reliable program.
- New program was written by technologists accustomed to programming in synchronous I/O update environment, so it must adapted to asynchronous environment.

First, we utilize the results presented in Table 1 for TAG-IOH and decide if we need to minimize mutual latencies for some signals in subset $\Sigma \cup \Omega$. For such I/O groups, we add their I/O buffering into the program. It is ordinary required for all numeric control algorithms or other parts which functionality depend on a proper sampling.

Finally, we consider dataraces. The I/O buffering exclude them. If we have buffered all $\Sigma$ inputs, then the program is surely datarace free. Otherwise, if I/O buffering concerns only some subset of $\Sigma$, we test all unbuffered inputs by either Proposition 8 or Lemma 7.

---

[4] The reduction of steps to half is possible, but it leads to less comprehensible form.

| No. | Size [kB] | Inputs | Possible dataraces | |
|-----|-----------|--------|---------|----------|
| | | | Lemma 7 | Prop. 8 |
| 1 | 3.2 | 5 | 5 | 5 |
| 2 | 3.5 | 4 | 4 | 3 |
| 3 | 6.7 | 4 | 4 | 3 |
| 4 | 13.4 | 94 | 63 | 54 |
| 5 | 28.2 | 132 | 102 | 88 |
| 6 | 143.8 | 419 | 25 | 2 |

Table 2. Tested PLC 5 Programs

We can try simple Algorithm 1 in Subsection 4.1. If it fails to give results for a $x \in \Sigma$, we apply I/O buffering to $x$, or we try any more exact and complex dataflow analysis.

### 5.1 Experimental Results

Algorithm 1 was tested on 6 program fragments extracted from different PLC programs that were written by several programmers for various industrial technologies. All fragments have only one possible trace (see Definition 5), which is common feature of many PLC programs.

Unfortunately, we have tested only PLC 5 programs because the import modules for another PLCs have not been finished yet, and we could analyze only fragments, since every PLC program contains some PLC dependent parts, for example special initializations of I/O modules.

The results are presented in Table 2. They are listed in size order measured in bytes occupied by programs in PLC internal memory. The table shows the number of their inputs, which must be tested, and possible input dataraces detected applying Lemma 7 and Proposition 8.

Lemma 7 does not require any additional equipment — we directly utilized the cross reference list in RSLogix 5 programming environment for PLC 5 processors. Testing dataraces according Proposition 8 was performed by an external program that processed PLC 5 programs exported into text files.

Proposition 8 gives more exact results. The biggest program (number 6) was nearly datarace-free due to copying many inputs into auxiliary variables to allow fast readdressing of I/O signals.

Much smaller programs 4 and 5 have many possible dataraces because they prioritized reducing auxiliary variables and preferred repeating inputs conditions to simplify troubleshooting of technologies.

Therefore, no statistically significant relation can be deduced. The number of inputs suspected for possible dataraces in a PLC program too depends on controlled technologies, employed programming styles, and additional requirements. On the other hand, the application of detailed tests according to Proposition 8 could save time in special cases.

## 6. RELATED WORKS

Dataraces are intensively studied and many papers deal with them, for example Choi et al. (2002); Ramanujam and Mathew (1994) cited in the previous sections. Unfortunately, all approaches that we have found assumed the knowledge of the source codes of all analyzed treads. No available publication studied a case similar to TAG-IOH, when some threads have random behavior and no synchronizations are available.

The optimization of compilers is the main domain of dataflow analysis used for tracking variable dependences. We already mentioned Sathyanathan (2001) and Zheng (2000), in which are long lists of publication dealing with this topic. The book Nielson et al. (1999) presents good overview of many methods.

Our simple method for testing variable dependences, presented in Subsection 4.1, utilizes the results of the transfer sets published in Šusta (2004) or Šusta (2003).

## REFERENCES

J.-D. Choi, K. Lee, A. Loginov, R. O'Callahan, V. Sarkar, and M. Sridharan. Efficient and precise datarace detection for multithreaded object-oriented prog. In *ACM SIGPLAN 2002 Conference on Programming Language Design and Implementation (PLDI), Berlin, Germany,* June 2002.

Flemming Nielson, Hanne Riis Nielson, and Chris Hankin. *Principles of Program Analysis.* Springer-Verlag, 1999. ISBN 3–540–65410–0.

J. Ramanujam and A. Mathew. Analysis of event synchronization in parallel programs. In *Languages and Compilers for Parallel Computing,* pages 300–315, 1994.

Patrick W. Sathyanathan. *Interprocedural Dataflow Analysis - Alias Analysis.* PhD thesis, Stanford University, Computer Systems Laboratory, June 2001.

Richard Šusta. *Verification of PLC Programs.* PhD thesis, CTU-FEE Prague, May 2003. avail. at http://dce.felk.cvut.cz/susta/.

Richard Šusta. Low cost simulation of PLC programs. In *7th IFAC Symposium on Cost Oriented Automation COA 2004, Gatineau (Québec) Canada,* pages 219–224. Université du Québec en Outaouais, 2004.

B. Zheng. *Integrating Scalar Analyses and Optimizations in a Parallelizing and Optimizing.* PhD thesis, Dept. of Computer Science and Engineering, University of Minnesota, 2000.

ELSEVIER
IFAC
PUBLICATIONS
www.elsevier.com/locate/ifac

# A ROBUST APPROACH TO THE PART LOADING PROBLEM IN FLEXIBLE MANUFACTURING SYSTEMS

**M. Monitto, T. Tolio[1]**

[1]*Dipartimento di Meccanica – Politecnico di Milano, Via Bonardi 9 – 20133 Milano, Italy*

The loading problem in flexible manufacturing systems (FMSs) concerns with the allocation, among the machine groups that constitute an FMS, of the operations and the required tools and fixtures to produce a given set of parts.

This problem has been deeply investigated in the past, but the turbulence and the strong competition in time and costs characterising the environment where manufacturing firms are operating nowadays, make the traditional approaches no more adequate, as they were developed considering "static" system conditions. What is now strongly needed by firms is a "robust loading", that means a loading that is "insensitive" to unforeseen FMS disturbances, as machine disruptions or uncertainty on arrival times of the jobs and so on.

The paper proposes a new approach to tackle the loading problem by means of stochastic programming techniques considering in a explicit way uncertain events. The output of the method is a loading program able to assure acceptable performance, within a reasonable range of variability of the parameter which define the problem.

Numerical results based on realistic test cases are provided to illustrate the efficacy and the robustness of the described approach. *Copyright © 2004 IFAC*

Keywords: Flexible manufacturing systems, Loading, Uncertainty, Robustness.

## 1. INTRODUCTION

In the current competitive environment, effective short term production planning has become a necessity to survive in the market place. However, available production planning tools generate accurate production plan that optimise specific performance measures, considering the shop floor as a static environment.

As a matter of fact, in real life production environments are subject to many sources of uncertainty, e.g. machine breakdowns, unexpected release of high-priority jobs, variability in job processing times, unexpected needs of reworking... Normally firms are well aware of these problems and tend to prefer robust solutions to production planning problems rather than optimal static solution. Many firms perform short term production planning manually or with the help of very basic tools but they tend to create plans where "robusteness" plays a crucial role. Indeed robustness allows to determine a short term production plan that makes system performance relative insensitive to the dynamic nature of the shop floor and of the environment where the firm operates. In these situation the introduction of short term production planning tools that do not consider the dynamic nature of the shop

floor my end up in a disastrous situation. Indeed with the available tools re-planning is frequently required, as the initial plan quickly becomes infeasible; however plan modifications have costs and tend to disrupt the normal operations of the firm. This is one of the reasons why short term production planning tools are not frequently adopted in practice.

The question therefore is whether it is possible to create short term production planning tools able to generate robust solutions. In the literature this problem has been addressed from a long time as far as regards the scheduling problem in single machine, flow shop and job shop environments. Little has been proposed for the problem of short term production planning in Flexible Manufacturing Systems (FMSs). In particular in the case of FMSs the most crucial step in the short term production planning activity is the loading phase defined by Stecke (1983) as follows:

"Allocate the operations and the required tools of the selected part types among the machine groups subject to technological and capacity constraints of the FMS".

The paper is organized as follows: a brief literature review regarding robust short term production planning is proposed in Section 2 while Section 3 introduces the problem statement and Section 4 provides a stochastic model to tackle the problem. Finally Section 5 presents some numerical results and Section 6 proposes some conclusions.

## 2. STATE-OF-THE-ART

Loading is one of the phases of operative production planning, and over the years after the pioneering work of K.E. Stecke a wealth of contributions appeared addressing every aspect of this problem (see Grieco, *et al.*, 2001, for an extensive review). At the moment however little has been done regarding the problem of robustness related to the loading problem even if many authors stress the importance of taking into account ways to react to unexpected events.

On the contrary, even though it is relatively self-contained and small, there is a body of literature coming from two decades of research in the field of robust scheduling, starting from the suggestions of Graves (Graves, 1981) regarding the need to understand schedule robustness through the explicit consideration of uncertain information. Therefore in the following we will look at the literature regarding robust scheduling in order to get some insight which can be helpful in addressing loading problems.

Looking at the literature regarding robust scheduling, no framework or classification scheme exist, but authors specify which representation of uncertainty is adopted, which type of system and which class of policies are considered, which formalisation for the robustness concept is defined. However, as a matter of fact, there is no an unambiguous definition and measure of robustness and each author suggests a different interpretation. Therefore, for each paper described in the following, we specify the system, the random variables considered and their description, the robustness measures adopted and the solution approach.

Leon, Wu and Storer (Leon, *et al.*, 1994) propose an approach to deal with robustness measures and robust scheduling for job shops, when the adopted control policy is "a right shift" policy: it means that, when a disruption occurs, the scheduling sequence is maintained while unfinished jobs are delayed as much as necessary to accommodate the disruption. The considered uncertainty sources are tied to the occurrences of machine disruption within the planning horizon. They are characterised by the distribution of the time to fail and time to repair. The robustness measure is obtained as a linear combination of both expected makespan (when disruptions are considered) and expected delay of the schedule (which is defined as the difference between the expected makespan in the presence of disruptions and the optimal and deterministic makespan obtained if no disruption occurs). An exact measure of schedule robustness is derived for the case in which only a single disruption occurs, while for the case in which multiple disruption may occur, approximate measures are developed.

Daniels and Kouvelis (Daniels and Kouvelis, 1995) adopt a different approach to the robustness concept as they consider "robust" a schedule with the best worst-case performance. Therefore, the robust scheduling concept is formalised for single-machine problems with variable processing times, where the performance of interest is the total flow time of the jobs. Variability is structured using scenario generation. Given a possible realisation of processing times $\lambda$ and the optimal schedule $S^*$ corresponding to $\lambda$, it is possible to evaluate the difference between the performance measure of a generic schedule S when scenario $\lambda$ occurs, and the performance measure of schedule $S^*$. Therefore, robustness of the generic schedule S is measured as the maximum difference over all the scenarios. The robust scheduling problem is then formalised using mathematical programming. A branch and bound algorithm and an heuristic procedure to solve the problem have been developed. Daniels, Kouvelis and Kairaktarakis (Daniels, *et al.*, 2000) consider the case of a two-machine flow shop in which the processing time of jobs are uncertain and the performance measure of interest is system makespan. To define the robustness measure, the concept of risk-adverse decision maker who is interested in hedging against poor system performance is adopted. In this case, given a possible realisation of the processing times $\lambda$, the optimal schedule $S^*$ corresponding to $\lambda$ is computed using the Johnson algorithm. The robust scheduling problem is formalised using mathematical programming while a branch and bound algorithm is used to solve the problem.

In Daniels and Carrillo, 1997, a single-machine scheduling problem where job processing times are uncertain is considered. In this case, schedule robustness is tied to the variance of system performance due to uncertain processing times. A decision-maker may prefer a schedule with sub-optimal average performance, but low performance variability, instead of an alternative schedule, with optimal average performance and significantly higher variance. In determining the optimal schedule, it is necessary to consider both average system performance and performance variability: therefore, the proposed robustness measure is based on the likelihood of achieving system performance not worse than a given target level. Within the paper, the robust scheduling problem is formalised considering, as performance measure of interest, the total flow time of the jobs. The authors propose a branch-and-bound algorithm and an heuristic approach to solve the robust scheduling problem.

Anglani, *et al.*, 2001, address problem of independent job scheduling on identical parallel machines with sequence dependent set-up costs in order to minimise the total set-up cost. Uncertainty about job processing times is modelled by means of fuzzy numbers. The scheduling problem is then formalised by means of a non-linear mixed integer-programming model, but an equivalent linear model is also proposed.

In literature, many different ways of interpreting scheduling robustness exist. Another concept strictly related to robustness, the schedule nervousness is also introduced (Kazan, *et al.*, 2000), (Kimms, 19). A schedule is nervous if it requires frequent updating due to the fact that the initial one has become unfeasible. A robust schedule is more than a non-nervous one: its stability is, in fact, combined by high performance.

Yellig and Mackulac (Yellig and Mackulac, 1997) propose a new scheduling strategy that is inspired on the anticipatory failure policy of inventory hedging points developed by Kimenia and Gershiwn (Kimenia and Gershiwn, 1983). Following the inventory hedging, failure rates and repair times of the machines are used to choose a production rate in a way that anticipates machine downtimes, meeting production demand without creating excessive inventory level. Clearly, because of the use of inventory hedge, the approach is not applicable in make-to-order and just in time environment. In Yellig and Mackulac, it is proposed a strategy that anticipates failures, based on historical machine performances, and maintains also a stable schedule by reserving machine capacity. This capacity reserve results in a reduction of throughput, but also in a hedge (a capacity hedge instead of an inventory hedge) that protects against stochastic failures.

In (Vidyarthi and Tiwari, 2001) a fuzzy-based solution methodology has been formulated to address the machine-loading problem in a flexible manufacturing system. The objectives considered are minimization of system unbalance and maximization of throughput, whereas the systems technological constraints are posed by availability of machining time and tool slots. The job ordering/job sequence determination before loading is carried out by evaluating the membership contribution of each job to its characteristics such as batch size, essential operation processing time and optional operation processing time.

Srinivas, Tiwari and Allada suggest to use the auction based heuristic control strategy to deal with the problem of distributed manufacturing systems, the next generation manufacturing system conceived to be intelligent enough to take decisions and automatically adjust itself to situations such as variations in production demand and machine breakdowns. In (Srinivas et al., 2004) this methodology has been applied to the flexible manufacturing system machine-loading problem where job selection and operation allocation on machines are to be performed such that there is a minimization of system unbalance and a maximization of throughput.

Another way to adapt to a rapid changes of manufacturing orders flexible manufacturing systems (FMS) is to go into the direction of dynamic rules. Machines become further versatile functionally and, if tools are controlled by fast tool delivery device, machines can perform a variety of operations when they are supplied with the required. In (Lee et al., 2003) it is presented an integrated model that performs operation sequence and tool selection simultaneously into the direction that minimizes tool waiting time when the tool is absent.

In the existing literature, little attention is given to the industrial applications of the approaches based on robustness concepts, even if the strategic importance of the topic is well recognised. Therefore, although the academic contribution to robust production

scheduling is quite abundant, its impact on practice has been so far minimal.

## 3. PAPER OBJECTIVES

The main goal of this paper is to understand and formalise loading issues in a FMS when uncertainty in system conditions is considered. This means in particular to define the robustness concept and to formalise and solve the loading problem when uncertainty sources are explicitly considered in the model (this means to specify which kinds of uncertainty factors are to be considered and how they influence system behaviour).

The majority of current loading strategies are often rigorous in determining an optimal solution while assuming a static environment. However, a production system is not a static environment: frequent re-assignments are then required, as the fixed resource (fixture, tools) assignment becomes quickly unfeasible, and this makes very difficult to achieve good performance.

It is therefore desirable to generate resource assignments that are robust within a reasonable range of conditions. If a disturbance occurs which makes impossible to proceed with the a priori defined resource assignment, then in response to the disturbance, a "control action", that is a set up, is taken to adjust the resource assignment to make it feasible.

With this aim, it seems reasonable to formalise the problem of robust resource assignment as a stochastic programming problem.

## 4. THE PROPOSED MODEL

Given the assumptions described here-before, the robust loading problem has been formulated as a two-stage stochastic integer non-linear problem. The planning horizon is divided into two stages: for the first one the decision maker has to define how to allocate fixtures to jobs and to cells in order to meet market demand with the minimum number of changes in existing configuration, while having at the same time a look to the future. Indeed, the objective function also considers the necessary reallocation of fixtures, to meet production requirements at the second stage of the problem.

The considered production system is a parallel machine FMS. In principle machines are able to perform every operation and therefore produce every part type of the part mix but they are organised in specialized cells (a cell is constituted by one or more machines) The output of the model will be the assignment of parts to cells and the specialization of the machines.

The model is detailed in the following.

### 4.1 Notation

$i$= scenario index

$j$= part type index (it is assumed that each fixture is dedicated to one single part type. Terms such as part type, pallet and fixture are interchangeable)

$k$= cell index

$p(i)$= probability of occurrence of scenario $i$

$f_0(j, k)$= number of pallets assigned to part types $j$ and cell $k$ at the initial time

$f_1(j, k)$= number of pallets assigned to part types $j$ and cell $k$ at the first stage

$f_2(i, j, k)$= number of pallets assigned in scenario $i$ to part types $j$ and cell $k$ at the second stage

$x_1(j, k)$= number of pallets assigned to part $j$ and processed by cell $k$ at the first stage of the problem

$x_2(i, j, k)$= number of pallets assigned in scenario $i$ to part types $j$ and processed by cell $k$ at the second stage

$fc_1(j, k)$= number of changes in pallet assignment of part $j$ at cell $k$ from the initial state to the first stage

$fc_2(i, j, k)$= number of changes in pallet assignment of part $j$ at cell $k$ from the first stage to the second stage in scenario $i$

$fcell(k)$= maximum number of pallets that it is possible to assign to cell $k$

$fpart(j)$= maximum number of fixtures, i.e. pallets, that can be assigned to part $j$

$H$= high constant value

$wt(j, k)$= working time of pallet type $j$ processed by cell $k$

$tt(k)$= time required by the transport unit to go from the load/unload station to cell $k$

$lut(j)$= time necessary for the load/unload phase of pallet $j$

$B_1(i,k)$= effective time of availability of cell $k$ at the first stage of the problem

$B_1(i,k)=\beta(i,k)B_1$ where $\beta(i,k)$ is the availability of cell $k$ in scenario $i$ and $B_1$ is the duration of the first stage

$B_2(i, k)=\beta(i, k)B_2$ where $\beta(i, k)$ is the availability of cell $k$ in scenario $i$ and $B_2$ is the duration of the second stage

$inv_0(j)$= initial inventory of part $j$, in terms of corresponding processed pallets.

$inv_1(i,j)$= inventory of part $j$ at the end of the first stage in scenario $i$ in terms of corresponding processed pallets.

$inv_{max}(j)$= maximum inventory of part $j$ in terms of corresponding processed pallets.

$D_1(i,j)$= demand of part $j$ in scenario $i$ at the the first stage in terms of corresponding pallets to be processed.

$D_2(i, j)$= demand of part $j$ in scenario $i$ at the second stage in terms of corresponding pallets to be processed.

$ls$= minimum service level.

### 4.2 Problem formulation

$$\min E(su) = \sum_i su(i) \cdot p(i) \qquad (1)$$

$$\sum_k x_1(j, k) + inv_0(j) \geq D_1(i, j) \cdot ls \quad \forall i, j \qquad (2)$$

$$\sum_k \left( inv_0(j) + x_1(j, k) + x_2(i, j, k) \right) \geq$$
$$\geq D_1(i,j) + D_2(i,j) \quad \forall i, j \qquad (3)$$

$$\sum_k x_1(j, k) + inv_0(j) - D_1(i,j) =$$
$$= inv_1(i,j) \quad \forall i,j \qquad (4)$$

$$inv_1(i,j) \leq inv_{max}(j) \quad \forall i,j \qquad (5)$$

$$x_1(j, k) \leq H \cdot f_1(j, k) \quad \forall j, k \qquad (6a)$$

$$x_2(i, j, k) \leq H \cdot f_2(i, j, k) \quad \forall i, j, k \qquad (6b)$$

$$\sum_j x_1(j, k) \leq$$
$$\leq \sum_j \left( \frac{f_1(j, k)}{wt(j, k) + 2tt + lut(j)} \right) \cdot B_1(i,k) \quad \forall i, k \qquad (7a)$$

$$\sum_j x_2(i, j, k)$$
$$\leq \sum_j \left( \frac{f_2(i, j, k)}{wt(j, k) + 2tt + lut(j)} \right) \cdot B_2(i,k) \quad \forall i, k \qquad (7b)$$

$$\sum_j f_1(j, k) \leq fcell(k) \quad \forall k \qquad (8a)$$

$$\sum_j f_2(i, j, k) \leq fcell(k) \quad \forall i, k \qquad (8b)$$

$$\sum_k f_1(j, k) \leq fpart(j) \quad \forall j \qquad (9a)$$

$$\sum_k f_2(i, j, k) \leq fpart(j) \quad \forall i, j \qquad (9b)$$

$$|f_0(j, k) - f_1(j, k)| = fc_1(j, k) \quad \forall j, k \qquad (10a)$$

$$|f_1(j, k) - f_2(i, j, k)| = fc_2(i, j, k) \quad \forall i, j, k \qquad (10b)$$

$$\sum_j \sum_k (fc_1(j, k) + fc_2(i, j, k)) = su(i) \quad \forall i \qquad (11)$$

$$x_1(j, k), x_2(i, j, k) \in R^+$$
$$f_1(j, k), f_2(i, j, k) \in N \qquad (12)$$

### 4.3 Problem description

The optimization problem above presented is a Non-Linear Mixed-Integer Problem with (i(3jk+j+1)+3jk) variables, of which ((i+1)jk) integer, and (i(5jk+4j+3k+1)+3jk+j+k) constraints, of which ((i+1)jk) non-linear.

Let description start from the constraints.

Equation (2) expresses the intention of satisfying at least the $ls$% of market demand at the first stage of the problem, using existing inventory and production. Since demand has to be completely satisfied it is possible to postpone satisfaction of the remaining (1-$ls$)% to the second stage. This is enforced by constraint (3), where the total demand of the planning horizon (that is given by the sum of demand at each stage) can be satisfied using inventory and production capacity of both the stages. Equation 4 defines inventory created at the end of stage 1, if an excess of production is planned, and equation 5 imposes a maximum to it.

For all the other constraints of the model, it is possible to recognize two families: in the first one there are constraints referred to the first stage of the problem ("a" constraints), while in the second one, those referred to the second stage ("b" constraints).

Equations 6 enforces the assignment of at least one pallet to a part and a cell if the considered cell has to produce that part. Equations 7 expresses capacity constraints tied to the cycle time of fixtures. The following equations 8 states that each cell cannot hold more than a maximum number of pallets. Analogously, equations 9, states that the number of pallets that can be assigned to each part type cannot exceed the maximum number of fixtures that are available for each part.

The initial configuration of the considered system is described through the set of parameters $f_0(j,k)$, that represent how many pallets are dedicated to each part type in each cell. Equation 10a evaluates how many changes in pallet assignment are to be made to pass from the initial configuration to the configuration of the first stage, while equation 10b, evaluates the number of changes in pallet assignment to be made to pass from the configuration of the first stage to each possible configuration, one per scenario, of the second stage. The total number of changes for each scenario is the value of the set-up variable (11).

The objective function of the problem is than the minimization of the expected value of the set-up variables, i.e. the sum of set-ups per scenario weighted by the probability of occurrence of that scenario.

Therefore the overall logic of the model is the selection of a resource assignment for the first stage of the planning horizon that can be re-adjusted with relative small changes in the second stage to satisfy scenarios with higher probability of occurrence, while less probable scenarios can require a larger number of changes in pallet assignment.

## 5. EXPERIMENTS

### 5.1 Performance measures

Stochastic programming allows considering the distribution of uncertain parameters within an optimisation problem and defining a solution that has good average performance in the considered range of variability of the parameters. The goodness of this solution can be measured using the Expected Value of Perfect Information (EVPI), that is the difference between the optimal objective function value of the stochastic programming problem and the optimal objective function value of the deterministic problem that assumes the perfect information about shop floor condition, and the Value of Stochastic Solution (VSS), that is the difference between the optimal objective function value of the traditional stochastic scheduling approach, that uses the expected values of uncertain parameters, and the optimal objective function value of the stochastic programming problem (Birge and Louveaux, 1997).

With respect to the proposed model, EVPI represents the fixture re-allocation that can be avoided if the decision maker knows in advance market demand and resource availability. VSS is a measure of the benefits obtained by considering uncertainty on market demand and resource availability instead of assuming deterministic values.

### 5.2 Reference system

The proposed model has been tested with reference to a real case. Starting from the real case, realistic cases have been generated by varying the following four characteristics of the problem:
1. number of products that constitute the production mix
2. standard deviation of uncertain parameters
3. correlation among product demands
4. autocorrelation among different stages of the problem.

The real case comes from a manufacturer of mechanical components for the automotive sector.

The manufacturing system considered is an FMS with four parallel CNC machines.

Collected data include, for each part type, daily market demand observed into six months, pallet configuration, tools and machining time. Moreover, for a period of two months, reasons and duration of idle times of productive resources has been collected, using some data sheets filled in by workers.

Using these data, uncertainty about market demand and system availability have been modelled in stochastic terms, using empirical distributions.

The problem, once formalized, entails 949 decision variables, (of which 288 integer), and 1078 constraints (of which 288 non linear). It has been implemented using Lingo 8.0® with a Pentium P4 2.6 GHertz workstation. The stochastic model is generally solved in 2 hours, while the expected value problem in 30 minutes and wait and see problem in 4 hours.

Here in after, some graphs that summarize the results of the experimental phase are presented and discussed. The real case is characterized by the presence of high variance, correlation among product demands, and autocorrelation of market demand. These parameters affecting the reference environment are addressed one per time. The basic case is that of 4 independent products, with a standard deviation equal to 10% of mean, and without autocorrelation.

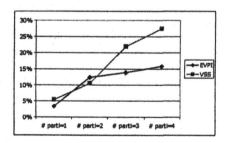

Fig. 1. EVPI and VSS in respect to the number of parts constituting the part mix.

As the graph in figure 1 shows, if the diversity of the part mix, in terms of number of produced parts, increases, the benefits of using the stochastic model increase as well. Moreover, VSS increases faster than EVPI: this means that it is more and more critical to ignore uncertainty as the diversity of the part mix grows.

Considering the standard deviation of market demand (figure 2), we can see that if the complexity and the turbulence of the environment increase, the performance of the stochastic model increases too; moreover EVPI is greater than VSS.

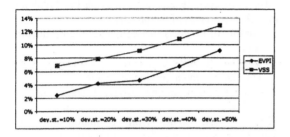

Fig. 2. EVPI and VSS respect to the standard deviation of market demand.

Results obtained considering the correlation among different products show that if there are many independent products both VSS and EVPI decrease because different sources of uncertainty average out.

Similar results are obtained considering autocorrelation among different periods of time.

Simulating the application of the described method to the real case (where all the considered aspects of the environment are present at the same time) in a period of three weeks, using a rolling approach, the total reduction of setup obtained using the stochastic model instead of the expected value problem (i.e. the VSS), is equal to 16%.. There is a further 11% reduction of setup if the decision maker knows the future in advance (EVPI). With the hardware and software resources previously described the computational time for the stochastic model is around 2 hours.

## 6. CONCLUSIONS

The proposed model allows obtaining better fixture assignment resulting in a reduction in the number of setups, in comparison with the traditional deterministic approaches. This is result is in line with the results obtained in the literature on other problems in the field of production planning.

Given the entity of the attainable advantages and considering at the same time the required high computational effort to obtain the optimum solution, it seems reasonable to devote future research to find heuristic approaches to the problem.

## REFERENCES

Anglani, A., Grieco, A., Guerriero, E., Semeraro, Q., Tolio, T. (2001). A fuzzy approach to a robust schedule of identical parallel machines with sequence dependent set-up costs. *Proceedings of the XVI ICPR, 29 July-3 August 2001, Prague (CR)*.

Birge, J.R., Louveaux F. (1997). *Introduction to Stochastic Programming*. Springer Series in Operations Research, Springer.

Daniels, R.L., Kouvelis, P. (1995). Robust Scheduling to Hedge Against Processing Time Uncertainty in Single-stage Production. *Management Science*, 41/2, pp. 363-376.

Daniels, R.L., Carrillo, J.E. (1995). β-Robust scheduling for single-machine systems with uncertain processing times. *IIE Transaction*, 29, pp. 977-985.

Daniels, R.L., Kouvelis, P., Kairaktarakis, G. (2000). Robust Scheduling for a two-machine flow shop with uncertain processing times. *IIE Transactions*, 32, pp. 421-432.

Grieco, A., Semeraro, Q., Tolio, T. (2001). A Review of Different Approaches to the FMS loading Problem. *The International Journal of Flexible Manufacturing Systems*, 13, pp. 361-384.

Graves, S.C. (1981). A Review of Production Schedule. *Operation Research*, 29, pp. 646-675.

Kazan, O., Nagi, R., Rump, C.M. (2000). New lot sizing formulation for less nervous production schedule. *Computers and Operation Research*, 27, pp. 1325-1345.

Kimenia, J., Gershwin, S.B. (1983). An algorithm for the computer control of a flexible manufacturing system. *IIE Transactions*, 15, pp. 353-362.

Kimms, A. (1998). Stability Measures for Rolling Schedules with Applications to Capacity Expansion Planning, Master Production Scheduling and Lot Sizing. *International Journal of Management Science*, 26/3, pp. 355-366.

Kuroda, M., Shin, H., Zinnohara, A. (2001). Robust scheduling in Advanced Planning and Scheduling environment. *Proceedings of the XVI ICPR*, 29 July-3 August 2001, Prague (CR).

Lee, C.S., Kim, S.S., Choi, J.S. (2003). Operation sequence and tool selection in flexible manufacturing systems under dynamic tool allocation. *Computers & Industrial Engineering*, 45, pp. 61–73

Leon, V.J., Wu, S.D., Storer, R.H. (1994). Robustness Measures and Robust scheduling for Job Shops. *IIE Transactions*, 26/5, pp. 32-43.

Srinivas, Tiwari, M. K., and ALLADA, V. (2004). Solving the machine-loading problem in a .exible manufacturing system using a combinatorial auction-based approach. *International Journal on Production Research*, 42/9, pp. 1879–1893.

Takriti, S., Ahmed, S. (2001). On Robust optimisation of Two-Stage Systems. *Optimisation On Line*, February 2001.

Vidyarthi, N. K., Tiwari, M. K. (2001). Machine loading problem of FMS: a fuzzy-based heuristic approach. *International Journal on Production Research*, 39/5, pp. 953–979

Weiss, G. (1995). *A tutorial in Stochastic Scheduling*, 1995, in P. Chretiénne, E. G. Coffman, J.K. Lenstra, Z. Liu, *Scheduling theory and its Applications*. John Wiley and Sons Ltd.

Yellig, E.G., Mackulak, G.T. (1997). Robust deterministic scheduling in stochastic environments: the method of capacity hedge points. *International Journal on Production Research*, 35/2, pp. 369-379.

ELSEVIER
IFAC
PUBLICATIONS
www.elsevier.com/locate/ifac

# HUMAN-ROBOT COLLABORATION IN AUTOMATED MANUFACTURING

Ulrich BERGER[1], Raffaello LEPRATTI[1], Heinz-H. ERBE[2]

[1] Brandenburg Technical University at Cottbus
Chair of Automation Technology

[2] Technical University of Berlin
Center of Human-Machine Systems

Abstract: Effective, efficient and reliable human work activities with automated manufacturing systems are only possible, if machine operators have both knowledge about effects of their inputs (*operating*) and comprehensive skills about system feedback (*understanding*). Consequently, features such as controllability and transparency prove to be relevant criteria for the intelligent system design. The paper stresses the importance of an innovative human-robot communication concept based on natural speech. A procedure for setting up ontological nets, which contribute in semantic filtering natural language instructions, represents the project emphasis. Thereby, a new way for the use of natural languages in production environments is proposed. *Copyright © 2004 IFAC*

Keywords: manufacturing systems, industrial robots, man-machine systems, robotics, natural languages, artificial intelligence, industry automation.

## 1. INTRODUCTION

The use of automation systems like industrial robots provides high flexibility in manufacturing plants (World Robotics, 2003). However, as shown in (Lepratti et. al, 2004), without an intelligent embedding of human knowledge and skills this could also represent an obstacle to the transition process of today's industry, evolving from a mass production towards a knowledge-based customer-oriented one. This concerns not only acceptance but also reliability of collaboration between humans and automation technologies, which have to be strictly seen as complementary elements within the same manufacturing system. Effective, efficient and reliable human work activities with automated systems are only possible, if operators have both knowledge about effects of their inputs (*operating*) and comprehensive skills about system feedback (*understanding*). Thus, features like controllability and transparency prove to be relevant criteria for the system design.

In the interaction between humans and automated manufacturing it is crucial to map functionality system properties and intentions pursued with its development, in order to close the gap between operating and understanding and make the human-machine feed-back loop more robust that way.

Experiences have already shown that machine operators are often confronted with uncertainties, which have to be correctly localized, interpreted and intercepted (Sheridan, 1992). On the other hand, in most cases, errors or even accidents are often attributed to human failures. However, results of investigations contradict this hypothesis and show, indeed, that humans are rarely causes of rule violations (Endsley, 1999). If human errors occur, they are mostly produced as consequences of momentary "obscurenesses" during automated operations. Under these circumstances, the use of the natural language could represent a step forward in harmonizing human work activities aiming at optimizing transparency and reliability and, thus, controllability of automation systems.

As underlined in (Lepratti and Berger, 2003), interoperation barriers are identifiable in each of the main interaction forms within a manufacturing environment, beginning from the Human-Human Communication and the Machine-Machine Data Exchange up to the Human-Machine Collaboration. In following, attention will exclusively be paid on Human-Robot Collaboration.

At this point, in order to avoid confusion in sharing terminology, it should be distinguished between the terms *Co-operation* and *Collaboration*. While former foresees only the sharing of same intentions with others; i. e. formation of partnerships and commitments among enterprises, latter requires also a deep involvement and commitment in a common design, production-process or service; i. e. to work jointly with others.

Erbe (Erbe, 2004) considers three different main research fields on Human-Robot Collaboration:

- hand guided robot (Cobot)
- tele manipulation, tele-operation
- robot guided by communication through gestic and speech recognition.

Hand guided robots are called Intelligent Assisting Devices (IAD), tele-manipulators / tele-operators are considered semi-intelligent devices, while robots of the third category as human co-workers.

The contribution focuses on the latter issue presenting an innovating concept for enhancing interaction through speech recognition based on natural language. A procedure for setting up ontological nets will be outlined. It contributes in semantic filtering meaning of natural language speech instructions trying to make human-robot interaction unequivocal.

In order to validate related conceptual approaches, a prototype platform has been developed. First trial experiments prove its robustness as well as suitability. Future research activities will focus on the extension of its application spectrum, beginning by adding further vocabulary terms up to the refinement of applied mechanisms.

## 2. HUMAN-ROBOT COLLABORATION

Recent research results presented in the literature regarding a robot as a human co-worker are focused mainly on humanoid robots. See for example the conference documentation of Humanoids 2003 (Knoll and Dillmann, 2003). These developments would have their application fields in offices and households as service robots and for assisting elderly and disabled persons.

One of the first developments for getting information of the human behavior when working was called "behavioral cloning" (Bratko, 1995). It is a process of reconstructing a skill from operator's behavioral traces by means of machine learning techniques. However, the research was aimed to replace humans with robots but not for collaboration purposes. This has changed in the recent past: Kimura et al. (Kimura et al., 1999) recognized the object and the human grasp by two vision systems for analyzing the tasks done by the operator. With these information the robot analyses the operators demonstration and generates a task model for assisting the operator. Sato (Sato et al., 2002) improved this development. Laengle and Woern (Laengle and Woern, 2001) consider a robot as an intelligent assistant for the worker. Instead of researching for solutions to achieve a complete autonomous execution of complex tasks in an uncertain environment, the authors enter into a compromise, wherein the operator helps the robot to finish the tasks correctly when uncertainties occur. The robot can switch from automatically to semi-automatically executed tasks with help of the human operator. For observing the tasks force and vision sensors are used. Stopp et al (Stopp et al., 2002) developed an assistant for an

order-picking task. The task is taught using a laser pointer and a hand-held computer. The interaction is supported by speech output of the user.

The application in the manufacturing industry is also gaining a large interest. While industrial robots at production lines are usually working in a well-known environment, the challenge is now to let them work together with human operators in an unstructured, uncertain environment. While sensor data permit robots to recognize the unknown surrounding environment, human speech instructions ensure their correct operation providing them with the necessary missing behavioral "know-how".

The following paragraphs consider ontological nets as essential core elements within the innovative human-robot communication concept. These allow to overcome possible semantic misunderstandings in instructions based on natural speech. Section 3 introduces the concept of ontological net, describing its structure using a mathematical model. In order to proceed natural languages a knowledge base with several syntactic and semantic commitments based on a specific domain of use is required. If needed, this can be enriched or adapted by means of two tools, whose structure is presented in section 4. Concluding remarks regarding further research activities are anticipated in section 5.

## 3. ONTOLOGICAL NETS

Although the use of natural languages still represents a hazard solution approach above all in connection with its employment in manufacturing due to possible misinterpretations, which could arise during the human-robot instruction process as consequence of syntactical, lexical and extensional ambiguities, they represent the most familiar and understandable communication form for human beings. Following Winograd's theory (Winograd, 1980), assuming that there is no difference between a formal and a natural language, one finds proper reasons for all the efforts to formalize instructions and therefore knowledge expressed by natural languages.

The solution approach has been called Ontological Filtering System (OFS). It consists of a Knowledge Base (KB), which encloses a variety of semantic and syntactic commitments defined for a specific domain of use but also primarily of ontological nets (ON), in which chosen terms of the natural language are chained together hierarchically per semantic relations. This network consists, on the one hand, of the set of terms used as key words to standardize information contents for the machine data processing. On the other hand, it foresees a set of additional terms, which could be used from different persons in their natural communication, since there are more ways to express the same instruction, i. e. knowledge meaning. These terms could have different abstraction degrees in their meaning (the so called granularity). Thus, some words are more general in their expression than others, while others can go very deep with their meaning. An example of this semantic network is proposed in Figure 1.

According to their specification levels (divided through dotted lines of Figure 1) all terms – key terms and additional terms - are linked together by means of semantic relations (see arrows in Figure 1). With support of this network and a syntax analyzer, i. e. a parser, the OFS main program ensures processing of semantic contents of natural language sentences leading back used terms meanings to these ones belonging to the set of pre-defined key words. Latter ones are used to build instructions for robot operations (see for instance in Figure 1 the term "conveyor" highlighted with fat circle line), In this way the OFS provides a so-called semantic filtering function.

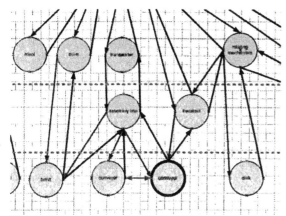

Fig. 1: Example of an Ontological Network

A mathematical description could help to explain its functionality. Figure 2 proposes a simplified example of an Ontological Network.

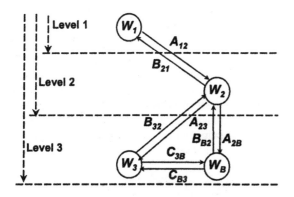

Fig. 2: Simplified Example of Ontological Network

Considering $W$ as set of chosen words belonging to the natural language and chosen for a specific domain of use:

$$W_{NL}=\{w_1, w_2, ...., w_n\} \qquad (1)$$

and a set of key words $W_B$, which represents the key terminology selected to formalize knowledge contents:

$$W_B=\{w_{1B}, w_{2B}, ...., w_{nB}\} \qquad (2)$$

using following set $R$ of semantic relations of natural languages such as: hypernymy ($A$), hyponymy ($B$), synonymy ($C$) and antonymy ($D$):

$$R=\{A, B, C, D\} \qquad (3)$$

one can define the ontological network as the following ordered triple:

$$ON=<W, R, S> \qquad (4)$$

where $W$ represents the addition set of terms $W_{NL} \cup W_B$ and $S$ takes the specification level of each element of $W$ into account. According to Figure 2, relations between the elements of $W$ can be included in a relation matrix $\Re$ :

$$\Re = \begin{bmatrix} 0 & A_{12} & 0 & 0 \\ B_{21} & 0 & A_{23} & A_{2B} \\ 0 & B_{32} & 0 & C_{3B} \\ 0 & B_{B2} & C_{B3} & 0 \end{bmatrix} \qquad (5)$$

Multiplying $\Re$ by the transposed vector $W^T$.

$$\Im = W^T \cdot \Re = \begin{bmatrix} W_1 \\ W_2 \\ W_3 \\ W_B \end{bmatrix} \cdot \begin{bmatrix} 0 & A_{12} & 0 & 0 \\ B_{21} & 0 & A_{23} & A_{2B} \\ 0 & B_{32} & 0 & C_{3B} \\ 0 & B_{B2} & C_{B3} & 0 \end{bmatrix} \qquad (6)$$

one attains the system of equations $\Im$, which reflexes the structure of the ontological net in turn:

$$\Im = \begin{cases} W_1 = A_{12} \cdot W_2 \\ W_2 = B_{21} \cdot W_1 + A_{23} \cdot W_3 + A_{2B} \cdot W_B \\ W_3 = B_{32} \cdot W_2 + C_{3B} \cdot W_B \\ W_B = B_{B2} \cdot W_2 + C_{B3} \cdot W_3 \end{cases} \qquad (7)$$

Furthermore, on the basis of the following figure 3

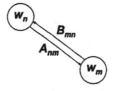

Fig. 3: Simple relation in the Ontological Network

one deduces the simple semantic relation (8)

$$\begin{cases} W_n = A_{nm} \cdot W_m \\ W_m = B_{mn} \cdot W_n \end{cases} \Rightarrow A_{nm} \cdot B_{mn} = \gamma \qquad (8)$$

where $\gamma$ represents an empty element, since paths from $W_n$ to $W_m$ and vice versa are equivalent over $A_{nm}$ and $B_{mn}$. Similarly, this counts also for:

$$C_{nm} \cdot C_{mn} = \gamma \qquad (9)$$

Resolving (7) as functions of $W_B$ using (8) and (9), one obtains following results:

$$\Im = \begin{cases} W_1 = A_{12} \cdot A_{23} \cdot C_{3B} \cdot W_B + A_{12} \cdot A_{2B} \cdot W_B \\ W_2 = A_{23} \cdot C_{3B} \cdot W_B + A_{2B} \cdot W_B \\ W_3 = C_{3B} \cdot W_B + B_{32} \cdot A_{2B} \cdot W_B \\ W_B = W_B \end{cases} \qquad (10)$$

In this way, every equation of (10) gives the number of different semantic paths, which lead a specific given element $w_i$ of $W$ to the corresponding key word $W_B$.

The structure of the Ontology Filtering System presented in this section should be easily extended to any other domain of use, i. e. manufacturing environment. When considering the natural language $L$ with a specific vocabulary of symbols $W$, one can rearrange the definition used above assigning elements of a further specific application domain $L'$ to symbols of $W$.

In the following section it will be shown, how to extend or modify the OFS Knowledge Base and in particular the ontological net.

## 4. THE PROTOTYPES STRUCTURES

As already mentioned, the mathematical model of the ontological network (ON) described in section 3 represents the core element of the Knowledge Base (KB) of the Ontological Filtering System (OFS). Conjoined with further information such as semantic classes and relations of terms as well as axioms and restrictions concerning the domain of use, i. e. the universe of discourse, it provides the OFS with the necessary "knowledge" for the processing of natural language contents. Likewise, it becomes possible for the OFS main program on the basis of the data stored in the knowledge base to process natural language sentences at both syntactic and at semantic level.

If applying the OFS in different domains of use, the knowledge base needs to be correctly managed and modified without a huge time effort. Thus, it should be possible for instance to introduce new terminologies, change or add semantic classes and/or relation as well as to remove or to define further restrictions in the domain of use every time. Thereby, the ontological network could be extended both horizontally, when specifying new terminologies in the same domain of use, or vertically, in case of adding new application fields.

Following these requirements, two tools have been developed. The first one is called OFS Knowledge Base Management Tool (KBMT). It enables the management to modify the entire knowledge base. The second one is called OFS Knowledge Base Visualisation Tool (KBVT) as it provides a graphical representation of the ontological network. However, while the former has been designed for permitting the access to all parts of the KB, the latter limits corrections, exclusively, in the ontological networks.

In the following two chapters (chapter 4.1 and 4.2), a general overview of corresponding tool structures is given.

### 4.1 The OFS Knowledge Base and the Knowledge Base Management Tool.

In order to correctly develop the KBMT, following fundamental questions have been posed:

- Which information must be provided to the OFS Main Program by this tool?

- In which way should information be stored into the OFS KB?

- Which kinds of connections are there between data in the OFS KB?

- How could modifications take place in the KB?

- How should the user interface look like under an ergonomic, i. e. user-friendly, point of view?

The information, which the KB provides to the OFS Main Program, can be divided into the following 5 groups:

*a) Semantic Class Definition.*
Each term in the KB belongs to a semantic class. Terms belonging to the same class have a shared meaning according to their assigned semantic role in the universe of discourse.

By means of the *Ontolingua* notation (Gruber, 1995) *Semantic Classes* such as *Humans*, *Machines*, *Processes*, *Failures*, *Work Tools*, *Work Pieces* could be defined. An example of the *Class* "machine" (?mac) is given below:

```
(define-class machine (?mac)
   "Any mechanical or electrical device,
   that transmits or modifies energy to
   perform or assist performances of human
   tasks"
   relation_to machine (?mac) = co-operation
   relation_to work piece (?wkp) = handling
   relation_to failure (?err) = restore
)
```

For instance terms like *Humanoid*, *Robot*, *Service Robot*, *Industrial Robot* etc. belong to this class.

*b) Basis Terms.*
A basis terminology $W_{NL}$ consisting of the addition of a set of basic terms $W_N$ has been fixed by recognizing principal elements playing a significant role within the regarded manufacturing environment as well as a set of possible actions $W_V$ among them. Figure 4 illustrates a simple exemplar scenario, where an industrial robot accomplishes an assembly and disassembly task.

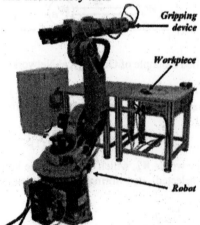

Figure 4: Example of fixing basis terminology.

The result in fixing terms is:

$$W_N = \{w_{N1}, w_{N2}, ..., w_{Nn}\} = \{robot, gripping\ device,...\}$$
$$W_V = \{w_{V1}, w_{V2}, ..., w_{Vn}\} = \{grip, open, move, ...\}$$

where the set $W_{NL}$ results from the association of $W_N$ and $W_V$. (Compare to equation (1)).

$$W_{NL} = W_N \cup W_V$$

As shown in section 3, the basis terminology is used to set up the *Meta Language* for the robot speech recognition. Therefore, it is very important to specify used term meanings exactly.

*c) Vocabulary.*

In the vocabulary all terms are archived, which can be recognized from the syntax analyser by parsing speech instructions within the OFS Main Program. Terms are hereby classified into nouns, verbs as well as adjectives, prepositions, articles and so on and are stored in the KB with additionally explaining information regarding their special meaning in the universe of discourse.

*d) Ontological Network.*

As explained in section 3, the ON contains both nouns belonging to the vocabulary (point *c*), included basis terms (point *a*) and all semantic relations between them. On this basis semantic connections between single terms and basis terms could be determined. (see Figure 1)

*e) Semantic and Syntactic Class Relations.*

The last group in the OFS is represented by semantic and syntactic classes, which are assigned to single vocabulary terms. Class and verb relations as well as relations between verbs and basis terms are necessary, in order to make the parser able to accomplish syntactic and content analyses of natural language and prevent possible misunderstandings. An example of semantic relation between the classes "machine" (?mac) and "workpiece" (?wrk) is given below:

```
(define-relation handling (?mac, ?wrp)
   "Manual or mechanical currying, moving,
   delivery or working with something"
   as_result: (?prc))
```

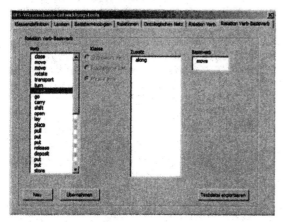

Figure 5: The KBMT Platform

For archiving data of all the five groups in the KB one has chosen the MS-EXCEL standard file as it offers a quite simple tabular structure.

Notwithstanding due to the quantity of information it falls still heavily, also for experts, to keep a clear overview of whole OFS KB structure. Thus, editing data within the KB requires also large time effort. For this reason, a KBMT has been programmed, in order to enable a systematic KB data management. Data modifications occur over the comfortable and user-friendly platform illustrated in Figure 5.

*4.2 The OFS Knowledge Base Visualisation Tool.*

As seen before modifications of the OFS KB can be easily carried out by means of the KBMT. However, this requires previous knowledge about the ON structure, such as e. g. by developers. Thus, one cannot expect that also someone, who has no knowledge about the KB structure, e. g. machine operators, can do it well. Since data sets become very complex and difficult to manage, already by few stored terms, a further possibility has to be given to make only essential changes also in the KB.

By means of a second tool called Knowledge Base Visualisation Tool (KBVT) the ON can be graphically visualized (see Fig. 1) and its structure adapted to specific requirements of every domain of use. This happens adding, removing or changing terms (circles) as well as semantic relations (arrows) through the platform of Figure 6 programmed within MS-VISIO.

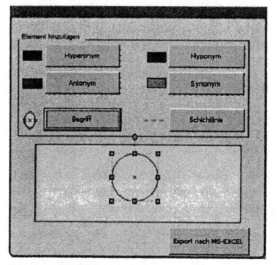

Figure 6: KBVT Input platform

It has to be guaranteed that every modification in the ON made by someone, without knowledge about the entire KB structure, is correct. In order to avoid possible errors a further functionality has been implemented. On the basis of the *Algorithm of Warshall and Floyd* (Gueting, 2003) the KBVT checks the consistency and accessibility of each ON element is checked automatically and gives out a feedback about its structure completeness.

Furthermore, after every modification resulting data must be updated into the KBMT and vice versa.

The necessary bi-directional data exchange between the two tools is ensured over a MS standard interface realized with help of macros of the *Visual Basic for Application* (VBA). The whole data exchange concept between both tools within the OFS is shown in the following figure (Figure 7).

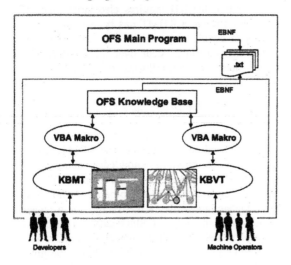

Figure 7: Data Exchange between both KB Tools

## 5. CONCLUSIONS

The contribution presented a concept for improving the human-robot collaboration based on the use of natural languages, where a procedure for setting up ontological nets represents the project emphasis. This provides functionalities for the semantic filtering of natural speech instructions.

For the validation of shown results several close-to-applications experiments – like the presented assembly scenario - have been carried out. The next steps aim, on the one hand, at extending the applicable vocabulary and, on the other hand, at refining the mechanisms used for the machine processing of natural speech. Additionally, a modularity and transmission of the interaction concept for further robotic controls are in progress.

Robot assistants for task sharing with human operators are still at the beginning but with promising outlook. The envisaged results have to be reflected, in order to proof principles of working science (e. g. experimental safety, motivation of shop-floor staff and so on). Also training and maintenance efforts and finally overall time and cost for implementation have to be regarded.

## ACKNOWLEDGMENT

The project described in this paper is performed in the laboratory of the Department of Automation Technology of the Brandenburg University of Technology. For the project tests a KUKA Robot KR15 has been employed. The authors are special indebted to Kai Henning, Ralf Kretzschmann and Christian Thies for their contribution to this project.

## REFERENCES

Bratko, I., T. Urbancic, C. Sammut (1995): Behavioral cloning: phenomina, results and problems. Proc. IFAC Symp. Automated Systems based on Human Skills, Elsevier Ltd. pp. 143-149.

Endsley, M. R. (1999). Situation Awareness in Aviation Systems. In: Handbook of Aviation Human Factors (Garland, D.J. et al (Eds)), pp. 257-276. Lawrence Erlbaum Assoc., London.

Erbe, H.-H. (2004): On Human-Robot Collaboration - A Survey on Cost Aspects. In *Proceedings of the 7th IFAC Symposium on Cost Oriented Automation*, Ottawa, Canada, June 7-9, 2004.

Gruber, T. R.: Toward Principles for the Design of Ontologies Used for Knowledge Sharing. In: *International Journal of Human-Computer Studies*, 43, 1995, p. 907-928.

Gueting, R. H.; Dieker, S.: "Datenstrukturen und Algorithmen", 2. Auflage, Teubner Verlag, Stuttgart-Leipzig-Wiesbaden, 2003.

Kimura, H., T. Horiuchi, K. Ikeuchi (1999). Human robot cooperation for mechanical assembly using cooperative vision systems. In: Proceedings IEEE & RSJ Intelligent Robotics and Systems 99, pp. 701-706.

Knoll, A.C., R. Dillmann (eds.) (2003). Int. Conf. On Humanoid Robots, VDI/VDE-GMA, Düsseldorf, Germany, ISBN 3-00-012047-5

Laengle,T., H. Woern (2001). Human-Robot-Co-operation using multi-agent-systems. Intelligent and Robotic Systems, pp. 143-159.

Lepratti, R.; Berger, U. (2003): Towards Ontology Solutions for Enabling Interoperability in Virtual Enterprises. In: *Processes and Foundations for Virtual Organisations* L. M. Camarinha-Matos and H. Afsarmanesh (ed.), Proceeding of the 4th IFIP Working Conference in Virtual Enterprises (PRO-VE 2003), Lugano, Switzerland, October 29-31, Kluwert, Boston-London, p. 307-314.

Lepratti, R.; Jing, C.; Berger, U.; Weyrich, M. (2004): Towards the knowledge-based Enterprise. In: *Proceedings of the International Conference on Enterprise Integration and Modelling Technology* (ICEIMT'04) Toronto, Canada, 9-11 October, 2004, (in print).

Sato, Y.K.B., H. Kimura, K. Ikeuchi (2002). Task analysis based on observing hands and objects by vision. Proc. IEEE/RSJ Intl. Conference on Intelligent Robots and Systems, pp. 1208-1213.

Sheridan, T. B. (1992). Telerobotics, Automation and Human Supervisory Control. The MIT Press, Cambridge, Massachusetts.

Stopp, A., S. Horstmann, S. Kristensen, F. Lohnert, (2002). Towards Interactive Learning for Manufacturing Assistants. Proc. of the 10th IEEE Inter. Workshop on Robot-Human Interactive Communication, Paris, France.

Winograd, T. (1980): What does it mean to understand language?. In: *Cognitive Science* 4, 1980, pp. 209-241.

World Robotics (2003): *Statistics, Market Analysis, Forecasts, Case Studies and Profitability of Robot Investment*, USA; 2003.

# HANDLING OF ALTERNATIVE PROCESSES FOR MACHINING OF AERONAUTICAL PARTS IN A CAPP SYSTEM

**V. Capponi, F. Villeneuve, H. Paris**

*Soils, Solids, Structures Laboratory, Grenoble - France,
phone: (33)4 76 82 51 44, fax : (33)4 76 82 70 43,
Vincent.Capponi@hmg.inpg.fr*

Abstract: This paper focuses on the management of alternatives machining operations, essentially from an access orientation point of view, during computer aided process planning of aeronautical parts. Characteristics of these parts entail to consider a larger amount of alternative machining operations for one single feature than considered in usual approaches. Using visibility map concept, taxonomy of "machining direction sets" is proposed to manage these alternatives. Then, procedure to handle theses alternatives sets through a step of the set-up planning stage is suggested. Copyright © 2004 IFAC

Keywords: CAPP, Process planning, Machining directions, Visibility map, Aircraft

## 1. INTRODUCTION

Computer assisted process planning (CAPP) is the last gap to jump in order to achieve the CAD/CAM integration challenge. In companies, process planning is still a task that requires a significant amount of both time and experience, and automated systems are not yet reliable, despite the research and industrial work done for the last twenty years. Results of these works can be roughly clustered in two main topics:

- Planning systems
- Cutting Path generation systems

The former are systems that really organize the manufacturing process. Historically, first developments were variant approaches, based on the retrieval of archived process plans (Group Technology concept). Then, generative systems were developed, using AI solving methods and knowledge base of generic manufacturing rules to construct the process plan. The domain targeted was mainly automotive-oriented where features were more or less isolated entities locally defined and treated, and where knowledge and know-how could be split into domain-, product-, enterprise-based. In such an approach, two generic major steps can be pointed out: matching a machining process to a machining feature and planning the set-ups.

The latter systems deal with the generation of the best tool path covering the whole machining of a feature, especially for free-form faces. It is mathematically difficult and optimized for cutting time, quality and form achievements. There is no organizational planning of the whole machining process of the part in these systems.

This paper presents an approach to handle alternatives machining operations during aided process planning of aeronautical parts. Aeronautical parts are machined from a parallelepipedic raw block, with a specific strategy that consider global mill roughing apart from finishing operations. Despite the complexity of these parts, major difficulties are rather due to planning consideration than due to tool paths generation. Indeed, free-form faces are not so numerous in considered parts. Prior constraints for process planning come from accessibility and precedence analysis. Visibility concept has been used to model machining directions, and a procedure is presented to handle alternative machining operations through a step of the planner module.

This work takes place in the context of a larger project developing a Computer Aided Process

Planning System for the aircraft industry. This French project includes aeronautical industrialists, software developers and vendors, and research laboratories.

The paper is organized as the following. Through a quick literature review, section 2 discusses on the management of alternative solutions in CAPP systems. Next section highlights the specificity of aeronautical parts proving particularities and the requirements needed for an adapted CAPP system. Section 4 presents the visibility concept as it has been used for workpiece orientation. Section 5 suggests using direction sets trough the visibility concept in order to meet aeronautical industrial needs. Definitions of different machining direction sets and their properties are proposed. To conclude, section 6 provides a simple example of our alternative sets management.

## 2. ALTERNATIVES MANAGEMENT IN CAPP SYSTEM

Computer aided process planning systems are waiting yet in research laboratories to be transferred to industry. To explain the fact that very few systems are used in companies, El Maraghy, *et al.* (1993) and Shen and Norrie (1999) pointed out the requirements a next generation CAPP system should have among others: *interoperability; open and dynamic structure; integration of human with software; support the decision making.* A way to integrate human planner decision in such system is to provide for him the relevant data to choose between a set of computerized alternatives rather than make him validate a unique rigid computerized solution.

Sormaz and Koshnevis (2003) states that the availability of alternatives process plans can speed up the incremental process plan generation as it allows more flexibility to fit with production schedules and resources availability. However, to compute alternatives process plans, the kind of data open to handle several candidates has to be identified. That is to say, one could consider and keep alternatives in many stages of the process plan generation, from the former one (for instance Raman and Marefat (2004) consider several feature descriptions for the same part) to the latter one (such as different operation sequencing in a same defined set-up).

Once a part is described in machining feature, two generic major steps can be highlighted in the literature survey on generative CAPP systems: *Operation planning* and *Set-up planning.* The former consists in matching a machining process to a machining feature, while the latter groups the process operations into set-ups and sequence operations into each set-up. Most of the CAPP systems reported in the literature select one unique machining process for an individual feature as an output of the *operation planning* step. Then, the *set-up planning* step generates all possible routes plans to combine the selected processes into set-ups, through constraints satisfaction. The best route plan is selected automatically from an objective function (often reduced to a lower-cost function or set-up minimization function). More rarely, the human planner through performance indicators chooses the best plan.

The reader should note here that final alternatives plans that really differ, that is to say involve different machining strategy choice, could only be generated through handling alternatives in the former stages. Indeed, alternatives in operation sequencing for instance often lead in local optimization on tool paths or on tool change, which can save money on the total cost for a mass production, but which cannot open the way to the optimal solution.

In operation planning, if manufacturing resources are adequate, several applicable processes may exist. It is possible that an optimal process, i.e. the most economical and efficient one, can be selected. However, an optimal machining process of an individual feature is not necessarily always optimal for a complete part. When more than one feature needs to be considered for a part at the same time, the optimal process for this individual feature may be changed. Therefore, alternative processes need to be generated and they are to be handled in order to reach a more optimal process plan. This approach guaranties the principle of least commitment.

. This report focuses on the way to handle a set of alternatives processes through the set-up planning stage. Alternatives can be between machining processes, or between machining directions in the same process. As shown in section 3, studies realized in an aircraft company on process planning activity point out the importance of this approach for aeronautical machined complex parts. Note that the methods to generate all the possible candidates processes for an individual feature is not the considered issue of this paper.

Although numerous machining CAPP systems have been reported in the literature, only a few have considered alternatives as an output of the operation planning step. Different methods have been employed to handle alternatives processes. Among others, methods use the Petri Nets modeling (Horvath, *et al.*, 1994; Kiritsis and Porchet, 1995), NLPP: Non Linear Process Plans (Kruth and Deland, 1992; Van Zeir, *et al.*, 1998), Process plan networks (Sormaz and Koshnevis, 2003), Weighted matrix (Wu, 2002), simultaneous fixture and set-up planning management (Paris, 2004).

However, most of theses studies consider alternatives processes for holes features, where the nature, the number and the sequence of operations changes, but where the machining direction is the same. Few studies take into account the alternatives between processes with different machining directions, like side milling and end milling for a plane. And when this is done, the problem is often simplified for 2-1/2 axes part, that is to say, at the most, only the 6 main directions of the parallelepiped prism will be evaluated to machine the feature. These simplifications are obvious on general mechanical parts because most of the time one can deduce the right process (peripheral or end milling) from the intrinsic specifications of the plane (roughness for instance). The complexity of aeronautical 5-axes parts entails to have a more detailed approach, in handling all possible machining directions in the space from the candidate processes.

## 3. AERONAUTICAL PART CONTEXT

### 3.1 Aeronautical parts specificity

Let us roughly demonstrates how specific are the aeronautical parts:

- General geometry: structural parts, elements for effort transmission and connection for example, are considered. They are 3D parts, meaning that the third direction is machined as well. General dimension could be 1000*500*150 mm. The stiffness/weight ratio is maximal.
- Material: mainly aluminum alloys, titanium and sometimes alloy steels.
- Form: 5 axis parts with many prismatic pocket recesses (i.e. bottom and flank are planes) and webs. Stiffeners are very often thin and sometimes free-form. The "complexity" of the form is rather due to the multiple different orientations of thin plane webs than to the free-form surfaces. Figure 1 shows an example of a simplified part.
- Quality: average level of 0.1 mm, with very few tight geometric specifications.
- Batch size: never more than 20 parts a year

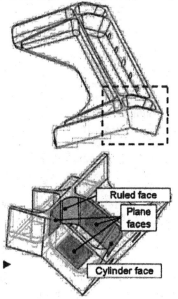

Figure 1. Part example and elementary features of a pocket recess.

### 3.2 Machining strategy

Studies made on the generation of process plans in an aircraft company (see Capponi 2002) can be summarized to the following points:

- General machining process: Unlike automotive parts where only functional surfaces are machined from a casted or forged blanks, the aeronautical parts are entirely milled from a prismatic raw block.
- Fixturing process: based most of the time on plane-centering-locating principle. Though, when automotive parts are positioned on their datum surface for their machining, the location and clamping surfaces of aeronautical parts are generally out of the functional surfaces of the part. It means that the raw material includes such elements along with the final part. These elements are classically in the remain of the blank after machining and must be managed throughout process planning and of course machining.
- Resources: machining processes are milling and drilling, 3 to 5 axes. Due to the lack of part stiffness, fixtures are specifically designed and realized; depression is sometimes used for clamping assistance.

In aircraft companies, these sculptured parts are milled through a 3 axes global roughing stage that removes the main volume of the initial raw block, followed by 5 axes finishing stage. That means that the pocket recesses are not milled by applying a global strategy (for instance, a typical generic pocket strategy could be composed of 3 stages, namely global roughing, bottom finishing and flank finishing) as in usual mechanical domains, but they are milled considering all the elementary features of the pocket recesses quasi independently (figure 1). This study deals with finishing operations.

### 3.3 Machining directions involved

Usually, an aeronautical part holds many pocket recesses versus few toleranced faces. These recesses have usually been designed to lighten the part, hence the surface quality (represented by the visual aspect, the intrinsic dimensional tolerance and the roughness) required for the part is usually poor. Indeed, peripheral milling or end milling modes are often technologically equal for achieving the required quality of a plane surface. On the other hand, ball-end milling, as it requires much more time, is only considered when the two formers modes are impossible. Thus, the machining directions for ball-end milling are not computed until it is necessary. Consequently, this leads to consider alternatives machining operation or/and alternatives machining directions for each elementary face. Note that the number of the alternatives could not be countable. As illustrated on figure 2, a plane can be peripheral milled, end milled or ball-end milled. For the peripheral mill and ball-end mill operations, infinite number of machining directions remains.

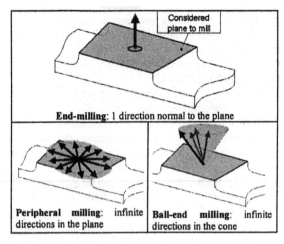

Figure 2. Alternatives directions for plane milling

A particular instance of a plane and its potential machining directions is shown on figure 3. This geometry is often found on aeronautical parts. An evident mode is to end-mill the plane as bottom pocket plane, with a fillet-end cutter. But, in order to save one new set-up orientation, sometimes the planner has to consider peripheral-milling modes to machine this plane. It leads to consider single operation with at least 2 machining directions.

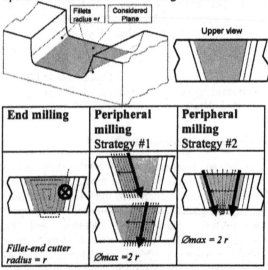

| End milling | Peripheral milling Strategy #1 | Peripheral milling Strategy #2 |
|---|---|---|
| Fillet-end cutter radius = r | Ømax = 2 r | Ømax = 2 r |

Figure 3. Alternatives directions for a particular plane

Last case encountered is when some feature holds only one possible machining direction (hole for instance), or when the part topology could forbid some machining directions of a feature to avoid tool/part collision. For instance, the bottom of a "closed" pocket recesses must be machined by an end-milling mode. As the axis tool must reach these directions, they are called "obligatory directions" (figure 4).

A feature model extending the initial machining feature model has been developed to match machining processes adapted to aeronautical

knowledge (Capponi, *et al.*, 2004). Another research[1] in progress is working on the extraction of these features and its associate machining directions from a CAD model.

Finally, once the possible machining directions have been computed for all features of a complex aeronautical part, one can states that alternative directions and obligatory directions are usually distributed in all the space. This paper focuses on the way to handle all theses directions, alternatives ones and obligatory ones are described in the next section.

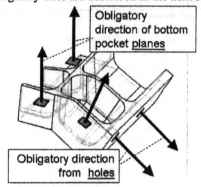

Figure 4. Obligatory directions in the example part

## 4. VISIBILITY CONCEPT TO HANDLE ALTERNATIVES PROCESS

This section describes the modeling of machining directions of an aeronautical part through the visibility map concept. First, visibility map concept will be explained. Then, a quick review of their utilization in machining will be presented.

### 4.1 The visibility concept

The visibility concept in manufacturing consists in mapping the relevant directions (for the considered manufacturing application) of a 3D workpiece onto the unit sphere, namely the gauss sphere $S^2$, in order to find the optimal workpiece orientation for the considered application. In various manufacturing domains, algorithms for this optimal orientation could be formulated as simple intersections on the sphere. This concept has been used with interesting applications in various manufacturing domains. Woo (1994) present applications, among others, in molding (finding the best parting plane to avoid auxiliaries), numerical control with probe (finding the best orientations to minimize the number of set-ups), layered manufacturing (finding the orientation minimizing stereolithography process time) and machining milling. A quick review of visibility approaches for milling is presented below.

[1] This research takes place in a laboratory of our partnership (see acknowledgements) and will not be tackled in this paper.

## 4.2 Visibility in machining

Among all works made on visibility in milling (Woo, 1994; Tang, *et al.*, 1998; Elber, 1994; Hascoët, *et al.*, 1996) the following same generic problem was studied: given a complex part, made of n sculptured surfaces milled by a ball-end cutter, find the set-up orientation that allows the maximum number of surfaces to be accessible and therefore milled by the tool on a 4- or 5-axis machine. Gupta *et al.* (1996) discuss the same problem with fillet-end cutters in their works. This approach is known as a "greedy" approach, that is to say that it has been proved that grouping the maximum number of surfaces in a set-up, in an iterative way, should not lead to the minimum number of set-ups. Different "elegant" algorithms have been used to solve this maximum intersection problem in the sphere (MIPS) in the less computing time using, among others, central projection on a plane or geometric duality mapping.
Cited approaches, through their works, have defining the following relevant visibility terms:

- *Real Visibility map of a surface*: spherical map on $S^2$, any points in a visibility map represents a tool orientation through which the entire surface can be machined. The Real Visibility map differs from the visibility map (Vmap) because the former considers both tool capabilities and tool/part collision while the latter considers only the tool capabilities. Constructions of the Vmap and their properties could be found in Gan *et al.* (1994).
- *Visibility map of a machine-tool (Machine Map)*: spherical map on $S^2$, represents the set of orientations the spindle of the machine-tool can access, due to the degrees of freedom of the machine. For instance, a generic 4-axis machine map is a great circle on the sphere (if the work-table can rotate a full 360° about the y axis, else it is only a part of the great circle), while a generic 5-axis machine map is a band surface on a sphere (considering that the spindle can also swivel partially about a second axis). This is illustrated on figure 5.

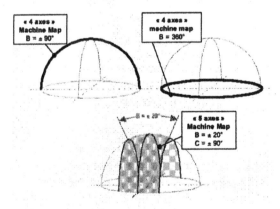

Figure 5. Machine map of a 5 axis machine-tool.

## 4.3 Approach suggested

The approach suggested in this paper use these visibility concepts to handle the « traditional » machining directions of alternatives processes of a feature, that is to say the directions of a flat end milling cutter for elementary plane or ruled faces instead of ball-end milling directions for a sculptured surface. First, this entails to differentiate the alternatives directions set from the so-called "obligatory" directions set. A taxonomy and graphic representation of these direction sets are provided in section 5. Then, an example of the utilization of these visibility maps through a step of the set-up planning stage will be explained in section 6.

## 5. TAXONOMY OF THE "DIRECTIONS SETS"

The different geometric nature of the elementary feature, that is to say plan, cylinder, ruled surface or free from surface, coupled with the availability of others candidates process, could lead to particular sets of machining directions for one feature, that have to be identified for a CAPP development. The proposed taxonomy below identifies two kinds of machining directions set for a feature. Taxonomy of elementary «accesses» is first presented, then the different access sets of a feature and their properties are detailed.

### 5.1 Access taxonomy.

A machining access is the generic term to define all the relevant possible combination of several machining directions to machine a feature in a determined machining mode. Five generic machining modes have been evaluated for typical aeronautical parts: peripheral milling (including 5-axis flank contouring), end milling, axial milling (i.e. drilling, boring, threading…), side and face milling, ball-end milling. This analysis leads to define 3 potentials "machining access" for one milling operation, each one illustrated for a real case in figure 6.

- **Single machining direction.** That is the simplest access, used in most of the milling operations. For instance, a single machining direction is required to machine a hole, a plane or a ball-ended surface (machined with a 3-axis strategy).
- **Multiple (non-continuous) machining directions.** Several discrete (i.e. denumerable set) directions define such an access. This access is required for the particular strategy to machine a plane as illustrated on figure 3.
- **Cone of machining directions.** Continuous directions (i.e. not denumerable set) on a spatial domain (typically a line on $S^2$) define this type of access. Any point of the domain is a machining direction to reach to machine the related feature. This access is mainly required for the flank contouring of ruled surface or any continuous 4 or 5-axis machining.

| Access Types | Single | Multiple | Cone |
|---|---|---|---|
| Matching Mode | PM, EM, AM, SFM, BEM | PM | PM, BEM |
| Example of configuration | 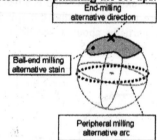 | | |
| Representation on $S^2$ | | Operation d match up with strategy#1 of figure 3. | |

PM : peripheral milling ; EM : End milling : AM : Axial milling ; SFM: side and face milling; BEM : Ball end milling

Figure 6. Illustration of access type

Note that the ball-end milling operation is always available to machine any kind of surface, but for a plane or a ruled-surface, it is known that it is not the optimal milling mode in term of surface quality or machining time. That is why the ball-end milling mode is only considering in our approach for free-form feature, or for the features for which no other milling mode is possible.

### 5.2 Access set

Now accesses for one single milling operation have been identified, set should be defined for a feature to take into account all the alternatives finishing operations available. Two kinds of set have been defined:

- **Obligatory set**: it is called obligatory because the tool axis must be oriented about all the machining directions define by the accesses in the set. Actually, a particular obligatory set is a non-single access obligatory set. All the alternatives operations lead to a set with the same access repeated x times. This kind of set is always composed of single repeated machining direction accesses, represented as a point on the unit sphere $S^2$. Most of the obligatory sets are single sets (i.e. with only one operation, one access), which could then be any kind of access listed above.
- **Alternative set**: alternatives sets group several accesses (corresponding to several operations) without same directions for a single determined feature of the part.

An interesting property to model not countable directions of the same milling mode is used in alternative sets. Particular groups can be identified on such alternative sets, because all alternatives operations belonging to the same milling mode draw typical shapes on $S^2$ for elementary face such as plane. These groups are:

- Alternative arcs (figure 7): the concerned mode is the peripheral milling mode for a plane. This mode leads to consider a succession of single machining directions, which corresponds to all

the possible tool axis direction. On $S^2$, this leads to a succession of points that draws a great circle or arcs of great circle. Obviously, 1 direction only among these lines has to be chosen between the "alternative arc".

- Alternative stain (figure 7): the concerned mode is the ball-end milling mode for any feature. This mode leads to consider a succession of single machining directions, which corresponds to all the possible tool axis directions compatible with such strategy. This leads to a succession of points that draws a surface or "stain" on the unit sphere. Depending on the ball end milling strategy chosen (3 axes or 5 axes), 1 or more direction(s) among this surface has to be chosen.

These particular groups are very useful because it enables to generate all the machining directions (that are uncountable because they are defined continuously on a domain) in a simple way. As it is shown in section 6, this taxonomy will be useful to take decision while planning the set-ups.

Figure 7. Alternative set of the plane of figure 2.

### 5.3 Visibility map of the part

The visibility map of the whole part is then an aggregation of the accesses set of each feature of the part. Due to the distinct nature of sets (i.e. obligatory or alternative), we can decompose the visibility map of the workpiece between

- The obligatory part of the visibility map
- The alternative part of the visibility map

The next section introduces the CAPP system architecture and illustrates the utilization of visibility map in the set-up planning stage.

## 6. HANDLING OF ALTERNATIVES THROUGH THE SET UP PLANNING

### 6.1 CAPP System specifications

In aeronautic industry, high capital cost machine-tools coupled with high level of workshop tasks automation leads to consider the number of set-ups as a crucial parameter to increase productivity. Since an added set-up requires the dismounting and setting-up of the workpiece by a human operator, the total number of orientations (to achieve the whole part) should be minimized. So to reduce costs in aeronautical parts workshop, main objective is to minimize set-ups.

An analysis of the characteristics of aeronautical parts production -such as: small batch size; high complexity of the parts; high hourly costs of the manual process planning activity - entails that the usual objectives such as lower cost or lower time, are not as relevant as in automotive industry. Prior constraints for process planning of aeronautical parts come from accessibility and operation precedence analysis. Indeed, as these parts hold very few tight tolerances, major constraints resolution can't be based on the tolerance charts of the part.

To meet the above industrial needs, human-centered specifications for a CAPP system have been developed. The three modules used for generating the process plan are illustrated in figure 8. The feature recognition module includes a machinability analysis stage that provides the obligatory sets and alternative sets of the part as an output. The planner module deals with set-up planning tasks and work holders definition. Then, the last module achieves operation sequencing in set-ups and microplanning tasks to compute the CN code and process planning documents. In the next sections, description of the way retained to handle the alternatives through a step of the planner module is proposed.

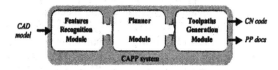

Figure 8. Modules of the CAPP System

### 6.2 The planner module

The set-up planning methodology is based on two analyses, one in the spatial domain, and the other on temporal domain.

Roughly, the spatial analysis deals with finding the best "partition" of the part between set-ups,

ensuring that each feature is accessible. This analysis provide as an output the minimum number of set-ups needed to achieve the whole part; the orientation to fix the part on the machine-tool for each set-up; the feasibility (i.e. from an accessibility point of view) of each feature in the considered set-ups.

The temporal analysis generate precedence constraints between machining operations to guaranty the respect of industrial manufacturing rules. In the aeronautical domain, these constraints could be triggered by generic manufacturing rules (due to the topology of the part) or more specific rules to ensure the rigidity of the thin parts of the workpiece during machining. Figure 9 illustrates a rule to ensure that the stiffener will not be vibrating or deformed during the machining of its top (rule #1) and a generic rule (rule #2).

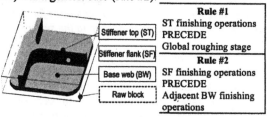

Figure 9. Instance of precedence rules.

To achieve the minimum set-ups objective, our approach gives priority to accessibility constraints. In other words, a sequential approach have been specified where the first stage is a spatial analysis and the second stage is a constraints satisfaction based on the temporal analysis. Note than in order to check with the particularity of the parts held by tabs fixture, a third stage is achieved to refine the macro planning with the new information computed while designing the work holders. The three stages are illustrated on figure 10.

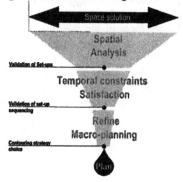

Figure 10. The three stages of the planner module

### 6.3 Handling of alternatives

The initial space of finishing solutions to machine a part represents all the possible reliable combinations sets to machine each feature. Explicit space of solutions cannot be computed for combinatory explosion problems. The strategy developed in our approach is based on a restriction of the alternatives solutions to mill a feature during the three stages listed above. In other words, each decision taken in each stage could restrict alternatives if it got the right information to. Conversely, alternatives could be kept at the end of each stage if these alternatives were not constrained by the decisions taken in the stage. This approach respects the least commitment principle. At the end of the third stage, a choice step (made by the planner or in an automatic way) will select only one solution for each feature if alternatives remain in order to get a unique final process planning of the part.

### 6.4 Spatial analysis detail

An overview of the spatial analysis is presented here. Two steps are computed in this stage: *defining the best workpiece orientations for the set-ups; associating a machine map to the set-up*s.

*Workpiece orientations for the set-ups.* The retained approach for this stage is based on a study that states that 90% of the parts are machined in a "double face" strategy. That is to say, only 2 opposed faces provided by the main planes of the raw block are used to locate the part on the worktable. The 10% remaining parts (usually more complexes ones) are machined on more than two location faces, but usually the two first set-ups are located onto the two opposed planes of the raw block as well. As raw material cost is not a relevant criteria in such context, the best-fit blank (usually parallelepipedic) is not the minimum material one but can be defined to ease the set up fixture by providing two (opposed) relevant and good quality location plans in a required orientation. Assuming this specificity, the concept of neutral plane has been highlighted (by analogy with a parting plane in molding). Neutral plane models the orientation of the planes faces of the raw block, therefore the two opposed workpiece orientations on the machine tool.

For "double face" parts, the best neutral plane is defined applying a heuristic; the plane is taken parallel at the larger pocket bottom(s) of the part. For more complex parts (where bottoms of pocket are not relevant considering the whole part), work is in progress to find an efficient algorithm based on the visibility map.

*Associating a machine map.* This step description is explained with parts machined using the "double face" strategy. The neutral plane will divide the visibility map of the part in 2 hemispheres. Each hemisphere is accessible to only one set-up orientation. When a machine map will be associated to a set-up, it will cover only a portion of the corresponding hemisphere. The problematic could be summarize as follows: find the best orientation of the two machine maps (each in one of the 2 hemispheres), ensuring that all obligatory access sets and at least one access of each alternative sets will be intersecting with the machine maps. Unfortunately, no elegant mathematic formal solution exists. To solve this problem efficiently, a technological hypothesis, relevant from what has been seen in industry, have been formulated and the following method have been retained:

Hypothesis: The machine map holds a preferential direction that corresponds to the most rigid configuration of the machine (for a serial machine-tool, this configuration is usually achieved when axes rotation are on their origin positions). To avoid oriented work holders that are too complex to manufacture, this preferential direction of the machine map is mapped onto the primary locating direction of the set-ups, that is to say the two opposed normals of the neutral plane. Hence, only one orientation parameter about the primary locating direction is needed to associate a machine map to a set-up.

Method: Considering the visibility map on each hemisphere, following distinction could be made:

- *Alternatives sets totally included in a hemisphere (ASI):* each access of the alternative set is totally included in a hemisphere. That means that the corresponding feature is only machinable on one given set-up.
- *Alternatives sets in between the two hemispheres (ASB):* this configuration is due to some accesses of the same set which are not included in the same hemisphere. Each access of the set, considered separately, must be totally included in a hemisphere to be used. Indeed, imagine the case where an access of the set is spanning on the two hemispheres. This "spanning" access is not viable, because it cannot be machined on a single set-up. This assumption leads to eliminate the spanning access from the set (restriction of the set). This restriction could lead to transform an initial ASB in an ASI.
- *Obligatory sets (OS):* each set is entirely included in a hemisphere, as it is an implicit constraint to define the initial hemisphere partition.

Method based on the visibility map information to find a relevant orientation of the machine map in the hemisphere is presented. This method is based on a augmented central projection algorithm (Chen and Woo, 1990), to transform spherical problem to planar problem (see figure 11). Applying this algorithm to a single hemisphere visibility map, associating the machine-map consists in finding the best orientation of its projection on the plane (see figure 12).

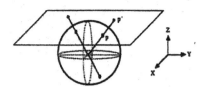

Figure 11. Central projection (from Chen and Woo, 1990).

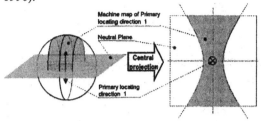

Figure 12. Machine map projected onto a plane.

Algorithm: Objective of the following algorithm is to find all possible orientations of the two machine-maps that allow at least one access for each feature.

---

1.Reasoning in each hemisphere i

   (a)  Find the range of possible orientation to match the OS of the corresponding hemisphere. This defines a congruent zone Oi to place the machine map (see figure 13).

   (b)  In the Oi domain, find the range of possible orientation to match at least one access of each ASI of the corresponding hemisphere. This defines a congruent zone Ii (Ii ⊆ Oi) to place the machine map.

   (c)  *for each ASBj* :In the Di domain, find the range of possible orientation to match at least one access (of the corresponding hemisphere). This defines j zones Bij (Bij ⊆ Di).

2.Reasoning on the two hemispheres ( i = {1;2} )
*for each ASBj*

   (d)  If B1j=B2j={∅}, consider a ball-end milling strategy or define new set-ups [GOTO 1]

   (e)  if B1j ={∅} or  B2j ={∅ }, a restriction of ASBj become an ASI. [GOTO b]

   (f)  if B1j ≠ {∅} AND B2j ≠{∅},

        i.  if B1j=D1 AND B2j=D2, then this ASBj is not taken in account for reasoning.

        ii.  if Bkj ⊂ Dk AND Bmj ⊆Dm, compute all explicit combinations of Bij zones on the 2 hemispheres.

---

Then, the planner would have to choose a solution if several ones exist, or to decide which action to take if any set is not reachable (that is to say no solution exists). The possible actions are defining a new neutral plane (hence defining new set-ups orientation), add a set-up orientation, or machine the inaccessible feature in a ball-end milling mode.

Once a viable solution have been chosen, the system will update the alternatives sets of visibility map with the surface covered with the oriented machine map. This could lead to restrict the alternative sets. Due to this restriction; some sets will become obligatory sets, some sets will lose one or several access, and some sets will not be changed.

Then, the temporal constraints satisfaction stage checks the feasibility of the workpiece according to the set-up orientations defined in the previous activity. It takes into account the machining precedence constraints to allocate the machining operations while computing set-ups, and by the way could restrict some ASB by allocating it to a considered set-up. In the Constrained Macro Plan (output of the stage), some "direction sets" may be sequenced in a set-up, some may be allocated but not sequenced, and the non-constrained ones may be "free", that is to say their machining could be made on any defined set-ups. Final decisions to compute the unique final plan will be made by the human planner through the "refining" stage.

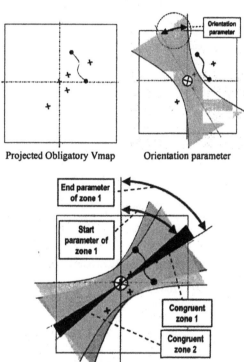

Congruent zone: range of possible orientation of the machine map that intersects with all obligatory sets.

Figure 13. Orientation parameter and congruent zone illustration

## 7. CONCLUSION

This paper focuses on an approach to handle alternatives machining operations during aided process planning for aeronautical parts.

Characteristics of these parts entail to take into account a larger amount of machining directions during the set-up planning activity. Our study suggests using the visibility map concept to handle the process alternatives. Alternatives can be between machining processes, or between machining directions in the same process. As the kind of machining operations modelled through the visibility concept differs from precedent works, taxonomy of the required accesses is defined and illustrated.

Retained approach is to keep alternatives processes for a single machining feature during the process planning, as all information to decide between these alternatives is not available for such aeronautical parts. General principles of the proposed CAPP system that will integrate this approach are exposed. Finally, procedure retained to handle the alternatives processes during a step of the planner module is described.

This procedure has to be further implemented in a CAD environment to be validated on a larger set of parts. By supporting the tasks of process planning activity, the main benefit expected by industrialists is a shorten and more reliable lead time.

To conclude besides the automation aspect, the reader should note that provided taxonomy and representation of alternative access on a sphere should be very useful and adapted to support human decision in interactive software. Indeed, our experience on CAPP system suggests us that real optimization for such difficult and opportunistic planning could not be achieved without human planner intervention. In the context of human decision required to choose between alternatives computed solutions, good and reliable indicators are necessary to support decision. Our study may contribute to set up indicators for spatial machining analysis of complex parts.

*Acknowledgments: This research is funded by the USIQUICK French Project whose partnership involves research laboratories, aeronautical and software industrialists.*

## REFERENCES

Capponi V. (2002). Proposition d'une architecture interactive pour l'élaboration de gammes d'usinage, guidée par l'étude d'une expertise industrielle. *Master of Research Thesis at Université J.Fourier,* Grenoble I, France

Capponi, V., O. Zirmi, D. Brissaud and F. Villeneuve (2004), Capp, Strategy And Models In Aircraft Industry. *IDMME'04 Conference Proceedings,* Bath, UK

Chen, L.L. and T. Woo (1990), Computational geometry on the sphere with application to automated machining, *Trans. ASME J. Mech. Design,* **114,** pp288-295.

Elber, G. (1994), Accessibility in 5-axis milling environment, *Computer-Aided Design,* **26,** 11, pp796-802

El Maraghy, H.A. (1993), Evolution and future perspective of CAPP, *Annals of the CIRP,* **42,** pp 1-13

Gan, J.G., T.C. Woo and K.Tang (1994), Spherical Maps : their construction, properties, and approximation, *Trans ASME Journal of Mech. Design,* **116,** pp357-363

Gupta, P., R. Janardan, J. Majhi and T. Woo (1996), Efficient geometric algorithms for workpiece orientation in 4- and 5-axis NC machining, *Computer-aided Design,* **28,** 8, pp577-587

Hascoët, J-Y, F. Bennis and P. Risacher (1996), Choix de configurations de machine-outil et détermination des visibilités réelles, *IDMME'96 Conference Proceedings,* Nantes, France, pp 395-404

Horvath, M., A. Markus and J. Vancza (1996), Process planning with genetic algorithms on results of knowledge-based reasoning, *Int. J. Computer Integrated Manufacturing,* **9,**2, pp145-166

Kiritsis, D. and M. Porchet (1996), A generic Petri-net model for dynamic process planning and sequence optimization. *Advanced Engineering Software,* **25,** pp61-71.

Kruth, J. P., and J. Detand (1992), A CAPP system for non-linear process plans. *Annals of the CIRP,* **41** (1), pp489-492.

Paris, H. and D. Brissaud (2004), Process Planning Strategy Based on Fixturing Indicators Evaluation. *Int. J. of Advanced Manufacturing Technology,* online available, to be published.

Raman, R. and M.M. Marefat (2004), Integrated Process Planning using tool/process capabilities and heuristic search, *Journal of intelligent manufacturing,* **15,** pp161-174

Shen, W. and D.H. Norrie (1999), Agent based systems for intelligent manufacturing: a state of the art survey, *Knowledge and Information Systems, an Int. Journal,* 1(**2**), pp129-156

Sormaz, D. and B. Koshnevis (2003), Generation of alternatives plans in integrated manufacturing systems, *Journal of Intelligent Manufacturing,* **14,** pp509-526

Tang, K., L.L. Chen and S-Y Chou (1998), Optimal workpiece setups for 4-axis numerical control machining based on machinability, *Computers in industry,* **37,** pp27-41

Van Zeir, G., J.-P. Kruth and J. Detand (1998), A conceptual framework for interactive and blackboard based CAPP. *Int. J. Production Research,* **36,** 6, pp1453-1473

Woo, T.C (1994), Visibility maps and spherical algorithms, *Computer Aided Design,* **26,** 1, pp6-16

Wu, R.R., L. Ma, J. Mathew and G.H. Duan (2002), Optimal operation planning using fuzzy Petri nets with resource constraints, *Int. J. Computer Integrated Manufacturing,* **15,** 1, pp28–36

ELSEVIER
IFAC
PUBLICATIONS
www.elsevier.com/locate/ifac

# AUTONOMOUS CONTROL OF SHOP FLOOR LOGISTICS: ANALYTIC MODELS

S. Dachkovski * F. Wirth * T. Jagalski **

\* Center for Technomathematics, University of Bremen,
28334 Bremen, Germany
\*\* Department of Planning and Control of Production
Systems, University of Bremen, 28334 Bremen, Germany

Abstract: In complicated shop floor environments of hundreds of machines it is difficult to organize an effective and robust central control strategy for all processing parts, buffers and machines. We consider a production line for several types of products produced in parallel lines of machines. For this problem we derive a continuous model using ordinary differential equations and discuss the corresponding optimal control problem. Here optimality criteria concern the idle time of machines, work-in-process as well as the total throughput of the production. To develop a robust and close to optimal control mechanism we study the case where the individual parts waiting for a processing step have the capability to choose a machine themselves. We discuss different strategies for this decision, depending on the desired optimization objective, and their advantages in comparison to simple parallel production and globally optimized solutions. These strategies may be interpreted as local autonomous control. The main advantages of the approach lie in the comparatively simple implementation of the control mechanism as well as in the added robustness, when compared to a pre-planned production schedule.

Keywords: Shop-floor systems, autonomous control, optimization

## 1. INTRODUCTION

In complicated shop floors consisting of hundreds of machines it is difficult to organize an effective and robust central control strategy for all processing parts, buffers and machines. In this paper we consider some simple production scenarios to demonstrate the idea of a possible autonomous production. Similar configurations have been simulated with stochastic models in (Scholz-Reiter *et al.*, 2004), where the advantages of an autonomous control are discussed.

We are interested in the following scenario. Consider a shop floor with $n$ production lines. Each of the lines is optimized to the processing of a specific type of parts. However, in case a certain line is idle, this line can be used to process another type of parts, albeit with reduced efficiency. Thus there are kinds of parts indexed by $i = 1, \ldots, n$ that each arrive with a time-dependent rate $a_i(t), i = 1, \ldots, n$. The $i$-th kind part can be processed by the $j$-th production line with a production rate of $b_{ij}, i, j = 1, \ldots, n$. We assume for all $i, j = 1, \ldots, n$ that $b_{ii} > b_{ij}, j \neq i$, and also that $b_{ii} > b_{ji}$. These two assumptions mean that the $i$-th line produces the $i$-th kind product more efficiently than any other line does and that also the $i$-th production line does not process parts of another kind faster than the $i$-th one.

In principle, we can think of production lines consisting of several stages, where the output of each stage becomes the input of the next stage. As the output of the previous stage may be interpreted as the arrival of parts to be processed, we concentrate here of a one-stage scenario. In a first step we describe a continuous model for the previous setup. For this an obvious optimal control strategy can be presented. We will argue that the implementation of this approach is impractical for large shop floors. It is the aim of the present paper to discuss distributed, local control strategies, that may be interpreted as autonomous regulation.

We now present a continuous model for this situation. Continuous models, also called fluid models, see e.g., (Kleinrock, 1975), (Armbruster, 2004). They allow to describe the material flows in terms of differential equations.

The arrival of the $i$-th raw material is governed by the time dependent function $a_i(t)$, which we assume to be a piecewise continuous, nonnegative function. The raw materials (parts) are stored in a single central buffer. The state vector of the buffer is given by $x(t) = (x_1(t), \ldots, x_n(t))$, where $x_i(t)$ denotes the amount of the $i$-th raw material stored in the buffer. The time derivatives of the $i$-th state is given by

$$\dot{x}_i(t) = a_i(t) - f_i(a_i(t), x_i(t)) - \sum_{i \neq j} g_{ij}(a(t), x(t)). \qquad (1)$$

Here the function $f_i$ describes the reduction in the $i$-th raw material due to the production process in the $i$-th production line, whereas the functions $g_{ij}$ denote the reduction in the stored raw material due to processing of the $i$-th part by the $j$-th line for $i \neq j$.

If we want to maximize the throughput, then as the $i$-th machine works in an optimal manner with the $i$-th part. Then as long as there is a supply of the $i$-th raw material it is optimal to use the $i$-th production line exclusively for this. That is we define

$$f_i(a_i, x_i) = \begin{cases} b_{ii} & , \text{ if } x_i > 0 \\ \min\{a_i, b_{ii}\} & , \text{ if } x_i = 0 \end{cases}. \qquad (2)$$

The terms $g_{ij}$ are used to distribute material for the $i$-th product to the $j$-th production line in case this line still has some capacity. We assume that $a, x$ are given and let $i_1, \ldots, i_m$ denote indices for which $x_{i_l} = 0, l = 1, \ldots, m$. (If there are no such indices, then all production lines work with their corresponding product, and there is nothing to discuss.) The optimal way to do this is given by the solution of the following linear program.

$$\text{Maximize } \sum_{i \neq j = 1}^{n} g_{ij} \qquad (3)$$

subject to

$$g_{ij} \geq 0, \qquad i, j = 1, \ldots, n, i \neq j, \qquad (4)$$

$$\sum_{j=1, j \neq i}^{n} g_{ij} \leq a_i - f_i(a_i, x_i), \ i = i_1, \ldots, i_m, \quad (5)$$

$$\sum_{i=1, i \neq j}^{n} b_{ij}^{-1} g_{ij} \leq \left(1 - \frac{f_j(a_j, x_j)}{b_{jj}}\right), \ j = 1, \ldots, n. (6)$$

The constraints represent to positivity constraint on the $g_{ij}$, the constraint that not more than $a_i - f_i(a_i, x_i)$ can be processed of the $i$-th product, in case that the buffer for the $i$-th product is empty, and finally, that the total capacity of the $j$-th machine may not be exceeded.

This optimization problem needs to be solved whenever some of the buffers become empty. In this way the amount of material delivered to a machine is exactly equal to the amount that can be processed, so that there is no build up of material other than in the buffer. Also it is easy to see that at each time instant the throughput is maximized and idle times are minimized. This is the content of the following statement.

*Proposition 1.* Consider the production line described by the equations (1), $i = 1, \ldots, n$ with a distribution according to (2) and the linear program (3)–(6). Then for all $T > 0$ this distribution maximizes the throughput, which is given by

$$\sum_{j=1}^{n} \int_{0}^{T} f_j(a_j(t), x_j(t)) + \sum_{i=1, i \neq j}^{n} g_{ij}(a(t), x(t)) dt.$$

Furthermore, the distribution procedure minimizes the total idle times of the production lines.

While the described approach yields an optimal solution, the described method has several drawbacks, so that it is not feasible in shop floors characterized by a large number of production lines. First of all, the solution of the linear program may be cumbersome and it requires full information concerning the states of all machines and the total work-in-process. Secondly, the common buffer may be impractical depending on the circumstances, or in case of a virtual buffer further communication needs may be undesirable.

In the following we will investigate decentralized control strategies. We present indications that solutions close to optimal may be obtained using autonomous control concepts. For ease of presentation we restrict ourselves to the case of two production lines. In the following Section 2 we present three different scenarios for this case. In Section 3 some discrete event simulations for these scenarios are discussed. Finally, in Section 4 we present continuous models for these situations and discuss some optimal control problems.

## 2. THE CASE OF TWO PRODUCTION LINES

Let us consider a production of two kinds of products in two lines. There are two kinds of parts to be processed coming from two sources (scenario 1, see Fig. 1) with rate $a_1$ and $a_2$ (parts per unit time), respectively. The processing rate of machines is denoted by $b_{11}$ and $b_{22}$ respectively.

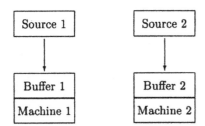

Fig. 1. Scenario 1

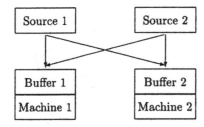

Fig. 2. Scenario 2

In the case that $a_i > b_{ii}, i = 1, 2$ the queues in the buffers grow linearly with time. If $a_i < b_{ii}$, the machines idle periodically. Now let $a_1 < b_{11}$ and $a_2 > b_{22}$ ($a_1 > b_{11}$ and $a_2 < b_{22}$ is symmetric). In this case the first machine idles periodically whereas the second has to proceed a growing queue.

To save the idle time of the machine 1 one can allow parts to choose another machine if this machine idles (scenario 2, Fig. 2) . Let us denote $b_{ij}$ the processing rate of the machine $i$ busy with part from the source $j$.

We consider also the following scenario 3. The parts arrive first to a common buffer and then decide to which machine to go (Fig. 3), whereby a part has a preference to go to the machine with a highest processing rate for it.

To demonstrate the advantages of the last two scenarios we perform a discrete event simulation.

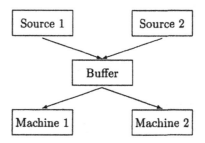

Fig. 3. Scenario 3

### 3. DISCRETE EVENT SIMULATION

We normalize the maximal arrival rate of source 2 to $a_2 = 1$. It is clear, that the interesting case is $a_1 < b_{11}$ and $a_2 > b_{22}$. The critical case occurs when the processing times are significantly smaller then possible arrival rates of one of the servers. Thus we set $a_1 = 1/24, b_{11} = b_{22} = 1/16, b_{12} = b_{21} = 1/20$ and vary $1/16 < a_2 < 1$, as smaller values of $a_2$ lead to scenarios similar to the previous one. The simulation time period is 500 units. The parts of the kind $i$ go to the machine $j \neq i$ if and only if it idles.

On following figures dash, solid and dotted lines correspond to scenario 1, 2 and 3 respectively. The total amount of parts processed by both machines is presented on the Fig. 4.

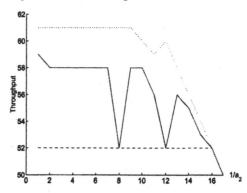

Fig. 4. Total throughput versus inter-arrival time $1/a_2$

As one can expect the three lines coincide if inter-arrival time $\frac{1}{a_2} > 16$, i.e, for low arrival rates. In this case all three scenarios work in the same parallel way. For higher arrival rates we see that with the exception of two points $a = \frac{1}{8}$ and $a = \frac{1}{12}$ the throughput in the second scenario is bigger as in the first one. The third scenario has a bigger throughput than in scenarios 1 and 2 or all $\frac{1}{a_2} < 16$ as we expect from Proposition 1.

The sum of the idle time over both machines is presented on the Fig. 5.

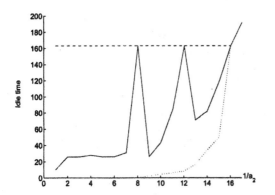

Fig. 5. Total idle time versus inter-arrival time $1/a_2$

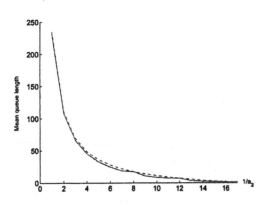

Fig. 6. Mean work-in-process versus inter-arrival time $1/a_2$

Again we see, that with the exception of the same two points the second scenario has less total idle time and the third scenario has less idle time than both scenarios 1 and 2 for all $\frac{1}{a_2} < 16$.

Figures 6 and 7 show the mean and the maximum work-in-process, i.e. the amount of parts in both buffers in the first two scenarios.

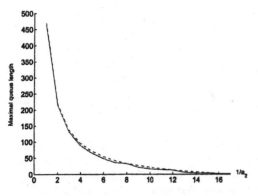

Fig. 7. Maximum work-in-process versus inter-arrival time $1/a_2$

We see that nearly the same buffer capacities are required in all three cases. However, the last two production scenarios have advantages in that the idle time is less and the total throughput is bigger.

## 4. CONTINUOUS MODELS

### 4.1 Scenario 1

As before let $a_i(t)$ be the time dependent rate of arrivals from the source $i$ and $b_{ii}$ be the processing rate of the machine $i$. Let $x_i(t), i = 1, 2$ be the amount of parts waiting in the first and second buffer. As there is no cross transfer of raw material (1) reduces to

$$\dot{x}_i(t) = a_i(t) - f_i(a_i(t), x_i(t)), \quad i = 1, 2, \quad (7)$$

with processing rates $f_i$ given by (2).

The total throughput and idle time as well as the queue length of the buffers can be calculated easily. There are no controls in this scenario. It is clear, that the non-interaction between the different production lines may lead to solutions that are far from optimal. In particular, idle times occur whenever a buffer becomes empty and these idle times cannot be used for other purposes, so that also the throughput is reduced. We do not discuss this simple case further.

### 4.2 Scenario 2

Let $a_i(t)$ and $b_{ij}$ be as above. Let $x_1(t), y_1(t)$ be the number of parts in the buffer 1 coming from the source 1 and 2 respectively and let $x_2(t), y_2(t)$ be the number of parts in the buffer 2 coming from source 2 and 1 respectively. Let $0 \leq \alpha_i(t) \leq 1$ be two time dependent parameters controlling the rate of parts arrived from the source $i$ and going to the machine 1. Then the evolution of the buffer queues can be described by the following system:

$$\dot{x}_1(t) = \alpha_1 a_1(t) - b_1(t)\frac{x_1(t)}{x_1(t) + y_1(t)}, \quad (8)$$

$$\dot{y}_1(t) = \alpha_2 a_2(t) - b_1(t)\frac{y_1(t)}{x_1(t) + y_1(t)}, \quad (9)$$

$$\dot{x}_2(t) = (1 - \alpha_2)a_2(t) - b_2(t)\frac{x_2(t)}{x_2(t) + y_2(t)} \quad (10)$$

$$\dot{y}_2(t) = (1 - \alpha_1)a_1(t) - b_2(t)\frac{y_2(t)}{x_2(t) + y_2(t)} \quad (11)$$

where $b_1(t)$ and $b_2(t)$ are the production rates of the machine 1 and 2 respectively. They can be calculated as follows: Consider the first machine and the queue $(x_1 + y_1)$ in the first buffer. If $(x_1 + y_1) > 0$, then let $\varepsilon(x_1 + y_1)$ be a small portion of the queue processed by this machine. The time spent for this portion is $\left(\frac{\varepsilon x}{b_{11}} + \frac{\varepsilon y}{b_{12}}\right)$. Then it follows

$$b_1(t) = \frac{\varepsilon(x_1(t) + y_1(t))}{\frac{\varepsilon x_1(t)}{b_{11}} + \frac{\varepsilon y_1(t)}{b_{12}}}. \quad (12)$$

With the same arguments for the second machine one can conclude:

$$b_i(t) = \frac{x_i(t) + y_i(t)}{\frac{x_i(t)}{b_{ii}(t)} + \frac{y_i(t)}{b_{ij}(t)}}, \quad i \neq j. \qquad (13)$$

For $x_1 + y_1 = 0$, i.e. $x_1(t) = y_1(t) = 0$ we conclude in the same way that

$$b_1 = \frac{\alpha_1 a_1 + \alpha_2 a_2}{\frac{\alpha_1 a_1}{b_{11}} + \frac{\alpha_2 a_2}{b_{12}}}$$

if the machine does not idle and $b_1 = \alpha_1 a_1 + \alpha_2 a_2$ otherwise (i.e., for small arrival rates). So we obtain

$$b_1(t) = \min\left(\alpha_1 a_1 + \alpha_2 a_2, \frac{\alpha_1 a_1 + \alpha_2 a_2}{\frac{\alpha_1 a_1}{b_{11}} + \frac{\alpha_2 a_2}{b_{12}}}\right), \quad (14)$$

for $x_1(t) = y_1(t) = 0$;

$$b_2(t) = \min\left((1-\alpha_1)a_1 + (1-\alpha_2)a_2,\right.$$
$$\left.\frac{(1-\alpha_1)a_1 + (1-\alpha_2)a_2}{\frac{(1-\alpha_1)a_1}{b_{12}} + \frac{(1-\alpha_1 a_2)}{b_{22}}}\right), \qquad (16)$$

for $x_2(t) = y_2(t) = 0$.

The control parameters $\alpha_i$ should be chosen to reach an optimal solution.

The criteria for the optimal solution can be for example:

• maximizing the total throughput over a time interval $[0, T]$

$$\int_0^T (b_1(s) + b_2(s))ds \rightarrow \max; \qquad (17)$$

• minimizing the total work-in-process

$$x_1(t) + y_1(t) + x_2(t) + y_2(t) \rightarrow \min; \qquad (18)$$

• minimizing the total idle time of the first machine:

$$T - \frac{\int_0^T \alpha_1(t)a_1(t)\,dt}{b_{11}} - \frac{\int_0^T \alpha_2(t)a_2(t)\,dt}{b_{21}} \rightarrow \min; \qquad (19)$$

Then an optimal control problem (Macki and Strauss, 1982), (Fleming and Rishel, 1975) can be formulated as follows:

Find $\alpha_1, \alpha_2$ such that the solution of (8-16) yields an optimal return for the functional (17), (18) or (19), respectively.

We note, that $b_i(t)$ is discontinuous at $x_i(t) = y_i(t) = 0$, hence the uniqueness of the solution of the system (8-11) is not clear at that point. To investigate this case we do the following transformation:

$$\begin{aligned} u &= x_1 + y_1 \\ v &= x_1 - y_1 \\ \xi &= x_2 + y_2 \\ \eta &= x_2 - y_2 \end{aligned} \quad \Rightarrow$$

$$\begin{aligned} \dot{u} &= \alpha_1 a_1 + \alpha_2 a_2 - b_1 \\ \dot{v} &= \alpha_1 a_1 - \alpha_2 a_2 - b_1 \frac{v}{u} \\ \dot{\xi} &= (1-\alpha_2)a_2 + (1-\alpha_1)a_1 - b_1 \\ \dot{\eta} &= (1-\alpha_2)a_2 - (1-\alpha_1)a_1 - b_2\frac{\eta}{\xi} \end{aligned}$$

where now $b_1$ may be discontinuous only if $u = v = 0$ and $b_2$ may be discontinuous only if $\xi = \eta = 0$. The initial conditions now are $u(0) = 0$, $v(0) = 0$, $\xi(0) = 0$, $\eta(0) = 0$.

Let $(u_1, v_1, \xi_1, \eta_1)$ and $(u_2, v_2, \xi_2, \eta_2)$ be two solutions of this system, then

$$(u_1 - u_2)^{\cdot} = 0 \qquad (20)$$

$$(v_1 - v_2)^{\cdot} = b_1\left(\frac{v_1}{u_1} - \frac{v_2}{u_2}\right) \qquad (21)$$

$$(\xi_1 - \xi_2)^{\cdot} = 0 \qquad (22)$$

$$(\eta_1 - \eta_2)^{\cdot} = b_2\left(\frac{\eta_1}{\xi_1} - \frac{\eta_2}{\xi_2}\right) \qquad (23)$$

with $(u_1 - u_2)(0) = 0$, $(v_1 - v_2)(0) = 0$, $(\xi_1 - \xi_2)(0) = 0$, $(\eta_1 - \eta_2)(0) = 0$. Firstly, it follows that $u_1 - u_2 \equiv 0$, $v_1 - v_2 \equiv 0$, then

$$\xi_1 - \xi_2 = C_1 \exp\left(\int \frac{b_1}{u_1}\,dt\right), \qquad (24)$$

$$\eta_1 - \eta_2 = C_2 \exp\left(\int \frac{b_2}{\xi_2}\,dt\right). \qquad (25)$$

With homogeneous initial conditions it follows $\xi_1 - \xi_2 \equiv 0$, $\eta_1 - \eta_2 \equiv 0$. The uniqueness is proved.

In order to maximize the throughput we now describe how to find $\alpha_i(t)$ that instanteneously maximize the production rate, i.e. $b_1(t) + b_2(t)$. We consider the system in different buffer states: Obviously, $\alpha_1(t) = 1$ and $\alpha_2(t) = 0$ for $x_1(t) + y_1(t) > 0$ and $x_2(t) + y_2(t) > 0$, i.e., the machine $i$ receives the parts only from the source $i$, $i = 1, 2$.

Let be $x_1(t) + y_1(t) = 0$ and $x_2(t) + y_2(t) > 0$. Then $\alpha_1(t) = 1$, since the second machine is busy. If $a_1 \geq b_{11}$, i.e., arrival rate is higher then it can be processed, then it follows $\alpha_2(t) = 0$. Otherwise if $a_1 < b_{11}$, the first machine can process some parts from the second source and having empty buffer. To find the appropriate $\alpha_2(t)$ we use the condition of empty buffer: let us consider a small time interval $\Delta t$. During that time $a_1\Delta t$ parts have arrived from the first source. The first machine has spent $\frac{a_1\Delta t}{b_{11}}$ time units processing them. The remaining time $\Delta t - \frac{a_1\Delta t}{b_{11}}$ can be used for the parts of the second machine which are processed with the rate $b_{12}$. It follows that

$$\left(\Delta t - \frac{a_1\Delta t}{b_{11}}\right)b_{12} = \alpha_2 a_2, \qquad (26)$$

if $a_2(t) > 0$ is big enough, such that there is no idle time. Finally

$$\alpha_2(t) = \min\left(\frac{(b_{11}(t) - a_1(t))_+ b_{12}(t)}{b_{11}(t)a_2(t)}, 1\right),$$

where we use the notation $a_+ = \max(0, a)$.

The case $x_1(t) + y_1(t) > 0$ and $x_2(t) + y_2(t) = 0$ can be treated similarly to obtain

$$\alpha_1(t) = \max\left(0, 1 - \frac{(b_{22}(t) - a_2(t))_+ b_{21}(t)}{b_{22}(t) a_1(t)}\right),$$

and $\alpha_2(t) = 0$.

The last possible state is that both buffers are empty, i.e. $x_1(t) + y_1(t) = 0$ and $x_2(t) + y_2(t) = 0$. Straightforward calculations yield in this case

$$\alpha_1(t) = \min\left(1, \left(1 - \frac{(b_{22}(t)-a_2(t))_+ b_{21}(t)}{b_{22}(t) a_{11}(t)}\right)_+\right),$$

$$\alpha_2(t) = \max\left(0, \frac{(b_{11}(t)-a_1(t))_+ b_{12}(t)}{b_{11}(t) a_2(t)}\right).$$

For this $\alpha_1(t), \alpha_2(t)$ the solution $x_1(t)$, $y_1(t)$, $x_2(t)$, $y_2(t)$ can be found solving the system (8-11).

*4.3 Scenario 3*

Now consider the third scenario. Again let $a_i(t)$, $b_{ij}(t)$ be as above. Let $x(t)$ and $y(t)$ denote the number of parts waiting in the buffer arrived from the source 1 and 2 respectively. Let $0 \le \alpha_i \le b_{1i}$, $0 \le \beta_i \le b_{2i}$ denote the rate of arrival of parts coming from the source 1 and 2 respectively and going to the machine $i$. The evolution of $x(t)$ and $y(t)$ is then given by:

$$\dot{x} = a_1(t) - \alpha_1(t) - \alpha_2(t), \qquad (29)$$

$$\dot{y} = a_2(t) - \beta_1(t) - \beta_2(t). \qquad (30)$$

Since the parts go to a machine from the buffer only if it becomes empty, i.e., the arrival rate is equal to the processing rate. Then

$$\alpha_i(t) + \beta_i(t) = b_i(t), \quad i = 1, 2 \qquad (31)$$

where $b_1(t)$ and $b_2(t)$ are the processing rates of the machine 1 and 2 respectively. Using (31) the system (29-31) is equivalent to

$$\dot{x}(t) = a_1(t) - \alpha_1(t) - (b_2(t) - \beta_2(t)) \qquad (32)$$

$$\dot{y}(t) = a_2(t) - (b_1(t) - \alpha_1(t)) - \beta_2(t) \qquad (33)$$

$$b_1^2(t) = \alpha_1(t) b_{11}(t) + (b_1(t) - \alpha_1(t)) b_{12}(t) \qquad (34)$$

$$b_2^2(t) = (b_2(t) - \beta_2(t)) b_{21}(t) + \beta_2(t) b_{22}(t) \qquad (35)$$

and one has only two independent control functions $\alpha_1(t)$ and $\beta_2(t)$.

Let us find $\alpha_1(t), \alpha_2(t), \beta_1(t), \beta_2(t)$ minimizing the work-in-process $(x + y)$: First we treat the case of empty buffers $x = 0$, $y = 0$. Analyzing the Fig. 8 and with the same arguments as in case of scenario 2 one the following solution is obtained

$$\alpha_1 = \min(b_{11}, a_1),$$

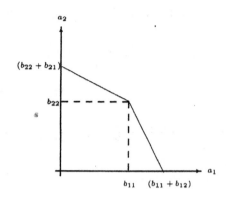

Fig. 8. Area of arrival rates variation

$$\alpha_2 = \min\left(\left(1 - \frac{a_2}{b_{22}}\right)_+ b_{21}, a_1 - b_{11}\right),$$

$$\beta_1 = \min\left(\left(1 - \frac{a_1}{b_{11}}\right)_+ b_{12}, a_2 - b_{22}\right),$$

$$\beta_2 = \min(b_{22}, a_2),$$

where we use the notation $a_+ = \max(0, a)$.

In case $x > 0$, $y > 0$ it follows that $\alpha_1 = b_{11}$, $\alpha_2 = 0$, $\beta_1 = 0$, $\beta_2 = b_{22}$.

For $x > 0$, $y = 0$ one has $\alpha_1 = b_{11}, \beta_1 = 0$, $\beta_2 = \min(b_{22}, a_2)$ and

$$\alpha_2 = \min\left(\left(1 - \frac{a_2}{b_{22}}\right)_+ b_{21}, a_1 - b_{11}\right),$$

and for $x = 0$, $y > 0$ we have $\alpha_1 = \min(b_{11}, a_1), \beta_1 = 0$, $\beta_2 = b_{22}$ and

$$\alpha_2 = \min\left(\left(1 - \frac{a_2}{b_{22}}\right)_+ b_{21}, a_1 - b_{11}\right).$$

We have found optimal controls $\alpha_i$ for all cases, with these data the evolution of the buffer queue $x(t)$, $y(t)$ is given by (29-31).

## 5. CONCLUSION

Several scenarios of shop floor control have been considered and it has been demonstrated that scenarios with autonomous control strategies, in which parts can decide locally which machine to go to, can be more effective than conventional pre-planned schedules where parts are handed over to the next machine in line according to a production plan.

The scenarios have been described by means of analytical models in form of differential equations and control functions in order to analyze the performance of distributed autonomous control systems. Several criteria of optimality, such as minimum idle times of machines, minimum work-in-process and maximum throughput have been stated. The optimal control functions can be found by solving the corresponding optimal control problem.

Moreover it has been shown that autonomous shop floor control is more robust in the case of machine breakdowns. Here, simple rules of local decision-making allow jobs to be transferred to another machine. These advantages have also been confirmed by simulations described in (Scholz-Reiter *et al.*, 2004) using a different modeling approach.

The presented research on decentralized and autonomous control scenarios has established a theoretical foundation for further research. The results have proven that it is promising to focus on more complex shop floor structures. In such a scenario the autonomy of a single part is more important as far as its influence on the performance in terms of less idle times of machines, higher throughput and less total work-in-process as well as the system's robustness is concerned.

## 6. ACKNOWLEDGMENTS

This research is funded by the German Research Foundation (DFG) as part of the Collaborative Research Centre 637 "Autonomous Cooperating Logistic Processes: A Paradigm Shift and its Limitations" (SFB 637).

## REFERENCES

Armbruster, D. (2004). Dynamical systems and production systems. In: *Nonlinear Dynamics of Production Systems* (G. Radons and R. Neugebauer, Eds.). pp. 5–24. Wiley-VCH Verlag, Weinheim.

Fleming, W.H. and R.W. Rishel (1975). *Determenistic and Stochastic Optimal Control.* Springer-Verlag. New York, Heidelberg, Berlin.

Kleinrock, L. (1975). *Queuing Systems.* John Wiley & Sons.

Macki, J. and A. Strauss (1982). *Introduction to Optimal Control Theory.* Springer-Verlag. New York, Heidelberg, Berlin.

Scholz-Reiter, B., C. de Beer and K. Peters (2004). Autonomous control of shop floor logistics. *IFAC Conference on Manufacturing, Modelling, Management and Control.*

www.elsevier.com/locate/ifac

# HEURISTIC ALGORITHMS FOR BALANCING TRANSFER LINES WITH SIMULTANEOUSLY ACTIVATED SPINDLES

Alexandre Dolgui[1], Nikolai Guschinsky[1,2], Genrikh Levin[2]


[1]*Ecole des Mines de Saint Etienne
158,Cours Fauriel 42033 Saint Etienne Cedex2 France
e-mail:{dolgui,guschinsky}@emse.fr*
[2]*United Institute of Informatics Problems
Surganov Str, 6, 220012 Minsk Belarus
e-mail:{gyshin,levin}@newman.bas-net.by*

Abstract: A balancing problem for transfer lines with workstations in series and simultaneously activated spindle heads is considered. The problem is to choose a block of operations for each spindle head from a given set of the blocks and to assign them to workstations while minimizing the line cost and satisfying technological constraints (precedence relation, inclusion constraints related to operations and exclusion constraints with regard to blocks). Two heuristic algorithms are developed. One of them assigns randomly step by step blocks to a current workstation. The second algorithm uses depth-first search technique. Experimental results are presented. *Copyright © 2004 IFAC*

Keywords: Computer-aided design, machining, optimization, heuristics, random search.

## 1. INTRODUCTION

Paced production lines with workstations in series and the possibility of simultaneous execution of several manufacturing operations in one block (by one spindle head) at the same workstation are widely used in mechanical industry for mass production (Groover, 1987; Hitomi, 1996). Concentration of operations in blocks allows to considerably decrease the total number of workstations and spindle heads for a required output. This results in decreasing the total cost of a line and the occupied area. However, searching for optimal assignment both of operations to blocks and of blocks to workstations is a very complex problem.

Dolgui et al. (2003a, 2004) consider the balancing problem for transfer lines with sequential execution

of blocks at workstations. In this paper, this problem is investigated for the case when all blocks (spindle heads) of the same workstation are executed simultaneously.

The line is equipped with a common transfer mechanism and a common control system. All stations of the line perform their operations simultaneously, and failure of one station (or the necessity to change tools) results in stoppage of the line. When a part is loaded on a workstation, it is fixed and all spindle heads of this workstation are activated. Each spindle head is equipped by several tools, which allow to execute corresponding operations at the same time (block of simultaneously executed operations). The workstation time is the maximal value among operation times of its spindle heads (block times) and the line cycle time is the maximum of workstation times.

It is supposed that a set **B** of all blocks, which can be used for line design, is known. For each block from **B**, its cost and execution time are given. Then, the problem is to choose blocks from **B** and to assign them to workstations in such a way that:
i) the total *line cost* is as small as possible,
ii) a given line *cycle time* is not exceeded,
iii) all operations from **N** are assigned and a *partial order relation* on the set **N** is satisfied,
iv) the constraints on the necessity to perform some operations at the same workstation (*inclusion constraints*) are respected;
v) the constraints on the impossibility to assign some blocks to the same workstation (*exclusion constraints*) are not violated.

If every block consists of one operation only, constraints iv) and v) are absent, and all blocks of the same workstation are executed in series, then the considered problem is a Simple Assembly Line Balancing Problem (SALBP). The problem is to assign all operations to workstations minimizing a specified criterion under a given partial order of operations and a fixed cycle time (Scholl, 1999). Usually, the criterion is the unbalance of the assembly line. The unbalance is minimal if and only if the number of the workstations is minimal.

Generally, for SALBP, integer linear programming models are formulated (Scholl, 1999). The exact methods are mainly based on branch and bound algorithms (Baybars, 1986). Since the problem is known to be NP-hard, exact solution of large problems is time-consuming and some heuristics (Helgenson and Birnie, 1961; Arcus, 1966; Sabuncuoglu *et al.*, 2000; Rekiek *et al.*, 2000, 2001) can be used to decrease the computational effort. Surveys of main publications on SALBP are given in (Baybars, 1986; Ghosh and Gagnon, 1989; Erel and Sarin, 1998; Scholl, A. 1999; Rekiek *et al.*, 2002 ).

The problem under consideration cannot be directly solved by the SALBP methods for the following reasons:
- each operation generally is presented in several blocks from **B** and it is not known in advance which block is preferable;
- the set **B** usually contains mutually exclusive blocks;
- the blocks from the same workstation are executed simultaneously;
- the line cost depends not only on the number of workstations but on the number and the cost of blocks.

For this problem, an exact method is proposed in (Dolgui *et al.*, 2003b). The method is based on transformation the initial problem into a constrained shortest path problem. Some dominance rules are used to reduce the size of the obtained graph. The results of computational experiments with heuristic relaxations of these rules are presented in (Dolgui *et al.*, 2003c). In this paper other heuristic approaches are developed.

The rest of the paper is organized as follows. Section 2 deals with the problem statement. Section 3 presents two heuristic algorithms. Section 4 is dedicated to experimental results.

## 2. PROBLEM STATEMENT

The following notation is used for modeling the design problem considered:
**N** is the set of all operations;
**B** is the set of blocks (spindle heads) which can be used for the line;
$m$ is the number of workstations in a design decision;
$n_k$ is the number of blocks of workstation $k$;
$C_1$ is the basic cost of one workstation;
$N_{kl}$ is the set of operations of block $l$ of workstation $k$;
$Pred(N_k)$ is the set of operations which must be executed before any operation from $N_k$;
$C_2(N_{kl})$ is the cost of the block $N_{kl}$;
$N_k = \{N_{k1}, ..., N_{kn_k}\}$ is the set of blocks from **B** which are executed at the workstation $k$;
$P = <N_1, ..., N_m>$ is a design decision.

It is assumed also that the line cannot involve more than $m_0$ workstations and each workstation at most $n_0$ blocks.

The line cost for design decision $P$ can be estimated

as: $C(P) = C_1 m + \sum\limits_{k=1}^{m} \sum\limits_{l=1}^{n_k} C_2(N_{kl})$ .

The constraints introduced in Section 1 can be represented in the following way:

i) A partial order relation over the set **N** is represented by the acyclic digraph $G^r = (\mathbf{N}, D^r)$. An arc $(i,j) \in \mathbf{N} \times \mathbf{N}$ belongs to the set $D^r$ if and only if the operation $j$ must be executed after the operation $i$.

ii) Since all blocks of the same workstation are executed simultaneously, the blocks with block time over the required line cycle time can be excluded from **B** before optimization. Therefore this constraint can be omitted after such transformation.

iii) Exclusion conditions for the blocks of the same workstation can be represented by the graph $\overline{G}^s = (\mathbf{B}, \overline{D}^s)$ in which a pair $(N', N'') \in \mathbf{B} \times \mathbf{B}$ belongs to the set $\overline{D}^s$ if and only if blocks $N'$ and $N''$ cannot be allocated to the same workstation.

iv) Inclusion conditions for the operations of the same workstation can be represented by the graph $G^s = (\mathbf{N}, D^s)$ such that a pair $(i,j) \in \mathbf{N} \times \mathbf{N}$ belongs to the set $D^s$ if and only if operation $i$ and $j$ must be allocated to the same workstation.

So, the design problem can be reduced to finding a collection $P=<\{N_{11},..., N_{1n_1}\}, ..., \{N_{m1},..., N_{mn_m}\}>, N_{kl} \in \mathbf{B}$, satisfying the conditions:

$$C(P)=C_1 m+ \sum_{k=1}^{m} \sum_{l=1}^{n_k} C_2(N_{kl}) \to min; \quad (1)$$

$$\bigcup_{k=1}^{m} \bigcup_{l=1}^{n_k} N_{kl} = \mathbf{N}; \quad (2)$$

$$N_{k'l'} \cap N_{k''l''} = \varnothing, \ k'l' \neq k''l'', \ k', k''=1,...,m, \ l'=1,...,n_{k'}, \ l''=1,...,n_{k''}; \quad (3)$$

$$Pred(N_k) \subseteq \bigcup_{r=1}^{k-1} \bigcup_{q=1}^{n_r} N_{rq}, \ k=1,...,m; \quad (4)$$

$$\bigcup_{l=1}^{n_k} N_{kl} \cap e \in \{\varnothing, e\}, \ e \in D^s, \ k=1,...m; \quad (5)$$

$$(N_{kl'}, N_{kl''}) \notin \overline{D}^s, \ k=1,...m, l',l''=1,...,n_k; \quad (6)$$

$$n_k \leq n_0, \ k=1,...m; \quad (7)$$

$$m \leq m_0. \quad (8)$$

The objective function (1) is the line cost; constraints (2)-(3) determine the condition of assigning all operations of the set $\mathbf{N}$ and including each operation into one block only; (4) define the precedence constraints over the set $\mathbf{N}$; (5) determine the necessity of executing the corresponding operations at the same workstation; (6) define the possibility of combining blocks at the same workstation; (7) – (8) provide the constraints on the number of workstations and blocks for each workstation.

## 3. HEURISTIC ALGORITHMS

The exact algorithm (Dolgui *et al.*, 2003b) is applicable for small and medium size. In this paper, two new heuristic algorithms are proposed. The first algorithm assigns randomly step by step possible blocks to a current workstation. It stops after the given number of trials or if the desired value of the line cost is achieved. The second algorithm uses depth-first search technique for finding a constrained shortest path in a special digraph. It stops if the first feasible solution is found.

### 3.1 RAB algorithm

The first algorithm, named RAB (Random Assignment of Blocks), is the following. As the heuristic algorithm (Dolgui *et al.*, 2003a, 2004), it is based on COMSOAL technique.

At each iteration, the algorithm creates stations step by step. First, it builds list $In$ of potentially assignable blocks to a current station taking into account precedence constraints. Then one block is chosen at random to be assigned to a current station. After the assignment, the list $In$ is modified in order to satisfy exclusion constraints for blocks and the assignment process is repeated. When the list $In$ is empty or $n_0$ blocks have been already assigned, the current station closed and inclusion constraints for operations are verified. If they are violated, assigned blocks are removed from the station and algorithm tries another attempt for creating a station. If the algorithm fails after a given number of attempts, it starts from the beginning (creation of the first station). The iteration is also unsuccessful if after creation of $m_0$ stations not all the operations from $\mathbf{N}$ are assigned.

Let $TR_{tot}$ be the current number of trials, $TR_{nimp}$ be the number of trials that do not improve the current solution, $TR_{st}$ be the current number of attempts to assign blocks to a station, $C$ be the cost of a current solution, and $C_{min}$ be the cost of the best solution.

*Algorithm* 1.
*Step* 1. Set $C_{min} = \infty$, $TR_{tot} = 0$, $TR_{nimp} = 0$.

*Step* 2. Set $v=\varnothing$, $k=1$.

*Step* 3. Put in list $In$ all blocks $B$ from $\mathbf{B}$ that satisfy precedence constraints for the set $v$, i.e. all the predecessors of operations from block $B$ are in the set $v$. If the list $In$ is empty then set $C=\infty$ and go to *Step* 10.

*Step* 4. Set $l=0$, $N_k = \varnothing$, $n_k = 0$, $C=C_1$, $TR_{st} = 0$.

*Step* 5. Choose block $B$ from the list $In$ randomly, set $l= l+1$, $n_k = n_k +1$, $N_{kl}= B$, $C=C+C_2(N_{kl})$. If $n_k = n_0$ go to *Step* 7.

*Step* 6. Remove from the list $In$ blocks $B$ if exclusion constraints are not valid for $N_k \cup B$. If the list $In$ is not empty then go to *Step* 5.

*Step* 7. If inclusion constraints are satisfied for $N_k$ then go to *Step* 9. Otherwise, set $TR_{st} = TR_{st}+1$. If $TR_{st}$ does not exceed a given value then go to *Step* 4. Otherwise, set $C=\infty$ and go to *Step* 10.

*Step* 8. Set $v=v \cup \bigcup_{l=1}^{n_k} N_{kl}$. If $v$ includes all the operations from $\mathbf{N}$ then go to *Step* 10.

*Step* 9. Set $k=k+1$. If $k > m_0$ then set $C=\infty$ and go to *Step* 10. Otherwise, go to *Step* 4.

*Step* 10. If $C_{min} > C$ then set $C_{min} = C$, $TR_{nimp} = 0$ and keep the current solution as the best, set $TR_{nimp} = TR_{nimp} + 1$ otherwise.

*Step* 11. Set $TR_{tot} = TR_{tot} + 1$.

*Step* 12. Stop if one of the following conditions holds:
- a given solution time is exceeded;
- $TR_{tot}$ is greater than the maximum number of iterations authorized;
- $TR_{nimp}$ is greater than a given value;
- $C_{min}$ is lower than a given cost value.

Go to *Step* 2 otherwise.

### 3.2 DFS algorithm

Let **P** be a set of collections $P = <N_1, ..., N_k ..., N_m>$, satisfying the constraints (2)-(7). The set $v_k = \bigcup_{r=1}^{k} \bigcup_{l=1}^{n_r} N_{rl}$ can be considered as a state of the part after machining it at the $k$-th workstation. Let $V$ be the set of all states for all $P \in \mathbf{P}$, including the initial state $v_0 = \varnothing$ and the final state $v_N = N$. A new acyclic directed multi-graph $G = (V,D)$ can be constructed, in which an arc $d$ between vertices $v'$ and $v''$ belongs to $D$ if and only if $v' \subset v''$, and there exists a collection $P$ that contains $(N_{k1}, ..., N_{kn_k})$ such that $\bigcup_{r=1}^{k-1} \bigcup_{l=1}^{n_r} N_{rl} = v'$ and $\bigcup_{l=1}^{n_k} N_{kl} = v'' \setminus v'$. The cost $Q(d) = \sum_{l=1}^{n_k} C_2(N_{kl}) + C_1$ is assigned to the arc $d \in D$ as well as a set $J(d) = \{j_1(d), ..., j_{n_k}(d)\}$ of block indices where $N_{kl} = N(j_l(d))$ for $l = 1, ..., n_k$. It is assumed that the set **B** is arbitrarily enumerated and $N(j)$ is a block from **B** with index $j$.

The path $x(P) = (d_1(x(P)), ..., d_{m(x(P))}(x(P)))$ from the set **X** of all paths $x = (d_1(x), ..., d_k(x), ..., d_{m(x)}(x))$ in $G$ from $v_0$ to $v_N$ can be associated with each design decision $P \in \mathbf{P}$. On the other hand, each path $x \in \mathbf{X}$ corresponding to a collection $P(x) = <N(d_1(x)), ..., N(d_k(x)), ..., N(d_{m(x)}(x))>$ satisfies (2)-(7) but may violate constraint (8). Here $N(d_k(x)) = \{N(j_1(d_k(x))), ..., N(j_{n_k}(d_k(x)))\}$.

Thus the initial problem (1)-(8) can be transformed to a problem of finding the shortest path in multigraph $G$ with at most $m_0$ arcs. This problem is stated as follows:

$$Q(x) = \sum_{k=1}^{m(x)} Q(d_k(x)) \rightarrow \min, \quad (9)$$

$$x \in \mathbf{X}, \quad (10)$$

$$m(x) \leq m_0. \quad (11)$$

The DFS (Depth-First Search) algorithm finds an approximate solution of the problem (9)-(11) or discovers the incompatibility of constraints (2) – (8) by generating a subgraph of $G$. The set $V$ is partitioned into subsets $V_i$, $i \in \{0,1,...,|N|\}$ in such a

way that $v \in V_i$ if $|v| = i$. Obviously, there are no arcs in $G$ between vertices of $V_i$ and vertices of $V_j$ for $i \geq j$. At each step, the algorithm generates all the possible arcs for selected vertex $v$ and simultaneously updates the minimal cost of paths for the current graph. For selection of a current vertex, it uses the depth-first technique. The algorithms stops if the vertex $v_N$ is reached or all the vertices from $V$ have already been explored.

In this algorithm, $last[i]$ is the current number of vertices in $V_i$, $cur[i]$ is the index of the selected vertex in $V_i$, $D(v)$ is the set of possible arcs from the vertex $v$, $COST(v,k)$ calculates the minimal cost path in current graph $G$ from $v_0$ to $v$ with $k$ arcs. Using $PREV(v,k)$ an optimal path in $G$ can be easily reconstructed, i.e. a solution of the problem is found.

*Algorithm* 2.
*Step* 1. Set $v_0 = \varnothing$, $COST(v_0, 1) = 0$, $v = v_0$, $cur[i] = 1$, $last[i] = 0$, $i = 1, 2, ..., |N|$.

*Step* 2. Put in list $In$ all blocks $B$ from **B** that satisfy precedence constraints for the set $v$. Construct $D(v)$ by generating feasible combinations of elements from $In$ taking into account inclusion constraints for operations and exclusion constraints for blocks.

*Step* 3. Perform *Steps* 3.1 – 3.3 for each $d \in D(v)$ .
- *Step* 3.1. Build vertex $w$ by combining vertex $v$ with operations from $d$.
- *Step* 3.2. If $w \notin V_{|w|}$ then set $last[|w|] = last[|w|]+1$, $V_{|w|}[last[|w|]] = w$ (include $w$ into $V_{|w|}$), and $COST(w,k) = \infty$, $k = 1, ..., m_0$.
- *Step* 3.3. If $v = v_0$ then set $COST(w,1) = Q(d)$, $PREV(w, 1) = v_0$. Otherwise, set $COST(w,k+1) = COST(v,k) + Q(d)$, $PREV(w,k+1) = v$ for $k = 1, ..., m_0-1$ if $COST(v, k) + Q(d) < COST(w,k+1)$ and $COST(v,k) < \infty$.

*Step* 4. Find maximal $i$ among $i = 1, 2, ..., |N|$ such that $cur[i] \leq last[i]$. If $i = |N|$ or $cur[i] > last[i]$ for all $i = 1, 2, ..., |N|$ (all the vertices from $V$ have already been selected), then go to *Step* 6.

*Step* 5. Set $v = V_i[cur[i]]$, $cur[i] = cur[i]+1$ and go to *Step* 2.

*Step* 6. If $last[|N|] = 1$ $(V_{|N|} \neq \varnothing)$ then set $C_{min} = \min\{COST(v_N, k)| \ k = 1, ..., m_0\}$. Otherwise, the problem has no feasible solutions.

## 4. EXPERIMENTAL STUDY OF ALGORITHMS

The purpose of this study is to compare the proposed algorithms on the quality of obtained solutions and the running time. Experiments were carried out on HP Omnibook x86 Family 6 Model 8. The results are presented in Tables 1 - 9. They correspond to 9 series of 10 test instances which were generated in random way for different values of $|N|$, $|B|$ and $p(G)$, $p(\overline{G}^s)$, $p(G^s)$, where $p(G)$ is the ratio between the number of generated edges (arcs) in $G'$ and the number of edges (arcs) in the complete graph (digraph) with the same number of vertices. In all examples $p(G^s) = 0.01$. Choosing $p(G^s) = 0.01$ is justified since the graph $G^s$ should be coordinated with graphs $G'$ and $\overline{G}^s$ to provide feasible solutions. So, edges from the graph $G^s$ are removed if they contradict graph $\overline{G}^s$ or belong to a path in $G'$.

In these tables, the following abbreviations are used:
- TC ($p(G')$, order strength and $p(\overline{G}^s)$ of test instances);
- GA (exact graph algorithm);
- RAB1 (RAB algorithm with $TR_{tot}$=10000, $TR_{st}$=2);
- RAB2 (RAB algorithm with $TR_{tot}$= 20000, $TR_{st}$=2);
- CA (combination of RAB1 and DFS algorithms);
- PM (performance measures);
- NSP (the number of solved problems);
- RT (running time in seconds);
- SD (the deviation of the obtained cost from the best known solution in percents).

Indices min, max, av for RT and SD mean the minimal, maximal and average values, respectively. The average value of the order strength (Scholl, 1999) is also calculated for test instances. The order strength is defined as the density of the transitive closure of the precedence graph.

#### Table 1 Results for |N|=50, |B|=150

| TC | PM | GA | RAB1 | RAB2 | DFS | CA |
|---|---|---|---|---|---|---|
| | $RT_{min}$ | | 3.68 | 7.26 | 0.15 | 4.19 |
| | $RT_{max}$ | | 5.78 | 11.46 | 1.31 | 5.88 |
| 0.05, | $RT_{av}$ | | 4.80 | 9.51 | 0.42 | 5.18 |
| 0.12, | $SD_{min}$ | | 2.12 | 0.04 | 0.00 | 0.00 |
| 0.05 | $SD_{max}$ | | 14.82 | 20.51 | 0.00 | 0.00 |
| | $SD_{av}$ | | 7.31 | 6.51 | 0.00 | 0.00 |
| | NSP | | 8 | 9 | 10 | 10 |
| | $RT_{min}$ | 1.17 | 4.47 | 8.87 | 0.11 | 3.98 |
| | $RT_{max}$ | 62.62 | 7.41 | 14.66 | 1.32 | 8.23 |
| 0.10, | $RT_{av}$ | 16.25 | 5.99 | 11.87 | 0.24 | 6.24 |
| 0.33, | $SD_{min}$ | 0.00 | 7.42 | 7.42 | 1.96 | 1.96 |
| 0.10 | $SD_{max}$ | 0.00 | 13.49 | 13.49 | 19.19 | 19.19 |
| | $SD_{av}$ | 0.00 | 10.17 | 10.17 | 9.67 | 8.88 |
| | NSP | 10 | 6 | 6 | 10 | 10 |
| | $RT_{min}$ | 0.51 | 5.04 | 9.91 | 0.09 | 5.06 |
| | $RT_{max}$ | 24.46 | 11.21 | 22.39 | 0.12 | 11.27 |
| 0.15, | $RT_{av}$ | 5.42 | 7.14 | 14.18 | 0.10 | 7.16 |
| 0.54, | $SD_{min}$ | 0.00 | 3.84 | 3.64 | 0.00 | 0.00 |
| 0.15 | $SD_{max}$ | 0.00 | 22.84 | 22.65 | 18.80 | 18.80 |
| | $SD_{av}$ | 0.00 | 11.47 | 10.29 | 8.74 | 8.42 |
| | NSP | 10 | 6 | 6 | 10 | 10 |

#### Table 2 Results for |N|=75, |B|=225

| TC | PM | GA | RAB1 | RAB2 | DFS | CA |
|---|---|---|---|---|---|---|
| | $RT_{min}$ | | 6.52 | 12.88 | 0.41 | 6.59 |
| | $RT_{max}$ | | 6.52 | 12.88 | 13.24 | 20.50 |
| 0.05, | $RT_{av}$ | | 6.52 | 12.88 | 3.31 | 10.18 |
| 0.14, | $SD_{min}$ | | 19.65 | 19.65 | 0.00 | 0.00 |
| 0.05 | $SD_{max}$ | | 19.65 | 19.65 | 0.00 | 0.00 |
| | $SD_{av}$ | | 19.65 | 19.65 | 0.00 | 0.00 |
| | NSP | | 1 | 1 | 10 | 10 |
| | $RT_{min}$ | 3.465 | 7.08 | 13.99 | 0.221 | 5.748 |
| | $RT_{max}$ | 275.3 | 11.77 | 23.263 | 0.57 | 11.83 |
| 0.10, | $RT_{av}$ | 73.76 | 9.93 | 19.69 | 0.29 | 8.84 |
| 0.45, | $SD_{min}$ | 0 | 18.13 | 16.6 | 4.1 | 4.1 |
| 0.10 | $SD_{max}$ | 0 | 29.1 | 29.1 | 24.69 | 24.69 |
| | $SD_{av}$ | 0 | 22.16 | 20.06 | 11.89 | 11.89 |
| | NSP | 10 | 4 | 4 | 10 | 10 |
| | $RT_{min}$ | 0.98 | 13.67 | 27.17 | 0.13 | 6.45 |
| | $RT_{max}$ | 48.04 | 17.81 | 37.11 | 0.30 | 18.02 |
| 0.15, | $RT_{av}$ | 13.06 | 15.89 | 32.41 | 0.25 | 11.93 |
| 0.63, | $SD_{min}$ | 0.00 | 4.43 | 4.43 | 3.99 | 3.99 |
| 0.15 | $SD_{max}$ | 0.00 | 14.02 | 13.73 | 13.89 | 13.89 |
| | $SD_{av}$ | 0.00 | 10.73 | 9.99 | 7.77 | 7.35 |
| | NSP | 10 | 3 | 3 | 10 | 10 |

#### Table 3 Results for |N|=100, |B|=300

| TC | PM | GA | RAB1 | RAB2 | DFS | CA |
|---|---|---|---|---|---|---|
| | $RT_{min}$ | | 9.78 | 19.24 | 0.64 | 8.49 |
| | $RT_{max}$ | | 16.54 | 32.53 | 338.90 | 340.3 |
| 0.05, | $RT_{av}$ | | 12.37 | 24.36 | 39.94 | 48.31 |
| 0.22, | $SD_{min}$ | | 0.00 | -4.99 | 0.00 | 0.00 |
| 0.05 | $SD_{max}$ | | 18.48 | 17.03 | 1.36 | 0.00 |
| | $SD_{av}$ | | 10.96 | 7.70 | 0.14 | 0.00 |
| | NSP | | 3 | 3 | 10 | 10 |
| | $RT_{min}$ | 11.79 | 11.24 | 21.99 | 0.47 | 10.13 |
| | $RT_{max}$ | 599.5 | 24.65 | 48.87 | 2.794 | 25.03 |
| 0.10, | $RT_{av}$ | 152.1 | 19.64 | 38.80 | 0.78 | 16.44 |
| 0.55, | $SD_{min}$ | 0 | 16.05 | 16.05 | 3.17 | 3.17 |
| 0.10 | $SD_{max}$ | 0 | 21.73 | 21.73 | 15.11 | 15.11 |
| | $SD_{av}$ | 0 | 19.37 | 19.37 | 10.11 | 10.11 |
| | NSP | 10 | 3 | 3 | 10 | 10 |
| | $RT_{min}$ | 5.56 | 18.72 | 34.06 | 0.48 | 11.37 |
| | $RT_{max}$ | 21.59 | 29.88 | 59.06 | 0.85 | 31.62 |
| 0.15, | $RT_{av}$ | 12.34 | 24.30 | 46.56 | 0.58 | 18.54 |
| 0.73, | $SD_{min}$ | 0.00 | 10.83 | 9.99 | 3.36 | 3.36 |
| 0.15 | $SD_{max}$ | 0.00 | 23.40 | 23.40 | 17.02 | 17.02 |
| | $SD_{av}$ | 0.00 | 17.11 | 16.69 | 11.06 | 11.06 |
| | NSP | 10 | 2 | 2 | 10 | 10 |

## 5. CONCLUSION

Two new heuristic algorithms have been presented to find a "good" design decision for balancing a transfer line with simultaneously activated spindle heads at workstations. It is supposed that a block of operations for each spindle head is to be chosen from a given set. The algorithms use different techniques to reach a "good" balance (random search and depth-first search).

The proposed algorithms are relatively efficient in terms of the quality of obtained solutions and the running time. This conclusion is based on their experimental comparison with an exact graph approach. For moderate size problems (less than 50 operations and 150 blocks) or when their order strength is relatively large, the exact graph approach is acceptable in terms of computation time, and thus the exact solutions have been obtained for this type of tests. The depth-first algorithm was able to find a relatively good solution for all test problems in an acceptable time (10 % worse in average than the optimal solution and with the average running time less than 1 min). The random search algorithm often fails to find a feasible solution, but for some instances it outperforms DFS algorithm in the computation time and the quality of obtained solutions. The experiments show that the heuristics performances depend on the characteristics of the problem (order strength, constraints of compatibility).

So far, it is impossible to conclude that one of the heuristic methods is better from the computation time point of view and the quality of the obtained decisions. Therefore, the use of combination of both techniques is preferable. This was confirmed by computational results.

Further investigations will be concern with mixed integer linear programming approach. The proposed algorithms can be used for finding "good" initial solutions and upper bounds for the value of an objective function.

## ACKNOWLEDGMENT

This work is financially supported by ISTC project B-986 and INTAS Project 03-51-5501.

## REFERENCES

Arcus, A.L. (1966). COMSOAL: A computer method of sequencing operations for assembly lines. *International Journal of Production Research*, 4, 259-277.

Baybars, I. (1986). A survey of exact algorithms for the simple assembly line balancing. *Management Science*, 32, 909-932.

Dolgui, A., B. Finel, N. Guschinsky, G. Levin and F. Vernadat (2003a). Some optimization approaches for transfer lines with blocks of parallel operations. In: *Preprints of the 7th IFAC Symposium on Intelligent Manufacturing Systems*, (L. Monostori, B. Kadar, G. Morel (Eds.)), Budapest, Hungary, pp. 261-266.

Dolgui, A., N. Guschinsky and G. Levin (2003b). Balancing production lines composed by series of workstations with parallel operations blocks. In: *Proceedings of the 2003 IEEE International Symposium on Assembly and Task Planning (ISATP'03)*, Besançon, July 10-11, 2003, IEEE, pp. 122-127.

Dolgui, A., N. Guschinsky and G. Levin (2003c). Graph approach for transfer lines balancing: exact and heuristic methods. In: *Proceedings of the 3rd International Conference Research and Development in Mechanical Industry* (RaDMI 2003), 19-23, September 2003, Herceg Novi, Serbia and Montenegro, CD-ROM edition (ISBN 86-83803-10-04), 9 pages.

Dolgui, A., B. Finel, N. Guschinsky, G. Levin and F. Vernadat (2004). A heuristic approach for transfer lines balancing. *Journal of Intelligent Manufacturing*, to appear.

Erel, E. and S.C. Sarin (1998). A survey of the assembly line balancing procedures. *Production Planning and Control*, 9(5), 414-34.

Ghosh, S. and R. Gagnon (1989). A comprehensive literature review and analysis of the design, balancing and scheduling of assembly lines. *International Journal Production Research*, 27(4), 637-670.

Groover, M.P. (1987). *Automation, Production Systems and Computer Integrated Manufacturing*, Prentice Hall, Englewood Cliffs, New Jersey.

Helgenson, W.B. and D.P. Birnie (1961). Assembly Line Balancing Using Ranked Positional Weight Technique. *Journal of Industrial Engineering*, 12, 394-398.

Hitomi, K. (1996). *Manufacturing Systems Engineering*, Taylor & Francis.

Rekiek, B., P. De Lit and A. Delchambre (2000). Designing Mixed-Product Assembly Lines. *IEEE Transactions on Robotics and Automation*, 16(3), 268-280.

Rekiek, B., P. De Lit, F. Pellichero, T. L'Eglise, P. Fouda, E. Falkenauer and A. Delchambre (2001). A multiple objective grouping genetic algorithm for assembly line design. *Journal of Intelligent Manufacturing*, 12, 467-485.

Rekiek, B., A. Dolgui, A. Delchambre and A. Bratcu (2002). State of art of assembly lines design optimisation. *Annual Reviews in Control*, 26(2), 163-174.

Sabuncuoglu, I., E. Erel, E. and M. Tanyer (2000). Assembly line balancing using genetic algorithms. *Journal of Intelligent Manufacturing*, 11 (3), 295-310.

Scholl, A. and R. Klein (1998). Balancing assembly lines effectively: a computational comparison. *European Journal of Operational Research*, 114, 51-60.

Scholl, A. (1999). *Balancing and sequencing of assembly lines*. Physica-Verlag, Heidelberg.

ELSEVIER

IFAC
PUBLICATIONS
www.elsevier.com/locate/ifac

# DATUM IDENTIFICATION IN REVERSE ENGINEERING

**G. J. Kaisarlis, S. C. Diplaris, M. M. Sfantsikopoulos**

*National Technical University of Athens (NTUA),
School of Mechanical Engineering,
Dept. of Mechanical Design & Control Systems*

Abstract: Reverse Engineering (RE) plays a important role in mechanical maintenance and plant equipment availability. To this date, RE-accuracy and tolerancing issues do not seem, nevertheless, to have been adequately addressed. Datum Reference Frame (DRF) is the core concept in Geometric Dimensioning and Tolerancing (GD&T), presently the dominant approach for mechanical accuracy specification. The paper presents the *Datum Reference Frame Establishment Module* of a currently under development knowledge-based computer aided system for the assignment of geometrical and dimensional tolerances in RE. *Copyright © 2004 IFAC*

Keywords: Manufacturing, Tolerance, CAD/CAM, Decision Support Systems, Knowledge – Based Systems.

## 1. INTRODUCTION

Since its introduction and standardization Geometric Dimensioning and Tolerancing (GD&T) is the dominant approach for the design and manufacture of mechanical parts due to its versatility and economic advantages. Appropriately used, it reduces for most applications the production cost, especially if it is effectively taken into consideration early in the process planning stage, (Diplaris and Sfantsikopoulos, 2003).

Datums and Datum Reference Frames (DRF) are fundamental concepts in GD&T. A datum indicates the origin of a geometric and/ or dimensional relationship between a toleranced feature and a designated feature or features of the part. The designated feature serves as a datum feature, whereas its true geometric counterpart establishes the datum. As measurements cannot be made from a true (theoretical) geometric counterpart, a datum is assumed to exist in and be simulated by the associated processing equipment. The main ISO standard for Datums, ISO 5459:1981, illustrates how datums are established by means of simulated datum features, e.g. machine tables, surface plates etc. Datum features are chosen to position the part in relation to a set of three mutually perpendicular planes, jointly called a *datum reference frame* (ASME, 1994) and restrict motion of the part in relation to it. Usually only one or two datums are required for orientation tolerances; positional tolerances, however, often require a three plane datum system, defined as primary, secondary and tertiary datum planes, Fig. 1. The total number of compartments in a DRF i.e. the number of datum features is established according to the necessary

restricted Degrees Of Freedom (DOF) of the toleranced feature. Any difference in the order of precedence of the datums referenced in the control frame requires different datum simulation methods and, consequently, establishes a different DRF.

In Reverse Engineering (RE) existing mechanical components for which technical data are not available or accessible or do not exist, have to be reconstructed and manufactured through a variety of techniques. A frequently met industrial need is that for spare parts for products that are out of production, whereas they may still have a considerable residual service life. Extension of equipment service life through re-engineering or spare parts remanufacturing by employing RE techniques obviously offers substantial economic benefits.

Tolerance design principles and advanced tolerancing techniques can be found in several publications (ElMaraghy, 1998; Zhang 1997). An extensive and recent review of the conducted research in the field of dimensioning and tolerancing is offered by Hong and Chang, (2002). The development and implementation of computer aided systems to assist designers in the allocation and validation of GD&T schemes that satisfy functional and assembly constraints represented in form of spatial DOF model has been treated by several different researchers (Shah *et al.*, 1998; Pandya *et al*, 2002.; Hu *et al.*, 2004). Of particular interest is the work of Tandler (1998) on the tools and rules for computer automated datum reference frame construction. Tolerance allocation in Reverse Engineering that takes into account the particularities emerging in such applications has only recently begun to appear as a research issue.

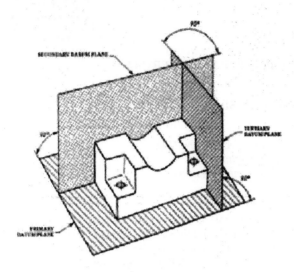

Fig. 1. Datum Reference Frame (from ISO 5459:1981)

Werghi *et al.*, (1998), give a general framework for the integration of geometric relationships aimed at constrained reconstruction of 3D geometric models of RE - objects from range data. A systematic approach through the development of a knowledge-based computer aided system (KBS) for RE-tolerancing has been proposed by Kaisarlis *et al.*, (2001). The KBS TORE (TOlerances for Reverse Engineering) was initially focused on the assignment of *nominal feature size, tolerance and type of ISO standard fit* in mating parts of assemblies. At its current stage of development *positional tolerances and basic dimensions* are also considered. Within this frame, the *Datum Reference Frame Establishment Module* of TORE that tackles the datums problem is presented in this paper. A brief overview of the KBS TORE concept and function precedes this presentation.

## 2. DRF IN REVERSE ENGINEERING

Manufacturing of RE-mechanical components which must fit and well perform in new or existing assemblies is a rather delicate job. The objective in such applications is the assignment of geometric and dimensional tolerances that match, as closely as possible, to the original (yet unknown) dimensional and geometrical accuracy specifications. RE tolerancing becomes even more sophisticated in case that CMM data are only available and a few or just only one of the original components to be reversibly engineered are accessible. Moreover, if operational use has led to considerable wear/ damage or one of the mating parts is missing, then the complexity of the problem becomes bigger.

Regarding DRF, RE–parts can be generally classified as *conventional mechanical* objects and as *stand-alone objects*. In a broad sense, the latter can be considered as "individual" items, whereas the former belong to assemblies and satisfy predetermined relationships with their mating counterparts. For the most frequently met RE-case of conventional parts,

the design guidelines request datum features to be readily discernible is a difficult task to be accomplished only by intuition and/ or experience. Datums sequence within DRF can considerably influence, on the other hand, the produced results, as illustrated by the example of Fig. 2 (a), originally presented by Chiabert *et al.*, (1998). The positioning sequence in Fig. 2 (b) (top) is the one defined by the DRF of the part drawing. In Fig. 2 (b) (bottom) the positioning sequence is different: A – C – B instead of A – B – C and the yielded results can also be respectively different. For RE mating parts, the assignment of different datum features or different sequence order of the datum features than the original ones will consequently put interchangeability at risk.

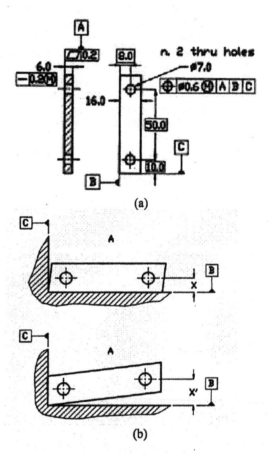

Fig. 2. Influence of Datum Sequence to the yielded results (Chiabert *et al.*, 1998).

## 3. KBS TORE

Mechanical engineering objects typically possess surfaces of simple geometries such as planes, cylinders, tori etc that justify their feature based description for RE. Furthermore, in a great range of applications the assessment of size and location tolerances of cylinders and/ or bosses constitutes one of the most critical tasks for a successful component reconstruction. The present status of the KBS TORE is developed in that context. Only parts with negligible or no wear are considered. The general concept of the method is shown in Fig. 3.

Feature nominal size and tolerance assessment is performed by the TORE–"S" (Size) module. Through a user-friendly interface, CMM obtained feature size and form error data and, as well as, surface roughness data are introduced into the system. The manufacturing process and measured, given or assumed functional clearances are also fed into the system. The KBS knowledge base incorporates independent modular "knowledge islands" such as the accuracy capabilities of the manufacturing processes, surface roughness–tolerance relationships etc. These are represented in the knowledge base either as sets of rules or as numerical functions and are manipulated by the KBS inference engine using both object-oriented and rule-based programming tools. The system then establishes the maximum possible tolerance range, taking also into account that the tightest reasonable tolerance should be, at least, five times larger than the accuracy threshold of the CMM in accordance with the ISO 10360-2.

The Size Tolerance Type Module performs a search towards either general ISO 2768 tolerances or ISO standard fits or non standard tolerances. Possible scenarios starting from "one only available original component & no mating component available" to "two or more original pairs of mating components available" are treated by the relevant algorithms of the *Nominal Size/ Dimensional Tolerance* module. Cost- tolerance relationships are used to "filter" the finally proposed alternatives. The present implementation is restricted within the 1-500 mm nominal size range and cylindrical (external or internal) features.

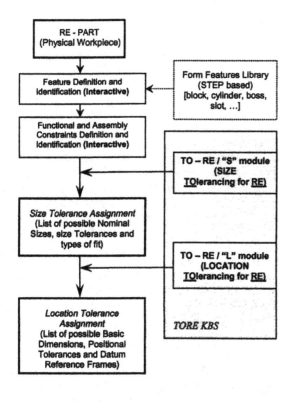

Fig. 3. TORE Concept.

For a *cylindrical feature location* the TORE-"L" (Location) module makes use of a search algorithm in order to generate possible alternatives for the basic (theoretical) dimensions and the positional tolerances. This algorithm is analytically presented by Kaisarlis *et al.*, (2004). In order to attend the required assembly clearance limits the calculation of the positional tolerances is effected using the formulas quoted in the ANSI/ASME Y 14.5M (1994) standard:

- "Fixed fastener case" (one of the parts to be assembled has restrained fasteners such as studs),

$$T_P = ½ (MMCD_H - MMCD_S) \qquad (1)$$

- "Floating fastener case" (two parts are to be fastened together with nuts and bolts),

$$T_P = MMCD_H - MMCD_S \qquad (2)$$

(symbols are explained in the end of the paper)

The maximum possible positional tolerances, $T_{PMAX\ S}$ and $T_{PMAX\ H}$, are respectively then calculated by adding the MMC "bonus tolerance",

$$T_{PMAX\ S} = T_P + MMCD_S - LMCD_S \qquad (3)$$

$$T_{PMAX\ H} = T_P + LMCD_H - MMCD_H \qquad (4)$$

Location and validation of candidate datum features is accomplished through the *Datum Reference Frame Establishment Module* the analysis of which follows in the next section. Feature location CMM data based on the established DRFs are then fed into the basic dimensions search algorithm to generate candidate X and Y basic dimensions. The thus obtained results are finally verified by the fundamental positional tolerance constraints, i.e. for the "fixed fastener case":

$$(X_{SM}-X_S)^2+(Y_{SM}-Y_S)^2 \leq ¼(T_P +MMCD_S - MD_S)^2 \qquad (5)$$
(in case of shaft)

$$(X_{HM}-X_H)^2+(Y_{HM}-Y_H)^2 \leq ¼(T_P +MD_H -MMCD_H)^2 \qquad (6)$$
(in case of hole)

## 4. DATUM REFERENCE FRAME ESTABLISHMENT MODULE

A datum feature is chosen on the basis of its geometric relationship to the toleranced feature and the particular characteristics of the design intent. To ensure proper RE-part interfacing, the datum features of the original part and the RE- part must apparently coincide, Fig. 2. In order to safeguard this RE-principle, DRF establishment has to provide an answer for the following three critical issues:

1. How many were the original datums? (i.e. 0, 1, 2 or 3)
2. Which features of the part were initially designated as datum features?
3. Which was their order of precedence?

In the adopted approach this is carried out in four sequential steps, Fig. 4. The required number of datums in the DRF is first associated with the number of restricted Degrees of Freedom (DOF) the RE feature has. Since RE features are considered as individual geometric entities they are treated like rigid bodies with three three independent translations (Tx, Ty, Tz) and three independent rotations (Rx, Ry, Rz), i.e. max. six DOFs before constraining. Due to geometrical symmetry, certain feature types may have less DOFs, for example a spherical ball seems to have only three degrees of freedom, (Tx, Ty, Tz), owing to its spherical symmetry that renders all rotations irrelevant as far as its geometry is concerned. Cylindrical (external or internal) RE features in TORE have a maximum of five DOFs, three translational (Tx, Ty, Tz) and two rotational (Rx, Ry) ones.

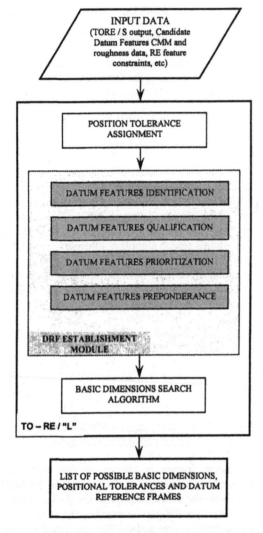

Fig. 4 Overview of the TORE/"L" and DRF Establishment Module function structure.

During Datum Feature Identification the user has to define the functional and assembly constraints and to decide for the number of the restricted DOFs. He is then prompted by the system to recognize the features of the part that come in contact with the mating part(s), the ones that have the most important impact to its function and the ones that are most probable to be used as location features during manufacturing and inspection. Functional and assembly constraints are interactively interrelated with candidate datum features by taking into account the DOFs that they may restrict them. In the general case that all five DOFs are restricted, primary datum can control three, secondary two and tertiary one DOF. Datum features must eliminate degrees of rotational freedom before attempting to eliminate degrees of translational freedom, therefore datum features which may constrain Rx and Ry are considered as primary datum candidates. Where two or more datum features can control the same DOF, the degree is constrained by the datum that is placed in the DRF highest order of precedence among them. In order to filter out non consistency with engineering practice and candidate all other possible datum features the rules given in Table 1 are used. This step leads to the configuration of two separate groups of potential datum features, the Candidate Primary Datum Feature Set and the Candidate Secondary and Tertiary Datum Feature Set.

Datum Feature Qualification is performed separately for the two datum sets. Qualified primary datum features should conform to both constraints below:

$$\text{DFFE} \leq T_P \text{ and } \text{REFperp\_to DF} \leq T_P \qquad (7)$$

Whereas for secondary and tertiary datum features the constraints that have to be satisfied are:

$$\text{DFFE} \leq T_P \text{ and } \text{DFperp\_to PDF} \leq T_P \qquad (8)$$

In any case, if the examined datum feature is a feature of size (cylinder, boss, tab etc) to the constraints (7) and (8) above another one is also added:

$$\text{DF\_T}_S \leq T_P \qquad (9)$$

## Table 1 Datum Feature Identification and Prioritisation Rules

1. A candidate datum feature should be discernible and easily accessible.
2. A candidate datum feature should be larger than the 1/5 of the largest flat surface of the part to permit subsequent location/ processing operations.
3. A candidate datum feature should be accurate and offer the best repeatability, (constraints (7), (8), (9)).
4. Real datum (e.g. datum plane) features have priority over derived (e.g. datum midplane or axis) or simulated (e.g. points in space) ones.
5. External datum features have priority over internal ones.
6. Datum prioritization and preponderance is based on the distribution density of the measured feature position data. (applicable in case of two or more RE-features in the same part)

For the prioritization of the Candidate Primary Datum Features the rules (1) to (5) of Table 1 are only applied in order to establish the primary datum. In the Candidate Secondary and Tertiary Datum Feature Set, features are grouped according to their orientation. A hierarchy on candidate datum features that belong in the same group is established in the prioritization phase. Finally, datum preponderance is determined. The highest ranked datum features of each orientation group are chosen. Criterion for the two last phases is Rule 6 of Table 1 that is applicable for parts with more than one RE features.

## 5. DISCUSSION AND CONCLUSIONS

Validation of the TORE RE-methodology has been so far performed on mating parts for which original engineering drawings are available. For the test run of the system CMM and surface roughness data of pairs of unused machined die components with one pattern per part of 4 equidistantly spaced holes with same diameter were used. Size and positional tolerances of the examined RE features were within the range of 0.02 to 0.08 mm Algorithm programming and graphical user interface of the system were effected through MATLAB and KAPPA – PC a commercially available KBS development shell. The produced to this date results were found to be in agreement with the drawings's requirements and supportive for the continuation of the research work.

Given the importance that RE applications have for the solution of a range of industrial engineering problems and the key role that manufacturing tolerances play for their successful implementation, a KBS that can be used as a decision support tool for the assignment of geometrical and dimensional tolerances in such applications proves to be a very useful tool. In that context, the TORE further development aims to respond to the majority of the cases that can be met in the industrial practice.

## LIST OF SYMBOLS

| | |
|---|---|
| $T_P$ | Positional Tolerance |
| S | Index for Shaft |
| H | Index for Hole |
| MD | Measured Diameter |
| $DF\_T_S$ | Datum Feature Size Tolerance |
| MMCD | MMC limit of Diameter |
| LMCD | LMC limit of Dimeter |
| $X_N, Y_N$ | Basic Dimensions |
| $X_M, Y_M$ | Measured Dimensions (of the coordinates that locate the center of the Measured Diameter) |
| DFFE | Datum Feature Form Error |
| REFperp_toDF | RE-Feature perpendicularity to Datum Feature. |
| DFperp_toPDF | Datum Feature perpendicularity to Prior Datum Feature |

## REFERENCES

ANSI/ASME Y14.5M–1994, (1994). *Dimensioning and Tolerancing*, ANSI, New York City.

Chiabert, P., F. Lombardi and M. Orlando (1998). Benefits of geometrical dimensioning and tolerancing, *Journal of Materials Processing Technology*, **78** , p. 29-35.

Diplaris S. C. and M. M. Sfantsikopoulos, (2003). Maximum Material Condition in Process Planning, *Proceedings of 6th SMESME International Conference*, Athens.

ElMaraghy, H. A., (1998). *Geometric Design Tolerancing: Theories, Standards and Applications*, Chapman &Hall, London.

Hong, Y. S. and T. C. Chang, (2002). A comprehensive review of tolerancing research, *International Journal of Production Research*, **40 (11)**, p. 2425-2459.

Hu, J., G. Xiong and Z. Wu, (2004). A variational geometric constraints network for a tolerance types specification, *International Journal of Advanced Manufacturing Technology*, **24**, p. 214-222.

Kaisarlis, G. J., S. C. Diplaris and M. M. Sfantsikopoulos, (2001). Tolerance Allocation in Reverse Engineering, *CD Proceedings of 16th International Conference on Production Research*, Prague.

Kaisarlis, G. J., S. C. Diplaris and M. M. Sfantsikopoulos, (2004). Positional Tolerance Allocation in Reverse Engineering, *Proceedings of XIV Metrology and Metrological Assurance Conference*, Sozopol, Bulgaria.

Pandya, G., E. A. Lehtihet and T.M. Cavalier, (2002). Tolerance Design of datum systems, *International Journal of Production Research*, **40 (4)**, p. 783-807.

Shah, J. J., Y. Yan, B. C. Zhang, (1998). Dimension and tolerance modelling and transformation in feature based design and manufacturing, *Journal of Intelligent Manufacturing*, **9**, p. 475-488.

Tandler, W., (1997). The tools and rules for computer automated datum reference frame construction. In: *Geometric Design Tolerancing: Theories, Standards and Applications*, (ElMaraghy, H. A. (Ed)), p. 100-111, Chapman &Hall, London.

Werghi, N., R. Fisher, C. Robertson, A. Ashbrook, (1999). Object reconstruction by incorporating geometric constraints in reverse engineering, *Computer-Aided Design*, **31**, p. 363-399.

Zhang, H. C., (1997). *Advanced Tolerancing Techniques*, John Wiley & Sons, New York City.

ELSEVIER
IFAC
PUBLICATIONS
www.elsevier.com/locate/ifac

# A CONTROLLED SYSTEM MODEL WITH PETRI NET FORMALISM FOR RECONFIGURATION

**DESCHAMPS, Eric – HENRY, Sébastien – ZAMAI, Eric – JACOMINO, Mireille**

*Laboratoire d'Automatique de Grenoble (LAG)*
*ENSIEG - Rue de la houille blanche, Domaine Universitaire, BP 46*
*38402 Saint Martin d'Hères Cedex, France*
*Tel : (33) 4-76-82-64-05, E-mail : eric.deschamps, henrys, zamai, jacomino*
*(@lag.ensieg.inpg.fr)*

Abstract: Usually, in an industrial context, the design of discrete control law to drive manufacturing system is assumed off line by several experts. Consequently, not only the design of all the control laws mobilizes many PLC program developers, but also, in case of unexpected resource failures, the reconfiguration process can be only considered from a manual point of view. This is mainly due to the lack of a generic method to model the controlled system abilities from which a control law can be automatically synthesized. So, to bring a solution in this field of research, a methodological approach to model a controlled system is first given. Second, to provide validation processing of the controlled system model, an automatic translation in Petri nets formalism is proposed. So, with such a structure, the model allows helping the decision-making for the reconfiguration processing. In this way, a generic algorithm to use it from a synthesis point of view is presented. *Copyright © 2004 IFAC*

Keywords: Control system Model, Reconfiguration, Manufacturing systems, Programmable Logic Controllers, Petri Net.

## 1. INTRODUCTION

In the automated production field, main interest expressed in the last twenty years is the automation of manufacturing systems. This automation looks for replacing shop-workers from hard and dangerous jobs transferring them to deciding and supervising places. Automated Manufacturing Systems (AMS) are commonly represented according three different elements (see. Fig.1):

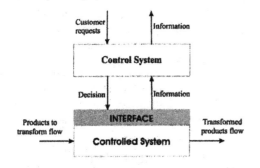

Fig. 1. Schema of an AMS

• a controlled part including the set of physical components (machines) actuating over the product flow to give it an added value,
• a control part containing the software, the hardware and the information charged of the process management (control law and controlled system model),
• an interface charged to insure the broadcasting between the controlled system and the control part.

The more the automation level of manufacturing systems grows, the more the problem of monitoring becomes crucial. Moreover, if constraints like security, ecology, productivity and quality, already imposed in nominal production, must be also satisfied even if process fails, or variability of the customer demand, then the problem can be considered as very difficult. A first solution to reduce this complexity can be found into the considered control architecture. So the control structure on which our approach is based is a hierarchical and modular structure. The second solution to face to disruptions in such a control structure leads to integrate monitoring and supervision abilities at each level of the structure. However, even if several works have been developed in this field of research (Kramer and Seheni, 1993), they give often solution in terms of monitoring and control but less on the decisional aspects. These decisions must define "what doing" after disruption, and so propose solution (control laws modelled using the IEC61131-3 norm) to come back in normal running satisfying constraints previously given. To do this, several information must be at least available as the initial state of the controlled system, the objectives to be reached and the abilities offered by the controlled system.

In this context, we propose to give our contribution to the supervision of complex industrial processes, especially in the reconfiguration field of research. So, a complete approach starting from the design of the controlled system model to the synthesis of a specific control law is dealt with. Consequently, this paper is organized as follows. In the first time, section 2 proposes a short review of the main works giving contribution in the (re)configuration field of research. Section 3 describes an application example that will be developed during the presentation. Based on this case study, section 4 proposes and details a methodological specification guide to help the engineers to build the data structure of the controlled system. From such a structure, section 5 proposes an automated processing able to formalize the controlled system model using Petri nets. Considering such a model, section 6 proposes an algorithm able to synthesize a specific control law. The conclusion and future works end this paper.

## 2. STATE OF THE ART AND MODELISATION REQUIREMENTS

To launch a (re)configuration process following failures occurrence (machines disruptions, product specification variation ...), an adequate knowledge of the controlled system is required. It means that the used model integrates all the required information to synthesize a control law. To quickly fix the main data that must be represented in the controlled system model, it is pertinent to considerate the main goal of the control system. A control system must transform the product flow (see Fig. 1) according to the product specifications. However, and from a practical point of view, the control system is not directly able to drive the product flow evolutions, but only the actuators (resources). So, to synthesize a control law respecting the product specifications, it is required to model not only the resources evolution from which the control system can act on the product flow but also the product flow itself.

In the (re)configuration field of research, several approaches have been proposed. However, the aim of this paper is not to do a complete review of all approaches; the only three main ones are presented. The two first approaches (Gouyon et al., 2004; Toguyeni et al., 2003) take place in the hierarchical context at the scheduling level of the CIM architecture. Indeed, their models represent accessibility and preceding relationships between operations. However constraints between operations and controlled system state are not taken into account. Consequently, only high-level operations can be modelled and not offered operations by an operative element like a cylinder for instance. So, these approaches cannot allow to synthesize a control law. The approach presented in (Holloway et al., 2000) take place at the lowest level of the control architecture. It allows to synthesize a control code for automated manufacturing systems. However, this approach has an actuator-sensor point of view without taking the product into account. So the control code is synthesised from control specifications but not from product ones. Moreover, there is a combinatory explosion problem with the construction of the model. So this approach is not well adapted in the complex industrial context.

To give our contribution in this field of research, we propose here to provide to the control law designer a method to model the controlled system for reconfiguration requirements.

Control law synthesis is mainly based on the controlled system model. So to model the controlled system, what point of view has to be used? In a control law the actions have effects on resources, which have next effects of changing the product flow from an initial state to a final state. These two states are defined according to the objectives. Therefore, the controlled system model must have a product flow point of view to make the link with the objectives. But with only this point of view, the resource evolutions without effect on the product flow cannot be found as the action "Retract a cylinder" (after the action "Extend the cylinder"). Finally to synthesize a control law with the resources point of view, the controlled system model must have a double point of view, resources and product flow. Then, the model must give the links between both points of view. The links define the causal relations between effects on the product flow and on the resources. These links are the operations offered by the resources of the controlled system. This double point of view to model the controlled system is also a necessity to know the executable operations for a state of the resources and a state of the product flow. Indeed, the executable operations depend on these two states. Finally, the operation model must represent the constraints on the states of the resources and the product flow but also the effects on the resource state and the product flow state.

## 3. SAPHIR

This section presents an application example based on the loading system (see Fig. 2) of the research platform Saphir of the Laboratoire d'Automatique de Grenoble, France. This platform is dedicated to the assembly of camshafts. A four places rotating storage is used to receive until six different kinds of products. These products are identified by a weight identification system. Once a product has been identified, a central conveyor drives it to a sorting device. Depending on the kind of the product, it is directed to a left or to a right conveyor. At the conveyors end, a position station allows to index the products. So, a robot takes the different products to assembly them. A shopworker is charged to fill the rotating storage and to empty the assembly station.

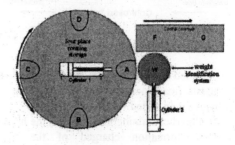

Fig. 2. Loading system of SAPHIR

Then, the next section presents the method to model the controlled system, which rests on the precedent principle described in section 2.

## 4. METHODOLOGICAL GUIDE FOR CONTROLLED SYSTEM SPECIFICATION

The method proposed here is mainly based on the specification of the effects on the resources and on the product flow with associated constraints (Henry et al., 2004). To explain this principle, let us take the above presented example, especially considering the "extend cylinder 1" (EC1) operation.

First, the operation modelling considers only the effect on the resource (C1) without studying the possible effects on the product. The beginning of the operation modifies the C1 state from the retracted position to the intermediate position and the end of the operation modifies the C1 state to the extended position. But to carry out this effect on the resource, constraints must be satisfied. There are constraints on other resources and on product flow. Indeed, to extend C1, C2 must be retracted before and during the operation else there is collision. And for product flow constraints there should not be product between A and B and between A and D for the same reason. All these constraints must be satisfied before and during the operation. This first part defines the basic behaviour, which is presented in the bottom array of the Fig. 3.

Then, the operation modelling considers the effects on the product flow. Indeed, if there is a product in the position A, the operation has an effect on this product. So, as the effect on the resource, to describe the effect on the product, the initial, intermediate and final states of the product are defined. In our example, the product must be in A, and during the operation, it is between A and W. The final product state is in W. Constraints must be also defined. For resources constraints, the rotating storage tray must be stopped and positioned before and during the operation. For the product flow constraints, the product that is transferred should not strike another product. This part defined the first extra behaviour (see Fig. 3). But the initial state of the product could be a product between A and W, after a failure for example. So, a second extra behaviour must be defined for this operation. The complete description of the operation "extend C1" is presented in Fig. 3.

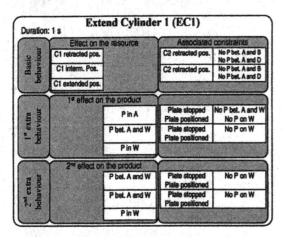

Fig. 3: Complete EC1 operation

## 5. PETRI NET MODELLING

This resulting specification allows to capitalize all the required data for synthesis use. However, before using this data base, several classical properties must checked. Obviously, because the normal behaviour of the flexible manufacturing controlled system must be represented by the resulting model, it must verify reversibility, boundedness, liveness and deadlock-freedom properties. For this reason, it is required to formalise the modelling. Among several formalisms proposed in the literature (Bucci et al., 95) to model manufacturing systems, the Petri net tool seems to cover all the requirements from a verifying properties point of view. Moreother, Petri nets allows to easily represent effects, flow, constraints and states as it is shown in Fig. 4. With such a representation for which mathematical matrix expression is systematic, and considering that the synthesis of an admissible control sequence amounts to looking for the existence of a path in a graph, it is possible to envisage an easily solution to be implemented.

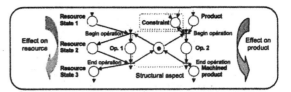

Fig. 4: modeling abilities with Petri net formalism

According to the resulting specification presented in the previously section, and considering the modelling abilities with Petri net formalism (see Fig. 4) six models have to be obtained: two state models and four constraint models.

### 5.1. Resource state model

First, the possible resource states must be known. This model is obtained from the part effects on the resources. Each state of a resource is modelled by a place and for each operation of this resource, two arcs are positioned between initial state and intermediate state and between intermediate state and final state. Transitions of each arc are respectively instantiated by the beginning and the end of the operation. In our example, for the resource C1, there are two operations: extend C1 (EC1) and retract C1 (RC1). The model of the resource obtained is presented in Fig. 5. All models of each resource give the resource state model.

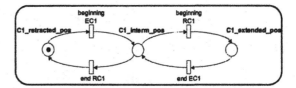

Fig. 5: Cylinder 1 model

### 5.2. Product flow state model

As for resources, the product flow state model must be defined. It is built in the same way than resource state model. The part of the model corresponding to the EC1 operation is presented in Fig. 6. These state models allow to follow the state of the controlled system, but they do not allow to know if an operation can be carried out. So models of constraints must be defined.

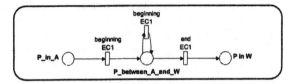

Fig. 6: A part of product flow states model

There are two kinds of constraints, constraints that must always be satisfied, and constraints, which must be satisfied only if the operation has an effect on the product. In considered example, the cylinder 1 can be extended if the rotating storage tray is stopped and positioned. This constraint must be satisfied only if there is a product in A or between A and W. Constraint models must not modify the marking of state place. To fire the transition "beginning operation", constraints for the start of operation must be satisfied. To simplify the notation, to test the lack of a token in a place (for example No P in W), the zero test is used (see Fig. 7). This notation is not restrictive because it can be replaced by the marking test of the complementary place. But the readability of the model will be worse.

### 5.3. Permanent pre-constraint

The first constraint model is the permanent pre-constraints model. It corresponds to constraints before operation of the basic behaviour. To fire the transition "beginning operation", resource constraints and product flow constraints must be satisfied. For the EC1 operation, the resulting permanent-constraint model is presented in Fig. 7. In this figure, places surrounded by a circle in dotted line are state model places.

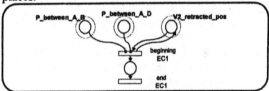

Fig. 7: permanent constraints model for EC1

### 5.4. Pre-constraints depending on effect model

The second pre-constraint model models the constraints depending on effects on the product flow. To build this model, associated constraints before operation of all extra behaviour are used. With each possible effect combination is associated a set of constraints which can be different. In our example, for EC1 operation two effects cannot be made at the same time because associated constraints are incompatible. So, there are three cases: no effect, the first effect or the second effect. An effect is made if the associated condition is true. The condition corresponds to the initial state of the product in this effect. So EC1 operation has no effect if conditions *Product_in_A* and *Product_between_A_and_W* are false. In this case there are not constraints (see transition 1 in Fig. 8). But if the condition *Product_in_A* or *Product_between_A_and_W* is true, to start the operation, the associated constraints *Tray_positioned* and *Tray_stopped* must be satisfied (see transition 1 and 2 in Fig. 8).

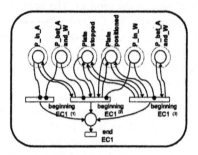

Fig. 8: second constraints model for EC1

So, two constraint models are obtained. One constraint model of basic behaviour, and the other constraint models of extra behaviours to start the operation. These models consider only constraints, which must be satisfied before the start of the operation. But constraints, which must be satisfied during the operation, are not satisfied. In our example, to extend C1, C2 must be retracted and to extend C2, C1 must be retracted. But if C1 and C2 are retracted, these constraints model authorize the start of operation extend C1 and extend C2 at the same time. Indeed, constraints *C1_retracted_pos*, and *C2_retracted_pos* are true. This simple example shows that these two constraint models are no sufficient. Constraints during the operation must also be considered.

### 5.5. Permanent-constraint

To model constraints during the operation, two new constraint models (constraints of basic behaviour and extra behaviour) must be defined. The first new model translates the constraints of basic behaviour, which will be satisfied during the operation. For each operation, the model prohibits the start of operations that modify constraints of considered operation. So this model is formalized with mutual exclusion between the start of the considered operation and the starts of prohibited operations.

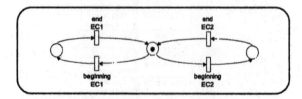

Fig. 9: first part of the third constraint model

In our example, in the basic behaviour of the EC1 operation, there are a constraint on resource states and two constraints on the product flow. Indeed, during the operation EC1, C2 must be retracted. So the start of operation EC2 is prohibited during the execution of EC1 operation. The first part of the model is presented in Fig. 9.

### 5.6. Constraints depending on effect model

Next, constraints on product flow must be translated. In this case, the beginning of EC1 operation is put in mutual exclusion with the beginning operation, which modifies the product flow. But an operation has or not an effect on the product flow. So, in this part of model, the transition "beginning EC1" is put in mutual exclusion with the only transition corresponding to the beginning of the operation with an effect that modifies the constraints. So a second part of the model is obtained (see Fig. 10).

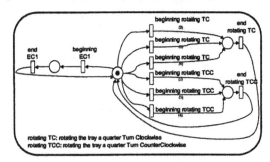

Fig. 10: second part of the third constraints model

The second new model translates the constraints of extra behaviours, which will be satisfied during the operation. But these constraints must be satisfied only if the associated effect is carried out. In this model the transitions corresponding to an effect of the considered operation is in mutual exclusion with the beginning of operations that modifies constraints. This model has also two parts. The first part translates resource constraints and the second part translates product flow constraints. In our example, if the EC1 operation has an effect on the product flow, so the constraints *Tray_positioned* and *Tray_stopped* must be satisfied. So transitions corresponding to the beginning of EC1 with effect are put in mutual exclusion with the beginning of two operations of tray rotating (clockwise and counter clockwise). Indeed these two operations modify constraints *Tray_positioned* and *Tray_stopped* (see Fig. 11).

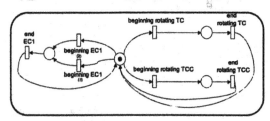

Fig. 11: first part of forth model

In our example, there is not a second part. Indeed, the only operation, which can modify the constraint *no Product_in_W*, is the considered operation.
These six models represent the whole controlled system model. The two state models allow to know

the states of the controlled system, indeed, when an operation is carried out, these models are updated. After, the two first constraint models allow knowing if an operation, which will be started, is compatible with the controlled system state. Finally, the two last constraint models allow knowing if two or more operations can be started at the same time. All these models are separated, but they are linked by state places and by transitions. So we can obtain one and single model merging these state places and transitions of beginning and end of operations.

With such a Petri net model, next section will propose a specific algorithm able to synthesize an admissible control sequence from product specifications.

## 6. OVERVIEW OF CONTROL LAWS SYNTHESIS

For concision reasons, we just present here an overview of the control law synthesis. The controlled system allows synthesizing a control law from product specifications. The presented process uses an algorithm given in (Zamaï, 1994). Considering that a control request amounts to fire the equivalent transition in the controlled system model, this algorithm allows finding using a backward technique, an acceptable sequence of firing transitions (i.e. an acceptable control sequence). It must be noted that this algorithm uses the mathematic aspect of the Petri nets model. So, the principle T-invariant, Pre, Post and C matrix are used. In the follow of the paper, this algorithm will be noted A1. Moreover, to synthesize a control law, (Henry et al., 2004) proposes to decompose the search of path into three steps (see Fig. 12), which correspond to different types of operations and constraints. The process which is next presented uses this principle, running the algorithm A1 in the two last steps.

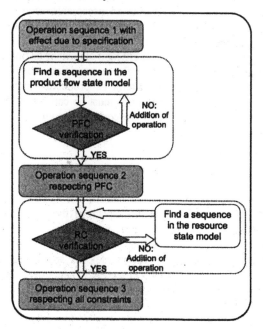

Fig. 12: Algorithm steps

Initially, product specifications are known. These specifications are constituted by a set of state variable

values of the product, which are modified by operations. The ends of these operations are modelled by transitions in the product flow model. So, a first sequence of transitions is known.

The second stage of the algorithm consists in running the algorithm A1 to find a sequence between each transition of the first sequence in the product flow state model. Then, the product flow constraint (PFC) must be checked. So, for each transition corresponding to the beginning of operations, PFC must be checked in the two first constraint models. If a constraint is not satisfied, it is required to fire a transition of the product flow model to validate this constraint. This transition is added in the sequence 1, and the algorithm A1 is run again to find a new sequence to fire all transitions of the sequence 1. PFC are again checked and so on. This step gives a second sequence of transitions, which respects all product flow constraints but not resources constraints (RC); this is the aim of the last step.

For each transition corresponding to the beginning of operations, RC must be checked in the two first constraint models. If a constraint is not satisfied, it is required to fire a transition of the resource state model to validate this constraint. Then, the algorithm A1 is run to find a sequence to fire this transition. Operations started by the founded transitions must be without effect on the product flow and must satisfied PFC. This obtained sequence is added in the sequence 2, and RC are again verified and so on. After this step, the algorithm gives an acceptable sequence that respects all the constraints.

Each *Begin operation* carries out this operation, and the *end operation* waits the end of this operation (sensor information for instance).

The algorithm does not use the two last constraint models. Indeed, it founds an acceptable sequence without optimisation. It does not try to carry out two operations at the same time. But, if we want to improve the solution, an efficient algorithm to find a path in the Petri net to fire a transition must be developed integrating criteria as production delay. Moreover, several operations must be carried out at the same time if the two last constraint models allow it.

## 7. CONCLUSION AND FUTURE WORKS

In this paper, problem of control law synthesis is dealt with. An innovative aspect of the proposed approach is to treat the entire problematic, from the controlled system modeling to the automated control law synthesis. So, a methodological specification guide is provided to help engineers to take out all the required data from the controlled system. The resulting data base allows to capitalize the knowledge on the considered controlled system. The advantage of this is to improve the reactivity in case of controlled system modification to easily update the data base, or in case of failures or customer request change, to prepare and to simplify the synthesis process of the new control laws. So, the resulting data base is then used to design automatically the whole control system model using here Petri nets formalism. Such formalism allows not only to verify the main properties of the model, but also to structure the adequate algorithm able to synthesize the adequate control law. From such a resulting control law models, it seems to be easy to translate them to one of the languages of the IEC61131-3 norm. However, it must be notify that the proposed algorithm does not yet guaranty performances as production delay, and quality criteria.

For this reason, our future works will first focus on the optimization of the proposed synthesis algorithm. Second, from these models, a study will be launch to consider the automatic translation of them into the languages of the IEC61131-3 norm. So, from such a possibility, the real time reconfiguration of PLC would be envisaged. Third, to integrate this approach in a real context, the proposed methodology to model the controlled system must be improved to allow its use with designers with a minimum level of expertise. In this way, the development of a software tool to guide first the designer to model the controlled system and second to implement the synthesis algorithm will be proposed.

## REFERENCES

G. Bucci, M. Campanai, P. Nesi (1995). Tools for Specifying Real-Time Systems, *The international Journal of Time-Critical Computing Systems*, vol. **8**, pp 117-172.

R. David, H. Alla (1992). "Du grafcet aux réseaux de Petri ", Hermes, Paris.

D. Gouyon, J.M. Simão, K. Alkassem, G. Morel (2004). Product-driven automation issues for B2M-control systems integration, *11th IFAC Symposium on Information Control Problems in Manufacturing (INCOM)*, Salvador, Brazil.

S. Henry, E. Zamaï, M. Jacomino (2004). Real Time Reconfiguration of Manufacturing Systems, *IEEE International Conference on System Man and Cybernetic (SMC'04)*, Hague, Netherlands.

L. E. Holloway, X. Guan, et al., (2000). Automated Synthesis and Composition of Taskbloks for Control of Manufacturing Systems, *IEEE Trans. On Systems, Man and Cybernetics*, **Part B**, Vol **30**, (5), pp 696-712.

T. Kramer, M. K. Seheni (1993). *Feasibility Study : Reference Architecture for Machine Control Systems Integration*, NISTIR 5297, National Institute of Standards and Technology, Gaithersburg, MD.

A.K.A. Toguyeni and P. Berruet and E. Craye (2003). Models and Algorithms for failure diagnosis and recovery in FMSs. *International Journal of Flexible Manufacturing System*, vol. **15**, pp 57-85.

E. Zamaï (1994). "Commande et surveillance des systèmes à évènements discrets complexes: Utilisation d'un modèle du procédé", *report of master*, LAG, Grenoble.

ELSEVIER
IFAC
PUBLICATIONS
www.elsevier.com/locate/ifac

# RESPONSIVE DEMAND MANAGEMENT WITHIN THE FOOD INDUSTRY

**Rob Darlington and Shahin Rahimifard**

*Advanced Manufacturing Systems and Technology Centre, Loughborough University, UK*

Abstract: There has been a proliferation of advances within the processed foods sectors, resulting in increasing availability of value-added and convenience foods. This is due to consumer demands for foods that can easily and quickly be prepared and has resulted in significant increase in product variety and their related preparation, cooking and packaging processes. The research reported in this paper investigates particular applications where long make-spans resulting from the increased processing and preparation requirements for convenience foods and short notification of reaction times demanded by retailers creates considerable wastage in the form of over-production to ensure order fulfilment. *Copyright ©
2004 IFAC*

Keywords: food processing, planning, production control, performance analysis, optimization

## 1. INTRODUCTION

The food industry is bounded by a number of constraining factors that have created a unique production ethos which make it distinct from many other manufacturing sectors. The volatility of demand for food products can be extreme; dependant on a diverse range of factors while the products themselves often have relatively short shelf life which limits the possibility of holding a safety stock to guard against demand fluctuations. In addition, food manufacturers are often given very short reaction times by increasingly powerful global food retailers while the software support which would otherwise aid manufacturing in such responsive environments is largely adapted from other sectors and is ill-equipped to deal with food industry requirements. The growth of convenience foods has been steady and well defined, with many new products being introduced at regular intervals providing labour saving meal accompaniment or solution. Product sectors such as fresh foods and ready-meals have grown from small beginnings to have strong prominent positions in supermarkets, as a result of popularity with consumers seeking time

saving products who are prepared to pay a premium for this convenience. The grocery supply chain in the UK is predominantly through a small number of large retailers, who thanks to the economies of scale offer "own label" products branded with the supermarket chains own name which has been manufactured on their behalf to their specifications.

In this paper, the authors discuss the issues involved in the development of a structured approach to combating the long make-span and short reaction time constraints in order that a responsive planning framework to reduce overproduction wastes may be realised. The following section of this paper provides a review of the previous research work relating to this aim. The remaining sections of the paper describe the research concepts related to the development of a responsive demand management framework for the food industry.

## 2. RESEARCH BACKGROUND

The structure of food retailing in the UK is based around a small number of retail chains, with a heavy bias for large out of town stores, where the majority of groceries are sold (Bell *et al.*, 1997). The relationships between these retailers and their suppliers need to be effective for consumer demands to be met, however it is considered that the retailers dominate in their dealings and inter-organisation co-operation is lacking (Robson and Rawnsley, 2001).

Demand for food products varies across the industry. Some foods have a fairly steady, easily predictable demand pattern, meaning that the consumer demand for the product can be met accurately, without wasteful overproduction, or disappointing consumers by not meeting their needs. Other products, for example prepared sandwiches, display highly volatile demand, for which there may be considerable wastage when demand is over-predicted or consumer dissatisfaction when stock-outs occur. Retailers may attempt to smooth large fluctuations by managing the demand, for example by running various promotional activities to maintain demand for products at a steadier level. The demand management efforts for products are merely an attempt to compensate for the external driving conditions such as weather conditions, holiday seasons and sporting events which contribute to the demand in highly volatile sectors.

In this context, the bullwhip effect was first identified by Forrester in the early 1960s, and has been investigated thoroughly since (Towill, 1996; Metters, 1997). Its basis is the amplification of demand across a supply chain, to the point of creating seemingly unpredictable demand volatility in suppliers. Additional problems occur when one companies actions are reflected across the whole supply chain, so organisations that create demand volatility can increase the complexities and costs across the whole of their supply network.

This is a clear indication that old data causes delay, amplifications of demand and overhead. Wal-mart (an American retailer) broke barriers with an innovative approach of introducing Point Of Sale (POS) data, made available to each level of supply chain, making actual demand much clearer across the companies (Mason-Jones, 2000). In this approach the inventory levels at retail point may be directly linked to suppliers for instant notification of product demand.

One of the primary considerations when dealing with food industry manufacturing is the Shelf Life of the ingredients and final products being processed. The times over which foods are fit for consumption may be accurately predicted, dependant on storage conditions.

In general food quality degrades with chemical changes and micro-organism growth over the time it is held until the shelf life expires (Pegg, 1999). Retailers aim to provide consumers with as much of the available shelf life possible, limiting the time that manufacturers have to hold produce.

When it comes to the scheduling of the production of convenience foods a number of considerations must be borne in mind; hygiene is of paramount importance and regular, intensive "clean-downs" of the processing equipment must be scheduled into the production schedule which may take as long as an hour per assembly line, typically occurring every 24 hours. In terms of planning production there are further restrictions as to the order of products to be processed dependant on flavour and allergen content (for nut-free products for example). Production schedules must take into account the order that products must follow in keeping with constraints such as flavours as described in detail by Nakhla (1995).

The 'rules of thumb' and constraints that must be imposed by the production planner are also poorly defined, rarely recorded, and it is often the case that it is only the planner or scheduler that knows when and where to apply these rules (Van Donk and Van Dam, 1998). This places great responsibility upon the production planner, and a formalised system of identifying these constraints will clearly be of benefit in times of staff illness or turnover, and will aid dissemination of vital production knowledge across the enterprise.

Further scheduling complications are evident based on the highly dynamic nature of the food industry (Gargouri *et al.*, 2002) such as when shipments of ingredients from suppliers arrive relative to when orders are placed and how this influences the way a schedule is created. Additionally, changes to the production plan from confirmed orders that had been forecasted create considerable complexity to the scheduling effort, particularly when the scheduling is being completed on such a compressed timeline, i.e. for orders which must be completed in a number of hours. Scheduling software has aided significantly the process of schedule generation, both speeding up the process, and allowing optimisation of the schedules produced to improve production planning. Some systems even potentially enable scheduling of jobs on the production plan to be integrated directly with the Manufacturing Execution System (MES) to take out all manual intervention, though the sophistication of such systems in dealing with food and drink industry problems is unclear, and certainly is at this stage unproven.

## 3. DEMAND MANAGEMENT IN THE FOOD INDUSTRY

Traditionally operational planning within the food industry has been based on a predictive make-to-stock approach, which utilised a number of forecasts for various product demands as a basis for production levels. However in recent years there has been greater pressure to adopt a more reactive make-to-order approach, similar to that adopted by the discrete parts manufacturing companies in the engineering sector. The make-to-order format of production is represented in Figure 1a, where the overall reaction time allowed for production is greater than the processing time of products. The implementation of such reactive customer order driven systems has proved to be more feasible in the engineering sector where such reaction times are allowed for the given make-spans. In this research, reaction time is taken to be the length of time from when customer orders are confirmed until the finished products are despatched. In convenience food sectors such as chilled ready-meals and sandwiches, there is greater complexity where the make span of products often represents a significantly longer time period than the reaction time required by the customer. This means that the production processes must be started before the exact levels of customer demands are known. This is represented diagrammatically in Figure 1b, where production starts before the orders have been placed, and thus order quantities must initially be predicted through some means of forecasting activity. This basing of production volumes upon forecasting methods inevitably introduces inaccuracies in the production plans, creating waste

due to the difficulty in accurately predicting the customer orders to be met. In addition, further complexities are created in this current state of manufacturing convenience foods by the volatile pattern of customer orders due to seasonal promotions/marketing activities and the limited shelf life of raw material and finished goods as mentioned previously which prevents products being made to stock. It may be considered that the problem of long make-span and short reaction time overlap may be resolved by separate focus on the two distinct dimensions of this problem, namely minimising the production make-span and maximising the reaction time as outlined in Figure 1c. While there are many potential improvements that may be made in tackling the problem, the changes required may take some time to implement, and in some cases may yield only slight improvements to the overlap of make-span and reaction time. In these cases there is a requirement for improved planning activities to reduce the impact of late notification of demand volatility.

The research reported in this paper has developed a structure for sandwich and ready-meal production to be improved with regard to the situation described earlier by adoption of a systematic approach as outlined in the IDEF0 diagram in Figure 2. This approach is based on four major activities namely health check, product/ process optimisation, supply network optimisation and a novel two stage production planning. The Health Check aims to provide the appropriate information related to production processes, supply chain and production planning activities through a structured modelling activity.

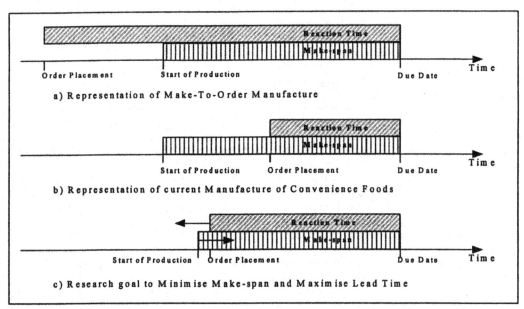

a) Representation of Make-To-Order Manufacture

b) Representation of current Manufacture of Convenience Foods

c) Research goal to Minimise Make-span and Maximise Lead Time

Fig. 1. Relationships between Make-span and Reaction Time for Food Industry Manufacture

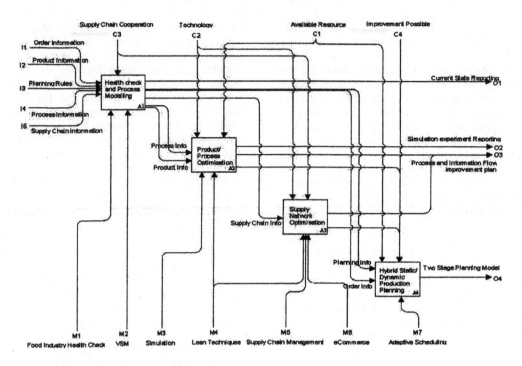

Fig. 2. IDEF0 Representation of the RDM Framework

The principal tool considered for this modelling has been Value Stream Mapping (VSM), which allows for simple and insightful mapping of both process and information flows along a supply chain. This mapping process provides the required information and knowledge to undertake the subsequent three activities as outlined in the following sections.

### 3.1 Product/ Process Optimisation

In focusing upon reducing the overall production lead-time for the manufacture of convenience foods, it may be seen that non value-adding steps in production may be quickly identified, leaving only the time required for the processing of tasks that are essential for manufacture. This requires the make-span of the products themselves be reduced, which may be achieved by only a limited number of means. Considering each of the essential processes in terms of the product's requirements, and the technology and resources reasonable to achieve those requirements, it may be infeasible to reduce the make-span of many products significantly without considerable technological development or resource investment. As such, the effort in "minimising make-span" encompasses considerable work in reducing the time-based wastes in the production system by the application of Lean techniques, and the careful consideration of how current practices may be improved upon to reduce the incumbent value-adding processes.

The Lean tools initially considered as being of most use when undertaking improvement of convenience food manufacture include Layout redesigns, Single

Minute Exchange of Dies (SMED) philosophies applied to changeovers and set-ups and Work In Progress (WIP) reduction. The preferred method of determining the potential impacts of changes to production processes is Simulation. Small scale models focused on particular aspects of the process being considered for improvement are used for initial data relating to the applied techniques. A commercial simulation software package, namely Simul8 has been used for this purpose. The details of this research work is the subject of further publication and is beyond the scope of this overview paper

### 3.2 Supply Network Optimisation

The aspects of the problem related to the reaction time are considered under Supply Network Optimisation. This includes order processing at both retailer and manufacturer, point of sale data processing and information flows between supply chain members. The significant improvements that may be made in reducing the demand notifications will not be limited to the manufacturer. In order that the best improvements may be made, it is likely that all members of the supply chain are involved and it is to that end in part the health-check was devised in order to promote collaboration between members of the chain. The positive dissemination of Point of Sale (POS) data across the supply chain has been successfully exploited by only a small number of retailers, and would substantially reduce the bullwhip effect described in section 2 which may be mapped across supply chains using the Demand Amplification Screen as applied in VSM.

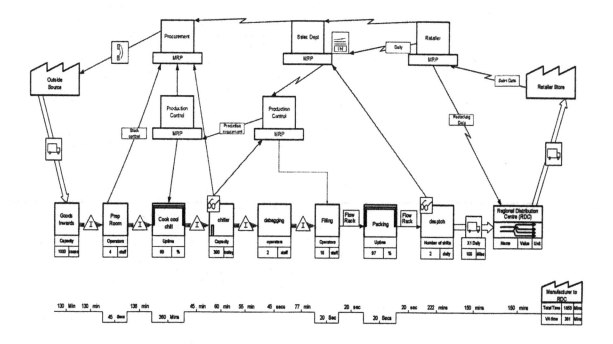

Fig. 3. Value Stream Map for ready-meal supply chain

The application of new e-commerce technologies to better enable information flows in this way will reduce the time spent in order placement and processing, and the mapping of such flows, as undertaken in VSM will outline where specific improvements may be made. A simple value stream mapping for a ready-meal production facility undertaken in electronic (eVSM) form is outlined in Figure 3. Further integration of planning activities via Advanced Planning Systems (APS) will also impact upon the overall lead time allowed for convenience foods.

### 3.3 Two Stage Planning

Once the activity concerning the optimisation of make-span and reaction time have been undertaken, focus must then be placed upon production planning activities. In this context, the research has developed a two stage planning framework based on a hybrid approach of utilisation of static and dynamic production scheduling rules. In this approach, operations are divided into two categories of standard and special operations. Standard operations are those which do not give the product identity and are shared among many products. Special operations are those that give identity to a product. The main principle of the two stage planning is to use static planning for the standard operations based on traditional forecasting approach in the first stage to generate a soft schedule, and to utilise a dynamic (real time) approach for special operations. The second stage is initiated when customer orders are confirmed. The confirmed production levels will be used to re-adjust the batch sizes for special operations to produce

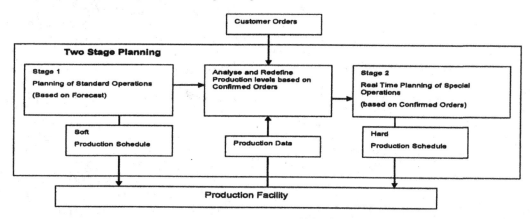

Fig. 4. Two Stage Planning of Production in the Food Sector

a hard schedule indicating that this final work-plan based on confirmed orders.

In this two stage planning framework, the processing of standard operations are initiated based on the soft schedule. The processing of special operations are however subject to change, dependant upon the confirmed orders and shop-floor data indicating the current state of production as shown in figure 4.

A commercial software scheduling system, namely PREACTOR has been adopted to implement the two stage planning. PREACTOR is a highly configurable finite capacity planning system, and utilises graphical user interfaces for ease of use and rapid access to information. It has modular structure of functionality, named PREACTOR 100, 200, 300 and APS, and among these PREACTOR 300 and APS enable users to create their own scheduling rules using Visual Basic programming. In order to develop the two stage planning system, the functionality of the standard PREACTOR software has been extensively enhanced to include custom scheduling routines, custom import and export scripts and specially designed user interfaces to support the real time planning of special operations. A simple schedule for a ready-meal production facility generated in PREACTOR is shown in Figure 5.

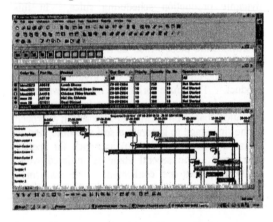

Fig. 5. PREACTOR scheduling software screen capture

## 4. CONCLUSION

Food industry manufacturers face production conditions that demand the application of specialist tools, techniques and software owing to the nature of, and the volatile demand for, these products. This highlights a need for increased use of IT tools, especially within small enterprises for effective production in the face of:

- Large volumes and varieties of product
- Relatively short production lead times
- Short shelf lives of product ingredients
- Short Product Life-cycles

The research reported in this paper considers a sector of products, namely convenience foods where the product make-spans exceed the reaction times that manufacturers are given to meet these demands. This has led to a situation where much waste is generated through overproduction to ensure retailer orders are met. As a result a responsive demand management framework has been proposed which utilises a number of contemporary process modelling and manufacturing planning tools to eliminate this wasteful overproduction. With the ever-increasing trend in variety of convenience foods and the large production volumes which are forecasted to rapidly increase, the adoption of such responsive demand management framework will not only provide significant financial benefit, but also support a sustainable approach to food production through reduction/elimination of waste.

## 5. REFERENCES

Bell, R., R. Davies and E. Howard (1997). The Changing Structure of Food Retailing in Europe: the Implications for Strategy. *Long Range Planning*, **30**, 853- 861.

Gargouri, E., S. Hammadi and P. Borne (2002). A study of Scheduling problem in agro-food manufacturing systems. *Mathematics and Computers in Simulation*, **60**, 277-291.

Mason-Jones, R., D. Towill (2000). Coping with Uncertainty: Reducing "Bullwhip" Behaviour in Global Supply Chains. *Supply Chain Forum* **1**, 40- 45.

Metters, R. (1997). Quantifying the bullwhip effect in supply chains. *Journal of Operations Management*, **15**, 89-100.

Nakhla, M. (1995). Production Control in the food industry. *International Journal of Operations & Production Management*, **15**, 73-88.

Pegg, A. (1999). Shelf Life. *Nutrition and Food Science*, **99**, 131-135.

PREACTOR (2002). User Manuals. Preactor International Limited, Chippenham.

Robson, I. and V. Rawnsley (2001). Cooperation or coercion? Supplier networks and relationships in the UK food industry. *Supply Chain Management: An International Journal*, **6**, 39-47.

Simul8 (2001). User Manuals. Simul8 corporation, Boston.

Towill, D. (1996). Industrial dynamics modelling of supply chains. *International Journal of Physical Distribution and Logistics Management*, **26**, 23-42.

Van Donk, D.P. and P Van Dam (1998). Structuring complexity in scheduling: a study in a food processing industry. *British Food Journal*, **100** 18-24.

ELSEVIER

IFAC

PUBLICATIONS
www.elsevier.com/locate/ifac

# AN EFFICIENT SIMULATOR FOR THE PROFIBUS-DP MAC LAYER PROTOCOL

**C. P. Antonopoulos**
*Research Fellow*
*Department of Electrical & Computer Engineering, University of Patras, Greece*
*cantonop@ee.upatras.gr*

**V. Kapsalis**
*Researcher*
*Industrial Systems Institute, University Campus, University of Patras, Greece*
*kapsalis@isi.gr*

**S. A. Koubias**
*Assoc. Professor*
*Department Of Electrical & Computer Engineering, University of Patras, Greece*
*koubias@ee.upatras.gr*

**Abstract**

The main goal of this paper is to develop an integrated general-purpose simulator for the Profibus MAC layer protocol in order to evaluate the steady-state behavior of this protocol, under a variety of possible operational conditions in a faulty communication channel. This protocol is part of the Profibus-DP system used widely as a standard and popular industrial communication system. This work examines deeply the performance of the protocol taking into account all critical system parameters and focusing on the main control (token/polling) and data communication (request/response) transactions. Using this simulator it can be shown how the data packet transfers, as well as the control packet transactions are strongly affected by the channel faults, especially for complex and expanded network topologies. However, under certain operational conditions the network performance can be acceptable. *Copyright © 2004 IFAC*

Keywords: Fieldbuses, Industrial networks, Wireless networks, Performance evaluation, Simulation

## 1. INTRODUCTION

The Profibus is the German standard and part of the European Standard CENELEC EN50170 for fieldbus networks capable to meet hard real time demands of a variety of industrial applications.

During the last years, a lot of work has been done, concerning mainly the Profibus-FMS system as well as the ring stability of the Profibus protocol under zero external load. (Willig, 2002), (Willig, 2003), (Willig, 1999b), (Willig, 2001), (Willig, 2002), (Tovar, 1999).

In this study we assume either a wireless medium (Gilbert-Elliot error model) or a wired channel (fixed error probability). Additionally, we are mainly concerned with the steady state functions, where external load is present and all user data are exchanged, while related research work is concentrated mainly at the control function. We also study the effect of parameters that have not been examined thoroughly, such as Slot Time, Length packet ratio and a wide rage of different network topologies.

The paper is structured as follows. In Section 2 the simulation model is described along with some of the most critical parameters, while the simulation results are presented and analyzed in Section 3. Finally, in Section 4 the main conclusions are presented.

## 2. THE SIMULATION MODEL

In order to examine the real-time performance of the Profibus MAC layer protocol we developed a detailed simulation model, taking into account all critical time parameters concerning the data transfer procedures. This model is developed using Microsoft Visual Basic 6.0 development environment so as to build an analyzing tool through which the user can control and change a great number of network parameters and derive estimates of network performance measures such as mean network and station throughput, mean network and station packet delay as well as packets dropped. Also the fluctuation of the above measures can be observed during the simulation.

### 2.1. The simulation algorithm

Wanting to focus on the data transfer procedures and how they are affected by a number of network parameters we assume that the token-ring formed by the master stations, is always stable and there are no stations leaving or entering that ring.

The most critical functions concerning the data transfer used by the Profibus MAC protocol is the polling cycles performed by the master stations towards their assigned slave and of course the token passing between the master stations.

Figure 1 presents the master station's life cycle, consisting of four states, in which the station enters periodically and with certain order within its life.

Inside each state the station must follow certain procedures, which are well defined by the Profibus protocol.

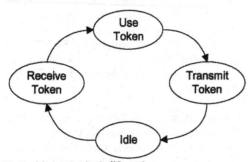

**Fig.1.** Master Station's life cycle

It is very important to mention two events that are independent of the algorithm and are checked every bit-time. These are: the creation of a packet at any of slave stations and the occurrence of an error.

### 2.2 Model Parameters

In order to achieve an accurate simulation of the data transmission part of the Profibus protocol we took into account all relevant network parameters, which can be configured by a user, through a friendly user interface, in order to simulate any possible network setting and to get a clear view of the performance affection due to these parameters. The most critical parameters the user can control are: Station Delay Time, Idle Time, SlotTime, Real Rotation Time, Target Rotation Time, Token Holding Time, Network workload, Packets length, Number of stations, Error model and Bit Error Rate.

## 3. SIMULATION RESULTS AND EVALUATION

### 3.1 Simulation Assumptions

The simulated performance of the Profibus protocol is evaluated through indicative Mean Throughput and Mean Normalized Packet Delay vs. Network Workload characteristic curves, taking into consideration a number the system's critical parameters.

The selected three simulation scenarios are the following.

Scenario 1: One master controls 10 slaves

Scenario 2: Two masters, each one controls 5 out of 10 slaves

Scenario 3: Five masters, each one controls 5 out of 25 slaves

The rest system parameters are defined as for a typical Profibus network.

In the next paragraphs indicative simulation results based on the previous simulation scenarios, showing the influence of important system parameters are presented.

### 3.2 The Effect of the Network Topology

Figure 2 presents Mean Throughput and Mean Delay curves respectively for B.E.R.=$10^{-5}$ and Polling/Reply=1/1. Figure 3 shows the same curves forPolling/Reply=1/5.

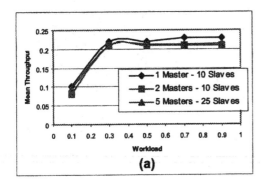

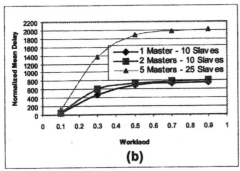

**Fig.2.** Mean Throughput and Delay for Polling/Reply=1/1, B.E.R.=$10^{-5}$

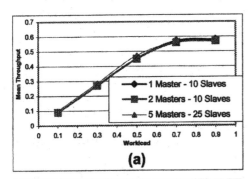

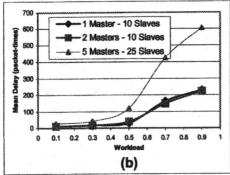

**Fig.3.** Mean Throughput and Delay for Polling/Reply=1/5, B.E.R.=$10^{-5}$

In order to examine the sole effect of network topology we considered a transmission medium

that is not affected by errors as the B.E.R. indicates. It is evident that network topology affects mainly the packet delay. Especially for Polling/Reply = 1/1, even for lightly loaded network (workload between 0.1 and 0.3) there is a vast difference in favor of the simpler network. Furthermore, this is the workload range where the highest increase rate is observed. After that range the increase rate drops indicating that the network is near the saturation point and finally reaches its highest value for heavily loaded network (workload above 0.7). The situation improves when we use polling/reply = 1/5. We can comment that mean delay difference between the three topologies is much smaller even for medium workload (up to 0.5). More interesting is that the increase rate for workload range 0.1-0.5 is very small as well, which is indicative of a stable network that handles all data transfer requests. However for workload above 0.5 there is a performance degradation which is more pronounced for the complex topology as the mean delay increase rate is, once more, much higher, indicating an unstable network. This conclusion is further supported from the respective mean throughput measurements, which show a much slower increase rate for workload above 0.5 reaching the highest value for workload equal to 0.7

Concerning the mean throughput, from these figures it is shown that in any case there is an upper limit for high values of the network workload. This means that there is a maximum number of data packets (max load) that the network can handle, under specific conditions (topology, B.E.R., Polling/Reply).

As it is shown in these figures, the system topology affects strongly the mean packet delay, especially if the topology is expanded and complex.

For simplicity reasons from now on we consider only two topologies. That's the cases of 1 master - 10 slaves and 5 master - 25 slaves , since the results for the case of 2 masters – 10 slaves are almost the same for the case of 1 master – 10 slaves.

### 3.3 The Effect of the Polling/Reply Ratio

In Figures 4, 5 simulation results concerning the mean Throughput and the mean packet delay vs. workload are presented for the selected scenarios (1 master-10 slaves, 5 masters-25 slaves) for Polling/Reply=1/1 and 1/5.

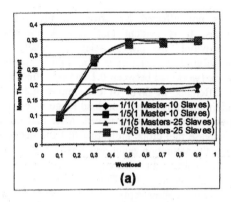

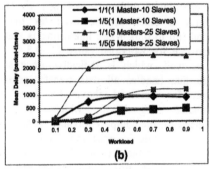

**Fig.4.** Mean Throughput and Delay for
Polling/Reply =1/1, 1/5 (B.E.R. = $10^{-3}$).

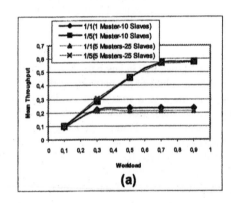

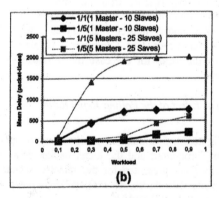

**Fig.5.** Mean Throughput and Delay for
Polling/Reply =1/1, 1/5 (B.E.R. = 0).

In these figures it is shown that the Polling/Reply ratio is a parameter that affects strongly the protocol performance for the whole range of the channel workload. Considering a transmission medium heavily affected by errors (B.E.R. = $10^{-3}$) and for all topologies there is a great performance difference for mean throughput even for 0.2 workload. At that point networks with polling/reply ratio equal to 1/1 reach saturation point which leads to the augmentation of the difference, for higher workloads, as it is shown at Figure 4a. However, even for a ratio equal to 1/5 the highest throughput is reached for a workload equal to 0.5, which is quite a poor performance. The effect is more intensive for mean delay where, especially for complex networks, there is a rapid increase even for a low workload when a ratio of 1/1 is assumed, while for a ratio equal to 1/5 the delay remains low for up to a workload equal to 0.4. An considerably better performance occurs when an ideal, from an error point of view, transmission medium is considered. Figure 5a shows that the mean throughput increases linearly up to a workload value of 0.55 and approaches 0.6, for a packet ratio = 1/5 while for a ratio of 1/1 the maximum throughput is 0.22, independent from the topology. In other words there is a performance improvement by a factor close to 3. From the respective delay measurements we can observe that for a workload up to 0.5, the mean delay remains fairly low for all topologies, while for a higher workload it is obvious that the mean delay is clearly favored by the simpler topology. At the same time when a packet ratio of 1/1 is chosen, the mean delay presents a vast degradation. Of course it must be pointed out that these are packet time delays. Another important observation is that when a packet ratio of 1/1 is used, a very high increase rate is experienced for low and medium workloads, which are the most representative working scenarios.

### 3.4 The Effect of the Error Model and B.E.R.

As it is already mentioned two error models are considered. The first one is a simple one providing a fixed B.E.R. value, while the second is the well-known Gilbert-Eliot model.

Figure 6 presents simulation results concerning these two error models for complex topologies and several values of B.E.R

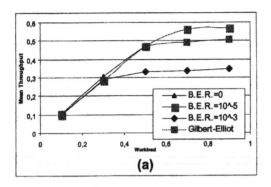

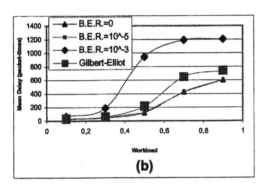

**Fig.6.** Mean Throughput and Delay for Polling/Reply=1/5 (5 Masters – 25 Slaves)

Simulations were conducted for simple topologies as well. However the results from both topologies were identical, meaning that B.E.R. is a network parameter that affects topologies the same way. Considering a wired medium (constant B.E.R.) we conclude that when that metric is equal to 0 or $10^{-5}$ the results are identical which leads to the conclusion that a B.E.R. $= 10^{-5}$ causes very few errors, at least for the duration of our simulations. However when B.E.R. increases to $10^{-3}$, which represents a typical error prone medium, the performance degrades considerably. This degradation is not obvious for low workloads because as the figures show all curves stay relatively close to each other and produce fairly good absolute measurements. When workload rises above 0.3, the degradation becomes apparent both in mean throughput, as it can't "follow" the network workload and in mean delay where it rises at a very high rate. Of course for high loads the rate drops but at that time the network is practically saturated.

In case of the Gilbert-Elliot model, a wireless channel is simulated. For the model parameters mentioned in Section 3, results are somewhere between the results for B.E.R.=$10^{-3}$ and B.E.R.=$10^{-5}$ but closer to $10^{-5}$. This is very important because according to (Willig, 2001) and

the parameters we considered, the respective mean B.E.R. is equal to $3.07*10^{-4}$ and thus someone could assume that it should be equally apart from $10^{-3}$ and $10^{-5}$, but that is not the case. An other important observation is that Gilbert-Elliot model curve has a form analogous to the form of the $10^{-5}$ and not $10^{-3}$, so there is much more "smoother" and linear increase, concerning both mean throughput and mean delay, which also being concluded by the fact that the difference compared to the $10^{-5}$ curve is always constant.

### 3.5 The Effect of Slot Time

Figure 7 presents simulation results concerning the effect of SlotTime parameter on the network performance. The throughput and delay performance is evaluated for low and high values for SlotTime (300, 900, 1500 and 2700 bit times), while errors are produced with B.E.R.=$10^{-3}$ as it was depicted by our simulations that SlotTime affects strongly the network performance under faulty conditions. In all cases the 5 master – 25 slaves topology is considered, assuming Polling/Reply=1/5.

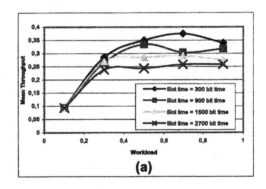

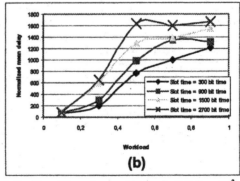

**Fig.7.** Mean Throughput and Delay for B.E.R. = $10^{-3}$

For B.E.R.= $10^{-3}$, which could represent an error prone wireless communication medium, there is a gradual delay performance degradation, that is a considerable extra packet delay, as SlotTime is increased. There are analogous results for mean throughput, which also exhibits gradual throughput degradation for high SlotTime values. Also it is expected that the effect of this parameter is going to be more apparent when more stations are present at a network.

According to the presented simulation results, it is obvious that for realistic cases, where a complex network topology exists and channel faults occur as in the case of a wireless medium, there is a serious delay and throughput performance degradation of the Profibus MAC protocol, especially for high values of the Polling/Reply ratio and high workloads. Therefore, the real-time response of this protocol is poor, under these operational conditions.

## 4. CONCLUSIONS

The main goal of this work is to present the development of a simulation model for the Profibus MAC layer protocol focusing on the user data transfer procedures. Additionally using this general purpose simulator the behavior of the Profibus MAC-layer protocol is evaluated. This evaluation is based on the study of the steady-state operation of the mentioned protocol, varying the critical system parameters. The overall outcome of this study is that this protocol behaves poorly when a faulty (wired or wireless) communication channel is considered. One of the main reasons is that the errors affect strongly both the control and data transfer procedures of this protocol. An equally important reason is proven to be the packet length ratio. A polling packet comparable with the data packet influences negatively the overall performance. Also, complex topologies contributed considerably at the performance degradation. However the detailed analysis of the graphs produced by the proposed simulator indicated that there are certain conditions where performance can be considered tolerable or even fairly good. This is very important as it leaves room to improve the performance of a Profibus network using a wireless transmission medium. This, of course requires certain changes to the Profibus MAC protocol, which is a very challenging research area.

## REFERENCES

Koubias, S., (1999) "Industrial Computer Networks", *Department of Electrical & Computer Engineering, University of Patras, Greece*

Monforte S. (19/5/2000), M. Alves, E. Tovar, F. Vasques, "Main Characteristics of the Profibus Data Link Layer", *Polytechnic Institute of Porto, School of Engineering*

Tovar Eduardo (1999), Francisco Vasques, "Real-Time Fieldbus Communications Using Profibus Networks", *Industrial Electronics, IEEE Transactions on, Volume 46, No 6, December*

Willig Andreas, (1999a) "Markov Modeling of PROFIBUS Ring Membership over Error Prone Links", *Technical University, Telecommunications Networks Group, Berlin May*

Willig Andreas, (1999b) "Analysis and Tuning of the PROFIBUS Token Passing Protocol for Use over Error Prone Links", *Technical University, Telecommunications Networks Group, Berlin March*

Willig Andreas, Adam Wolisz, (2001) "Ring stability of the PROFIBUS token-passing protocol over error-prone links" *Industrial Electronics, IEEE Transactions on , Volume: 48 Issue: 5 , Oct., Page(s): 1025 -1033*

Willig Andreas, (2002a), "Analysis of the PROFIBUS token passing protocol over wireless links", *Industrial Electronics, ISIE 2002. Proceedings of the 2002 IEEE International Symposium on , Volume: 1 , 8-11 July 2002, Page(s): 56 -60, vol.1*

Willig Andreas, Andreas Kopke, (2002b) "The adaptive-intervals MAC protocol for a wireless PROFIBUS", Industrial Electronics, 2002. ISIE 2002. Proceedings of the 2002 IEEE International Symposium on , Volume: 1 , 8-11 July Page(s): 61 -66 vol.1

Willig Andreas, (2003) "Polling-based MAC protocols for improving real-time performance in a wireless PROFIBUS", *Industrial Electronics, IEEE Transactions on , Volume: 50 Issue: 4 , Aug., Page(s): 806 -817*

European Standard EN 50 170 Volume 2, (1998) "PROFIBUS Specifications- Normative Parts of Profibus –FMS, -DP, -PA", March

PROFIBUS, (1999) "Technical Description", PROFIBUS Brochure, Order No. 4.002, September

ELSEVIER
IFAC
PUBLICATIONS
www.elsevier.com/locate/ifac

# PROGRAMMMING MICROASSEMBLY PROCESSES IN VIRTUAL REALITY

**Marc Seckner, Jochen Schlick, Detlef Zuehlke**

*Institute for Production Automation (pak)*
*Kaiserslautern University of Technology*
*P.O. Box 3049, 67653 Kaiserslautern*
*Germany*

Abstract: The generation and programming of reliable automated microassembly processes is still a major challenge. In this paper a new method for the off-line programming of microassembly processes using a virtual model of the microassembly cell is introduced. Off-line programming with the help of three-dimensional virtual models is a well known technique in macroscale manufacturing. Due to the specific attributes of the microscopic scale off-line programming techniques known from automated assembly processes cannot be applied in this field. The major aspect in microassembly is the use of sensor based compensation methods. In order to use off-line programming in microassembly those sensor techniques have to be integrated into the assembly program. Therefore a common off-line programming tool is enhanced for microassembly purposes. An assembly of a miniature gear is used to demonstrate the capabilities of the realized system. *Copyright © 2004 IFAC*

Keywords: Manufacturing, microsystems, off-line programming, process automation, production, simulation.

## 1. INTRODUCTION

Daily life becomes infused stronger and stronger by micro system products. The spectrum reaches from air bag sensors in vehicles over print heads for inkjet printers up to medical devices for minimalinvasive surgery. According to a market analysis of the NEXUS Task Force is the micro system technology subject to a growth of about 20 percent per year and will reach a worldwide volume of 68 bn. US dollars in 2005 (Wechsung, 2002).

Up to now most micro system products are designed monolithically, i.e. one part contains all important components for the system function. Since micro system products become continuously more complicated and higher demands are made on them it is inevitable – from engineering as well as from economic view – to unitize the construction. Thereby different materials can be integrated into the micro system, different manufacturing methods can be used for the functional parts, variants are supported and complicated 3D structured components can be implemented (Hesselbach, J., *et al.*, 2004). However,

a so-called hybrid construction demands an assembly which involves differentiated problems in the micro world (Zuehlke, D. *et al.*, 1997). By the low part dimensions and weights micro assembly is not comparable with assembly in macroscopic scale. Adhesive forces exceed weight forces and manufacturing as well as position tolerances are significant. The contamination of a part with slightest dirt particles has a lasting effect to its behavior. The structural fragility of micro parts is hardly to estimate and can lead to nearly invisible damages.

## 2. DIFFICULTIES IN THE PROGRAMMING OF MICROASSEMBLY PROCESSES

While the technical equipment available nowadays is – particularly by the integration of sensors – on the edge of fulfilling all requirements for a precise microassembly, the generation and programming of a reliable automated assembly process is a major challenge. Teach-In methods are approved in macroscopic assembly. They reach their limits in the microassembly on account of the tiny size of assembly objects, their fragility and the special constraints of the micro world. While positions with a tolerance frame of some 100 micrometers cause no problems when teaching, high-precision joining operations with tolerances of some micrometers need positioning procedures based on sensor information. Most of these are based on industrial image processing (Hoehn, 2001; Reinhardt, *et al.*, 2001; Hankes, 1997), on the evaluation of joining forces (Hankes, 1997) or on the internal function of the micro system. An example for function based positioning is the positioning of optical fibres (Hummelt, 2003) on a diode where the coupling maximum marks the optimum position.

Programming of such processes is characterized by numerous test runs with which position servo loops are parameterized, open loop procedures are tested and the complex program code is debugged. The frequently restricted number of parts which are available for test purposes often causes contamination and damages during the test runs. Thereby the object attributes change in a way that no reproduceable behavior exists any more. The attuned parameters are not optimal for a process with new objects. Thus programming an assembly process with high-precise positions is nearly impossible.

Another problem of on-line programming methods is the production outage of the assembly line for a long time when a new or changed assembly process is implemented. During programming and testing assembly tasks on the production line it cannot be used for other purposes. This causes high idle time costs by the unused capacity. Testing positioning methods based on sensor information requires additional expenditure. This leads to longer down-times. Besides, the handling devices can get damaged or decalibrated by unexpected collisions during test runs.

## 3. OFF-LINE PROGRAMMING AS A SOLUTION OF THE DIMENSION AND THE IDLE TIME PROBLEM

Off-line programming with the help of three-dimensional virtual models is a technology used commonly in a lot of robotics applications. With off-line programming the mentioned problems can be avoided. Positions can be acquired and defined with arbitrary precision by the free scalability of the simulation world. A hundred percent observableness of the assembly cell can be reached by a free view choice and by the possibility of fading out periphery and assembly cell components. Assembly processes can be repeated arbitrarily often without changing the physical attributes of the handling devices or parts to be assembled by unexpected contaminations or collisions. Besides, in the programming phase the assembly line is not needed and can therefore be used up to the implementation of the new assembly program.

Test runs as well as optimization cycles run completely in the virtual world. This provides an optimized assembly process which minimizes the risk of a periphery and component damage. In terms of temporal and economic aspects the best possible clock speed can be adjusted.

The off-line programming is applicable in microassembly by considering specific aspects.

## 4. SENSOR APPLICATION IN OFF-LINE PROGRAMMING OF MICROASSEMBLY PROCESSES

The essential difference between off-line programming in macro and in micro scale is the fact that the virtual model of the assembly cell does not comply precisely with reality in the micro world. In customary off-line programming methods divergences of some 10 micrometers do not play a role, because these are far below the permissible assembly tolerance. In microassembly the demanded tolerance is mostly lower than the divergence between the model and reality. Many narrowly tolerated positions must be determined by sensors during the real assembly process. However, these sensors must be implemented in the simulation model for off-line programming. Examples of sensors commonly used in microassembly are image processing systems for the localization of components and reference marks, distance and force sensors for relative and test signals for function based positioning.

Present off-line programming software is not designed for the application of sensors. Hence, the institute of production automation (pak) at the Kaiserslautern University of Technology enhanced the simulation software IGRIP® of the Delmia Corp. for microassembly purposes. Tools and GSL[1] macros were built which enable the simulation of sensors with their specific parameters and also receiving data of real sensors. The latter serves the synchronous cycle run of simulation and the real assembly line as well as process monitoring.

In order to simulate a sensor a GSL routine is generated which returns the expected result of a measurement plus a measuring error. In this routine the measurement range, the statistical distribution of the return values and the superposed statistical measuring errors are encoded. When starting a sensor routine it returns random values considering the physical properties of the corresponding real sensor. Therefore no physical model of the sensor is required, but the tolerances detectable with the sensor and the expected return values are modeled. Thereby the programming of sensor hardware is completely avoided in the simulation. Of course this is required

---

[1] GSL (Graphical Simulation Language): IGRIP® specific programming language

on the real assembly line with the compilation of the assembly program. Consequently the sensor in the real process can be exchanged with another one without adapting the simulation system as long as the exchange device behaves equally. Besides, this simplifies the translation of the assembly program to the programming language of the real robot control.

## 5. IMAGE PROCESSING SYSTEMS AS SENSORS

In the field of sensors used in microassembly image processing systems have a special importance. They deliver information about the position of components, geometrical figures or the surface structure. Due to the predominant Pick and Place procedures in microassembly image processing is used mainly to find certain component structures, as for example the outside contour or drillings. Therewith approximately known positions and orientations of components are specified.

Hence, in a simulation model image processing systems are simulated by three-dimensional sensors. Such a sensor returns a vector which contains the x- and y-coordinate of the center of the detected structure as well as its orientation. This vector refers to the field of view of the camera. So the values must be transformed into the robot coordinate system. To accomplish this transformation, the pixel to mm-ratio, the precise position of the field of view in the robot coordinate system and the perspective distortion of the camera must be known. Therefor the image processing system has to be calibrated with purpose-built objects. The calibration data is stored in a file where the image processing unit and the developed simulation tools have access to.

Whether the camera is mounted on an axis of the robot or is connected statically with the robot base the position vector of the field of view saved in the calibration file belongs to the associated axis or to the absolute robot coordinate system. The absolute position of a camera mounted on the robot is calculated after every robot movement.

To have the possibility to monitor and align the virtual with the real world, image processing systems

are implemented in the simulation model not only fictitiously by several routines. The field of view of the simulated camera is modeled by the physical properties of the camera and the lens read out from the calibration data. Besides, a window is opened which shows the virtual environment from camera view. Fig. 1 shows the simulation environment with the modeled field of view and the camera view window.

## 6. REALIZATION OF THE ASSEMBLY OF A MINIATURE GEAR BOX

Considering as example the assembly of a miniature gear box the concept of using off-line programming in microassembly is realized and validated. The gear box to be assembled is a miniature planetary gear with a diameter of 6 millimeters (Fig. 2). The assembly of one stage of the gear consists of mounting three planet wheels on their respective bearing pins on the planet carrier. For the further assembly two possibilities exist. First, the planet carrier of the following stage can be mounted on the completed stage. Then the planet wheels are put on this stage. Second, the next stage can be assembled separately. Then the complete stage is mounted onto the previous one. Mounting a planet carrier means to insert the sun wheel into the toothing of the planet wheels. The sun wheel is part of the planet carrier and is located in its center. Using these methods several stages can be assembled. The final stage consists of a planet carrier with an output shaft instead of a sun wheel as well as the planet wheels that are supported by the bearing pins of the planet carrier. In order to realize this assembly a tong gripper is used. The one that is used here is developed at the institute of production automation (Schlick and Zuehlke, 2003).

The position tolerance when assembling the planet wheels onto the bearing pins is 10 micrometers at maximum. This is due to the fact that there are no insertion chamfers either on the planet wheels or on the bearing pins. The lack of insertion chamfers results from the fabrication methods of those parts. The assembly situation represents a classical peg-in-hole problem. Neither the planetary wheels nor the planet carrier can be fed with this accuracy. Additional tolerances result from the handling device

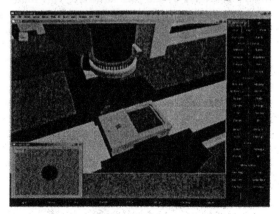

Fig. 1. Virtual image processing system in the simulation tool IGRIP®

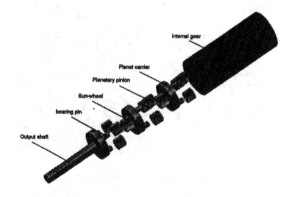

Fig. 2. Miniature planetary gear of Maxon Motor AG

and the gripper where positioning tolerances of several micrometers are inevitable. So the gear box has to be assembled using sensor based positioning strategies. The exact positions of the planet wheels and the bearing pins are unknown. The position of the bearing pins results from the orientation of the planet carrier. So the orientation of the planet carrier, the position of its center as well as the position of the planet wheels have to be acquired by analyzing sensor data. In order to acquire the position of a planet wheel the camera is first moved to a picture taking position that is approximately known by teaching. The picture is then searched for a circular blob that identifies the center bore hole of the planet wheel. The exact center position is calculated by identifying the center of gravity of the blob. The orientation of the planet carrier is calculated by focusing on the outer contour and taking advantage of the symmetry of the bearing pins (Fig. 3).

In the simulation environment the examination of the pictures features isn't necessary. Positions and orientations of all simulated objects are known. The tolerances are simulated by varying the positions and orientations of the parts randomly. The simulated parts change their position and their orientation when the simulated sensors are scanned. The magnitude of these changes depends on the ranges of the corresponding sensor. This behavior is comparable with a real system where the exact position of a part can only be estimated before the sensor is scanned. The position of the part is introduced to the robot control by scanning the sensor. After this the position is still afflicted with an uncertainty that corresponds to the measurement error of the sensor. So using sensors only helps to precise the information about parts positions.

In order to realize this assembly using off-line programming the complete assembly cell is modeled in IGRIP®. The kinematics of the robot and the gripper are implemented. Precise modeling of the kinematics is essential for a realistic simulation. This model is able to do every action the real assembly cell is able to.

The simulation of the sensors is realized using GSL routines. In the realized prototype these routines are generated by special software called "Sensor Routine Simulator" (SRS). The software generates the GSL

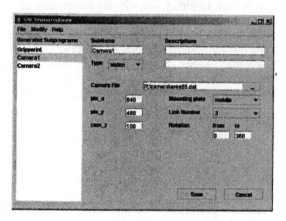

Fig. 4. Software tool for generating sensor routines in GSL

source code using the number of values in the return vector, the ranges and the distribution function of each value in the return vector of the sensor simulation routine. Fig. 4 shows a screenshot of the SRS software.

The simulated sensors are adapted to the real ones by modifying the above mentioned parameters. So the simulated sensor and the real one behave similar in statistical means. The software tool is able to deliver sensor routines for the image processing system, multidimensional force-torque sensors, distance sensors and other peripheral sensors integrated in the assembly cell.

In the realized assembly the image processing is the only sensor. The assembly program consists of the procedures putting the planet carrier into a holding device and assembling the three planet wheels onto the bearing pins. The gripper's maximum opening width is about two millimeters. However the planet carrier's outer diameter is 4.6 millimeters. So the planet carrier is gripped at a bearing pin instead of the lower disc. This is an eccentric gripping situation. In order to position and to orient the planet carrier correctly the exact position of the bearing pin and the orientation of the planet carrier are necessary. The planet wheels can be gripped at their outer boundary.

The simulation of the image processing is done by a routine generated by the SRS software. This routine returns the position and the orientation of parts when called by the assembly program. Using this routine the planet wheel's location and the location of the planet carrier are determined. The part to be located has to be visible in the camera window. The position of the camera window is determined by the mounting of the camera on the robot and the parameters of the camera's lens. These parameters are assessed by the camera calibration procedure. In order to have a precise simulation model these parameters are read prior to the simulation and the model is adjusted accordingly. The robot has to move to a certain position to make sure that the parts are in the camera window. Inaccurate positions of the bearing pins and the planet wheels are rendered more precisely by the image processing routine.

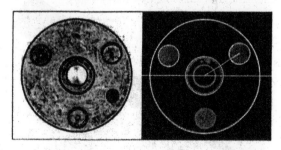

Fig. 3. Orientation calculation of a planet carrier with an image processing system (left side: original grabbed image, right side: calculated binary with graphic overlay)

The completed assembly program can be run several times to debug and to optimize it. During this testing phase in the simulation environment the parameters of the positioning methods can be easily adjusted.

In order to transfer the working assembly process from the simulation environment to reality a special postprocessor is used. This software translates the assembly program written in GSL to the V+ language used by Adept controller. The translation starts with the main program. During the translation the main program is scanned for subroutine calls. Some of these subroutines have to be translated, too. Some of them mustn't be translated. It concerns especially sensor routines and routines controlling the simulation system. Sensor routines are exchanged with associated V+ routines that scan the real word sensors. These real world routines control the sensor hardware. Thus they are very specific to a certain setup of the assembly cell. They are contained in a runtime library that is written for the unique setup of the cell. In case of the prototype the sensor and analysis routines for the image processing have to be exchanged.

The initial operation is done using a synchronous application flow of the simulation and the real word assembly program. Thereby the real sensor data is fed back to the simulation. So the correctness of the assumptions about the sensors behavior can be verified and parameters can be adjusted. In the last step of the transfer the process is decoupled from the simulation.

## 7. CONCLUSION AND OUTLOOK

Hybrid micro systems open up a variety of applications by combining different materials, fabrication methods and complete subcomponents. However combining different components means assembly. Thus a complex and error prone process is in the back end of the production line. While nowadays commercial assembly devices are available and sensor based positioning methods are known, programming the handling devices and sensors is still a complex, time-consuming and thus cost intensive task. However programming devices is a major part when automating processes. Off-line programming using three dimensional virtual models offers a lot of advantages considering the implementation of the control software.

Major components of automated microassembly processes are sensor based positioning methods. They are used when high demands to the positioning accuracy exist. Using off-line programming in microassembly, these sensors have to be included in the simulation. In this concept the integration of sensors into the simulation is done using a qualitative model of the respective sensors. In this model the statistical measurement error and the measurement range are included. So sensor based positioning methods can be simulated.

At the institute for production automation at the Kaiserslautern University of Technology an existing graphical off-line programming system has been enhanced for the simulation of microassembly processes. The assembly of a miniature gear box has been implemented using this programming system and has been translated to work at the real assembly cell. The usability of the system has been verified.

The system showed up to be very useful to implement microassembly processes. By using the developed tools the integration of sensors into off-line generated processes needs not much more time than programming a process without sensors. The sensor routines can be handled like other subprograms.

Due to the usefulness of the system several enhancements are worked out at present. Current work packages include the implementation of an active force based positioning method and a positioning method based on test signals. The latter one is often used to adjust lenses and fibers in the assembly of micro optical systems. These work packages enhance the system and open up a broad range of processes that can be set up using off-line programming with three dimensional virtual models.

## REFERENCES

Hankes, J. (1997). *Sensoreinsatz in der automatisierten Mikromontage*. PhD Thesis, University of Kaiserslautern, Fortschr. Ber. VDI **2 Nr. 459**, Düsseldorf, VDI

Hesselbach, J.; *et al.* (2004). International State of the Art of Micro Production Technology. In: *Production Engineering* **Nr. 1** (2004), Berlin, WGP

Hoehn, M. (2001). *Sensorgeführte Montage hybrider Mikrosysteme*. PhD Thesis, Forschungsberichte iwb **Nr. 149**. München, Utz

Hummelt, C. (2003). Making Connections: Real World Telecom Component Production. In: *Photonics Spectra*, **Nr. 9**, pp. 82 - 86

Reinhart, G., Jacob, D. and Fouchier, M. (2001). Automated Assembly of Holder Chips to AFM probes. In: *Proceedings of SPIE: Microrobotics and Microsystem Fabrication*, **Vol. 4568**, Newton MA, USA, 29.-30.10.2001

Schlick, J. and Zuehlke, D. (2003). Development of adaptive grippers for automated microassembly. In: *Proceedings MIRCO.tec 2003*, Munich, Germany, 13.-15.10.2003. Berlin: VDE

Wechsung, R. (2002). Market Analysis for Microsystems 2002-2005 – A Report from the NEXUS task force. In: *mstnews* **No. 2/02**, pp. 43-44

Zuehlke, D.; Hankes, J.; Fischer, R. (1997). Stepwise into microassembly, In: *Proceedings of the 8th International Conference on Advanced Robotics*, IEEE, Monterey CA, USA, 7.-9.7.1997

ELSEVIER
**IFAC**
PUBLICATIONS
www.elsevier.com/locate/ifac

# A FRAMEWORK TO REVIEW CONTINGENT INFORMATION SYSTEM DESIGN METHODS

**Virginie Goepp, François Kiefer**

*LICIA*
*Institut National des Sciences Appliquées de Strasbourg*
*24, Boulevard de la Victoire*
*67084 Strasbourg Cedex - France*
*Tel : 0033 (0) 3 88 14 47 48*
*Fax : 0033 (0) 3 88 14 47 99*
*Email : goeppvirginie@mail.insa-strasbourg.fr, francois.kiefer@insa-strasbourg.fr*

Abstract: Today, information systems (IS) and related information technologies occupy a prime position in our organisations. However, over 60% of IS projects represent a failure. In this boarder, the contingent and modular design methods, which enable to adapt the steps of the project to the context and to use several methods and techniques in a combined manner seem promising. Existing propositions emphasise on the description of such approaches, but there is no means to understand the underlying logic of them. Thus, a framework to do this is proposed. It enables to review eight existing methods and to propose and discuss evolution perspectives. *Copyright 2004 IFAC*

Keywords: information systems, information technology, design, models, project management.

## 1. INTRODUCTION

Today, information systems (IS) and related information technologies (IT) occupy a prime position at all levels in our organisations. Through up the years, a lot of methods and tools were proposed in the IS design field. In (Martin, 1984) they are defined as "a methodological jungle". In spite of this variety and number of propositions, over 60 % of IS projects represent a failure in terms of exceeding the budget, the deadlines and also in terms of unsatisfied requirements.

In this boarder, the contingent and modular design methods, which enable to adapt the steps of the project to the context and to use several methods and techniques in a combined manner seem promising. Existing propositions emphasise on the description of such approaches, there is no means to understand the underlying logic of them. However such means are essential to analyse the strengths and lacks of the existing approaches and to propose ways to improve them. The target of this communication is to propose a framework, which enables to characterise the contingency mechanism.

The second section deals with the notion of contingency in the IS design field and with a brief description of eight related methods. In the third section a framework based on this state of the art is proposed. This framework focuses on the underlying logic of contingency methods in the IS field. In the fourth section the framework is applied to review the previously presented methods. Then the last section is dedicated to analyse them. This analysis enables to suggest and discuss two directions for further research.

## 2. STATE OF THE ART OF CONTINGENT INFORMATION SYSTEM DESIGN METHODS

Before describing existing contingent methods in the IS design field, it is interesting to come back up to the contingency notion. This paper is not particularly concerned with the wider debate on the contingency approaches to organisational management in general; rather, it focuses on the contingency notion for IS design. Its definition in this particular field explains partially the existing research directions.

### 2.1. Contingency notion for information system design

In the IS design field different definitions for the contingency notion exist. In (McFarlan, 1981) the contingency principle emphasises on the fact that there is no better method in isolation but methods more or less adapted to a project or a context. Similarly in (Zhu, 2002) it is stated that the search for contingency approaches to IS design began when it was recognised that (1) there is no single best methodology for all IS design projects/situations and (2) there exists a variety of methodologies to select from.

In other words, the contingent IS design methods will focus on the selection of the most appropriate methodology or a set of methodologies to fit a particular case. These methods deal with the ongoing search for flexible, rigorous and workable design approaches. Even if the method adaptation is advocated, a detailed analysis will show that sometimes the existing contingent approaches propose also to adapt the sequence of the project activities.

After clarifying the main objective of contingent IS design method, an overview of the principal propositions is made. It shows the variety of such approaches and the need for means to analyse their underlying working principles. The first propositions like (Gorry et al., 1971), (Naumann et al., 1978) or (Naumann et al., 1978) are not described here. Indeed, they remain quite "basic". They classify the kind of system to design and the project context according its complexity and uncertainty. Their main assumption is then to match these situations with corresponding macro development strategies like prototyping, for example.

### 2.2. An overview of the main contingent information system design methods

This overview consists of the brief description of eight contingent approaches. Each description follows the following structure: objective, proposed approach and synthesis.

*Method engineering:* method engineering is a discipline to conceptualise, build and adapt methods, techniques and tools for the development of IS (Brinkkemper, 1996).

Method engineering, according to (Ralyté, 2001), proposes three main strategies to build a particular method:
- "ad hoc": the method construction is based on experience,

- reference meta-model instantiation: the method construction is based on the reference meta-model of existing methods,
- method fragment assembly: the method construction results from the assembly of method fragments. This strategy supposes to describe the method fragments and to store them in a so-called method base.

Method engineering as it was defined previously focuses on method construction with proposition of method fragment models, for example. The interaction with organisational project aspects of IS design is reduced here to the transformation of the built method model into an operational method and its integration into the design project. This interaction remains limited and explains the apparition of researches, which are defined here as "extended method engineering". These works like (van Slooten et al., 1996) or (Punter et al., 1996) deal, in parallel, with the strictly method part and with the so called "project strategy". Because of this difference, these kind of approaches will be presented separately.

*Euromethod:* the European project Euromethod (General Direction III of European Commission, 1996) has been designed to help organisations with the acquisition of effective IS and related services in a variety of situations. It encourages customers and suppliers to control costs and timescales, to manage risks and helps to improve mutual understanding.

For Euromethod an acquisition process consists in a set of decision points. A decision point is a milestone where the customer, possibly together with the supplier, is making decisions on the acquisitions. A decision point is characterised by the decision that are made and the deliverables that are exchanged. The following elements to build an adapted acquisition process are provided:
- 40 evaluation criteria of the project situation in terms of complexity and uncertainty
- 8 development strategies options: each strategy consists of a set of typical decision points
- a set of heuristics to choose the strategies, which are best adapted to the project.

Euromethod takes place in a customer/supplier situation. The choice between the proposed development strategies is based on a quite complete qualitative analysis of the project. This approach recommends no particular method. The recommendations are confined to the general principles of description, construction and control of the IS project.

*V-Model:* the V-Model (General Directive 250, 1997) is a development standard for IT systems which uniformly and bindingly lays down what has to be done, how the tasks are to be performed and what is to be used to carry this out.

The V-Model encompasses:
- the lifecycle process model: it contains binding regulations concerning work steps to be performed (activities) and results (products). This model is structured around 4 sub-models: system development, quality assurance, configuration management and project management. The tailoring process enables to adapt the process model to a specific project.
- the allocation of methods: this process enables to determine what methods are to be used to perform the activities established through tailoring. The V-Model includes 47 basic methods.
- the functional tool requirements: at this level the functional characteristics of the tools, which are to be used during system development are recommended.

The V-Model includes two adaptation mechanisms. On the one hand, the tailoring through a succinct project characterisation enables to select the activities, which have to be performed. The related activity sequence is selected according to an advantage/disadvantage analysis, which remains fuzzy. On the other hand, the method allocation matches one or several methods to each activity of the process model. Here, method selection is based on activity selection.

*Kiefer:* the approach proposed in (Kiefer *et al.*, 1995) provides four design scenarios for IS of manufacturing system.

The proposed scenarios include the project members to involve, the activity sequences to be performed and the methods to be used. The scenario selection is here not based on a "classical" project characterisation but on evolution modes. An evolution mode describes the situation in which the IS under study is. To each mode corresponds a specific scenario.

The main interest of this approach lies in the use of the evolution modes to guide the scenario selection. This concept is interesting because it takes the dynamics of the system to design into account. However, contrary to the previous methods, there no way to choose between the different modes is provided.

*Morley 1:* the approach described in (Morley, 1998) proposes to select a development model according to the risks attached to the project.

This approach encompasses:
- an evaluation framework: it consists of 6 risk factors or criteria to evaluate a given project situation
- a risk profile: it is a graphical representation of the level of each risk factor for a given project.
- development strategies: these strategies consist of a set of measures and development models to treat the identified risks.

This approach is similar to the project situation evaluation in Euromethod. The main differences lie in the proposed adaptations and the number of criteria. Here, they are less numerous but enable to select a development model and some countermeasures. Moreover, the evaluation framework takes partially the dynamics of the system to design into account.

*Morley 2:* the approach described in (Morley, 1999) is not really a contingent IS design method. Indeed, it is only a set of criteria to analyse the key-features of a given project. In further research this framework should guide the methodological choices through the links between criterion values and method elements.

The proposed framework consists of 13 criteria classified into four "points of view":
- teleological point of view: it represents the purpose and stakes of the project
- sociological point of view: it enables to assess the project in its human environment
- metrological point of view: it evaluates the quantitative aspects of the project
- informatics point of view: it assesses the informatics general features of the project.

This framework is similar to the criteria of Euromethod, even if the teleological point of view is not proposed in Euromethod. Some other points are more detailed in Euromethod. The main interest of this framework is the proposed research perspective which consists in linking directly methods and project evaluation.

*Van Slooten:* the contingent IS design method proposed in (van Slooten et al., 1996) provides a formal procedure to configure development scenarios from project characterization. A complete development scenario is an association of a route map and method fragments. Route maps are development strategy plans consisting of development activities and products. Methods fragments are parts of methods, techniques and tools that can be incorporated in a route map forming a complete project approach.

Scenario configuration consists of the following stages:

- Project characterisation: it is determined by assigning values and weights to contingency factors, which are characteristics of the development situation. These factors are categorised into four "domain groups".
- Based on the project characterization, the aspects, levels, constraints, and development strategy are determined. This is related to a set of possible route maps and method.
- A first project scenario is composed by selecting the most appropriate route map and a corresponding number of method fragments. These are finely tuned to the project characterization.
- The actual project performance is started using the scenario as guide line. The scenario is further refined during the course of the project, and possibly adapted in the case of unforeseen contingencies.

This approach is interesting because the scenario configuration procedure is clear and encompasses method fragments and route maps. Moreover the method fragment selection is based on the aspects and level notions and not on the project activities. However, the number of method fragments is low and their selection is related to the development expertise.

*Zhu WSR:* This approach called Wuli-Shili-Renli or WSR for short presented in (Zhu, 2001) stands out from the previous because it admits that various issues that shape IS design interact continuously with each other in an unpredictable manner. To WSR, all real-world situations and hence IS projects can be pragmatically seen as conditioned by the continuous interplay of wuli (relations in the world), shili (relations with the mind) and renli (relations with others).

For IS projects WSR encompasses the following elements:

- wuli-shili-renli interplay: the project is based on a so called "bubble management"
- IS project operational stages: activities are arranged according to six stages with regard to a few key questions. These stages can be further sub-divided into several more manageable steps, which can be intentionally geared towards incorporating

available activity elements from various methods

- Li-stage matrix: this matrix constituted by the WSR interplay and the six operational stages guides the developers through the IS design process in a spiral manner.

This WSR approach emphasises on the unpredictable dynamic interplay of multiple dimensions, perspectives and actions and does not adopt the established "social-technical" or "soft-hard" sequence. The IS project is structured around six main stages which can be sub-divided. The methods to be used are not stipulated but participants should ideally, be able to access whatever methods, techniques and tools they consider appropriate.

## 2.3. *From the overview to the necessity of a framework*

The previous section emphasises different points. The first is linked to the employed terms: the limits between the recurrent notions like methods, project and development strategy remain fuzzy. This is harmful for the good understanding of these methods. The second concerns the main target of them. Indeed according to the definitions (cf. 2.1) contingency should enable method adaptation. However Van Slooten (1996), focuses on methods fragments and route maps, which concern the main activities to be performed during IS design. Morley 1 only tackles the project activities and not the method aspect.

Moreover, all these researches focus on the description of the ways those methods and project activities should be adapted. Apart from the work done in (Zhu, 2002), where contingent IS design methods are classified and assessed from an operational point of view, there is no means to understand their underlying logic. However such means are essential to analyse them and better understand their strengths and lacks. So in the next section a framework to describe these methods from a theoretical point of view is first presented. Then in a second time the proposed framework is compared with the classification of Zhu (2002).

## 3. A FRAMEWORK TO REVIEW CONTINGENT INFORMATION SYSTEM DESIGN METHODS

This communication focuses on flexible, rigorous and workable IS design approaches. To tackle the analysis of the contingent IS design methods a framework concerning theoretical contingency elements for the IS field is presented. It is based on the previous definition of contingency and on the method overview.

### 3.1. "Level" and "type" or the Information System/ Methods/ Project triptych

The contingency principle deals with guiding method selection and their adaptation during IS design. Even if the first objective is method adaptation, the previous in depth analysis tends to show that it is not always the case. Beyond method adaptation contingent approaches emphasises the two following questions:

- What is adapted?
- Which are the elements that guide this adaptation?

To each question will correspond an element of the framework. Each element constitutes an elementary stone required to better understand the underlying contingency mechanisms. The framework elements corresponding respectively to each questions are:

- The "level" of contingency
- The "type" of characterisation

*"Level" of contingency:* this notion refers to the adapted element during IS design by the contingent approach.

Three kind of contingency "levels" are proposed:

- The information system (IS)
- The methods to be used
- The project, that is to say the sequence of the project activities

*"Type" of characterisation:* this notion refers to contingent factors or project characterisation criteria proposed by the contingent approach.

Two kind of characterisation "types" are highlighted:

- The "unique" characterisation type: this type implies a single contingency level.
- The "various" characterisation type: this type implies several contingency level.

*Related graphical representation of the framework:* in order to make the analysis easier a diagrammatically representation of the framework is proposed in fig. 1. The levels are represented as circle and the levels as arrows. The arrow points from the characterised level towards the adapted level.

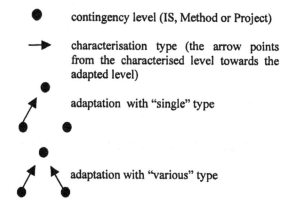

Fig. 1. Framework elements to review contingency approaches.

### 3.2. Use of the proposed framework

The proposed framework is applied to the eight methods described in section 2.2. This observation is shown in table 1. The dotted arrow concerns recommended adaptations, which are not guided and dependent only on the users. For example, the WSR approach (Zhu, 2001) recommends the use of methods whenever it seems necessary to the users but no particular method is specified.

Moreover adaptation sequences are highlighted. Sometimes the adaptation of a given level depends on the adaptation of an other level. This is the case for the V-Model (General Directive 251, 1997), where method allocation is based on project activities, whose adaptation results from tailoring. This tailoring is based on IS and project characterisation. In other words, the contingency "levels" are linked through the characterisation "type". These adaptation sequences are called "mode".

Two "modes" have been observed (cf. fig.2):

- "simple" mode corresponding to the adaptation of one level
- "composed" mode corresponding to the adaptation of a level through an other level

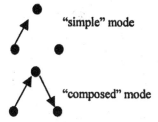

Fig. 2. Two adaptation "modes".

213

Table 1: Application of the framework to the eight studied contingency methods.

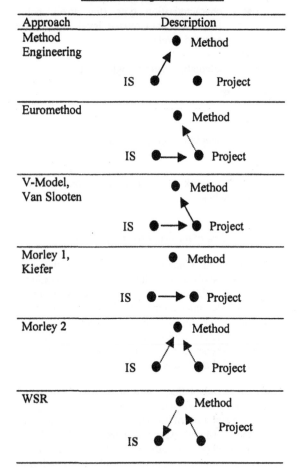

| Approach | Description |
|---|---|
| Method Engineering | |
| Euromethod | |
| V-Model, Van Slooten | |
| Morley 1, Kiefer | |
| Morley 2 | |
| WSR | |

The application of the proposed framework emphasises the six following "contingency mechanisms".

*Method contingency with IS characterisation*: this is generally the case for method engineering, where methods are adapted to a specific kind of IS. For example in (Gzara *et al.*, 2000) patterns are used to adapt the methods in product IS field. The project contingency level is not tackled.

*Project contingency with IS and project characterisation and possibility to adapt the method level:* Euromethod (General Direction III of European Commission, 1996) is classified in this category. Indeed, Euromethod provides different development and project control strategies selected according a in-depth characterisation of the situation. The proposed criteria concern the IS to design (complexity of data, complexity of target technology, …) and the project (heterogeneity of actors, size of the project team, …). After strategy selection there are two alternatives. The first consists in using the provided decision point structure. The second consists in choosing a method, which fits the selected strategies. For this process no particular method is recommended.

*Method contingency through project contingency with IS and project characterisation:* the V-Model, described in (General Directive 251, 1997) is representative of this contingency mechanism. Indeed, the tailoring enables to build an adapted lifecycle process (activities and products) thanks to a brief characterisation of the project and the IS to design. After tailoring, method allocation enables to match method fragments to the selected project activities.

The procedure proposed in (van Slooten et al., 1996) is quite the same as in the V-Model. A in depth project characterisation enables to choose a first route map. Then methods fragments can be incorporated in the route map. In this approach the method fragments are selected according to the tackled aspects (process, information, behaviour, organisation and problem).

*Project contingency with IS and project characterisation:* this contingency mechanism is proposed in (Morley, 1998) and (Kiefer *et al.*, 1995). Indeed, these approaches consist in choosing development models or development scenarios according to a situation analysis. In (Morley, 1998) this analysis is a brief qualitative analysis of the project size, technical difficulties. In (Kiefer *et al.*, 1995) the scenario selection is based on the evolution mode notion. This notion describes the situation of the IS according to its dynamics (strategic objectives, technical difficulties, people involved, …).

*Possible method contingency with IS and project characterisation:* this alternative is exposed in (Morley, 1999). It makes the assumption that it should be possible to link directly method fragments and a complete situation characterisation. However this path is only a research perspective that is not further detailed.

*Possible method contingency through project contingency with project and IS characterisation:* this mechanism is highlighted in (Zhu, 2001) with the WSR approach. Method and project contingency are recommended but there is no operational adaptation proposition. The involved project team has to sub-divide the six proposed development stages or to choose from their own experience the required methods.

# 4. ANALYSIS OF THE DESCRIBED CONTINGENT INFORMATION SYSTEM DESIGN METHODS

The use of the three framework elements (contingency level, characterisation type and adaptation mode) enables to describe contingent approaches from a theoretical point of view. A first analysis of the previous "contingency graphs" (cf. table 1) shows that even if "method contingency" is advocated "project contingency" is often proposed. This kind of adaptation takes several forms: route map, set of decision points, development model, development strategy, lifecycle process model. The following in-depth analysis begins with the simple adaptation modes and then deals with the composed modes.

## 4.1. Analysis of the "simple" adaptation modes

In the case of simple adaptation modes the characterisation type has a deep impact on the quality of the adaptation and the effective use of them. Indeed, for such modes the starting points are often restrictive: an IS type in (Gzara et al., 2000) or a particular design situation in (General Direction III of European Commission, 1996). Even if the proposed stepping seems clear it is not easy to fit perfectly the recommended limitation.

This difficulty is linked to the "single" characterisation. In this case the "various" characterisation type brings interesting solutions since the provided starting points are more realistic. Indeed, the "various" characterisation are based on a set of IS and project features. However, the adaptations are often limited to the project level like Morley (1998), who proposes different development models. Moreover this type of characterisation is generally based on factors assessing complexity and uncertainty of the situation. In this boarder, the approach of Kiefer et al. (1995) is interesting because the provided characterisation based on the evolution modes of the studied system takes the dynamics of the system into account.

## 4.2. Analysis of the composed adaptation modes

Concerning the composed modes, their analysis can be liken to the analysis of two simple modes achieved sequentially. Similarly the characterisation problem exists. Existing approaches, providing this mode type, are based on criteria whose quantification is difficult at project beginning like in the V-Model (General Directive 251, 1997). But from this quantification depends the adaptation at the project

and method levels since the levels are achieved in series.

This specificity emphasises the possible adaptation sequences. In other words which are the different adaptation orders. The "contingency graphs" analysis shows that the existing methods do not exploit all potential adaptation modes. Indeed the single existing composed mode consists in adapting the method level through the project level. This is in part linked to the predominance of the method level. This contingency level is an answer to the variety of existing IS design methods and the necessity and difficulty to select them consistently. The underlying assumption is the possibility to match project activities and products to at least one method fragments. In other words, methods fragments are selected according their potential use during specific design stages.

## 4.3. Comparison with Zhu's classification

The three elements of the proposed framework (contingency level, characterisation type and adaptation mode) provide a complete description of the underlying logic of contingent approaches. They enable to present an overview of the variety of such approaches, to surface their assumptions and to analyse similarities and differences. The proposed framework emphasises the IS / Methods/ Project triptych. These elements are essential to better understand contingency mechanisms and are complementary to the classification proposed in (Zhu, 2002).

Indeed, Zhu assesses contingency in the IS field from an operational point of view and proposes the following categories:

- Contingency at the outset: selecting a single methodology or a fixed combination of methodologies before the beginning of the project and following the same selected methodology or combination of methodologies throughout the whole project process
- Contingency with a fixed pattern: choosing techniques and tools at each individual stage of a project according to a fixed working sequence
- Contingency along development dynamics: using various methods and tools in an unpredictable manner.

This classification refers to ways to use the different contingency mechanisms during IS projects. In other words it emphasises the iteration number of the adaptation modes.

Indeed with contingency at the outset the contingency modes are applied once before beginning the project. For this class of approach method and project contingency depend generally on the type of IS to design.

For contingency with a fixed pattern method contingency is reiterated at each project stage according to the particular situation. However there is no contingency at the method level since the working sequence is linear and fixed.

Contingency along development dynamics make the assumption that various issues shape IS design. These issues interact continuously in an unpredictable manner and can only be tackled at each unique development moment rather than in a prescribed sequence. In other words this category suggest to use the different contingency adaptations whenever it seems necessary.

This classification refers particularly to the adaptation mechanisms during project progress and the difficulty to use such approaches. Even if existing contingency methods have generally to be performed at the out set, such mechanisms exist. However these are insufficient because they are not guided and remain in charge of the project team.

From the contingency point of the view the highest flexibility seems to be the best for the IS design. However if the relating processes are not guided and therefore unreliable, flexibility becomes harmful to the project. Indeed one of the risk is to focus on contingency to the detriment of the IS design progress. In other words contingency mechanisms must be improved from a reliability point of view in order to find a balance between the efforts dedicated to contingency and its efficiency.

## 5. PERSPECTIVES AND FURTHER RESEARCHES

In this communication after a brief overview of eight contingent IS design methods a framework to review them is proposed. The three proposed elements (contingency level, characterisation type, adaptation mode) enable to emphasise the underlying logic of such approaches. Such means are essential to analyse and better understand them.

Six different contingency mechanisms are highlighted through the use of the framework. Then they are analysed according to the contingency levels and the adaptation modes. Concerning the contingency level whereas the main target of

contingency is method adaptation some existing methods adapt the project activities too.

Concerning the adaptation modes the characterisation plays an important role. Actually characterisation is generally based on factors related to organisational and technical features of the situation. These features are difficult to assess at the beginning of the project and their values are likely to change during project progress. To improve contingent IS design methods it is essential to improve their characterisation. This could be done by using other characterisation concepts. These concepts must be relevant during all the project progress and enable to gather the required information for assessing existing factors. Such complementary concepts are provided in (Goepp *et al.*, 2003), where a semiotic based analysis leads to a set of key-problems. These key-problems are not linked with organisational and technical aspects of the situation but represents a set of problems which has to be solved during IS projects in general. The use of the proposed key-problems enables to analyse the situation and to build progressively the knowledge required by existing characterisations. However these links remain too loose and have to be further developed.

One other point of improvement is linked with existing adaptation modes. Indeed, actually the following composed mode predominates: IS → project → method. This mode is based on a classical characterisation of project situation and on a matching between method fragments and project activities. In other words, according to existing approaches, the full development of contingency requires to link the characterisation (criteria values, kind of representative subset) to selected project activities and then to potential method fragments. However some adaptation modes remain unexplored. This is the case of the mode which starts with method contingency and leads to project contingency trough method contingency. The corresponding "contingency graph" is shown in fig. 3. The operational use of this mode implies to link directly methods and characterisation. So method fragments can no longer be classified according to the project activities, and an other criteria must be provided. The key-problems proposed in (Goepp et al., 2003) could be useful to define classes of problems that method could help to solve.

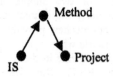

Fig. 3. "Contingency graph" of an unexploited mode.

216

Then, the classification criteria would be: "The method is useful to solve a class of problem", and the corresponding adaptation sequence would be: to focus on key-problems, then to select adapted methods according to classes of problems to solve, and finally to insert the selected method fragments into particular project activities. This potential change in the contingency mechanisms requires further researches like the identification of relevant problem classes and the efficiency evaluation of method fragments according to these classes.

## REFERENCES

Brinkkemper, S. (1996). Method engineering: engineering of information systems development methods and tools. *Information and Software Technology*, **38** (4), 275-280.

General Direction III of European Commission. (1996) "Euromethod, User Book Version 1." Euromethod Project France, Paris.

General Directive. (1997) "Life Cycle Process Model." V-Model - Development Standard for IT-Systems of the Federal Republic of Germany.

General Directive. (1997) "Method Allocation." V-Model - Development Standard for IT-Systems of the Federal Republic of Germany.

Goepp, V. and F. Kiefer. (2003). Towards a definition of the key-problems in information system evolution: Formulating problems to better address information system projects, In proceedings of *3rd International Conference on Enterprise Information Systems - ICEIS'03*, Angers, France, pp. 586-590.

Gorry, G. A. and M. S. Scott Morton. (1971). A framework for management information systems. *Sloan Management Review*, **13** (1), 55-70.

Gzara, L., D. Rieu and M. Tollenaere. (2000). Patterns approach to product information system. *Requirements Engineering Journal*, **5** (3), 157-179.

Kiefer, F., A. Michez, C. Le Guirrec and J.-M. Gazzo. (1995). Les enjeux de la conception des systèmes intégrés de production, In proceedings of *First International Congress of Industrial Engineering*, Montreal, Quebec.

Martin, J. (1984). *An information system manifest*, Prentice Hall International, United Kingdom.

McFarlan, F. W. (1981). Portfolio approach to informtaion systems. *Harvard Business Review*, **59** (5), 142-150.

Morley, C. (1998). La gestion des risques dans les projets système d'information. *La Cible* (10\98).

Morley, C. (1999). L'analyse a priori d'un projet système d'information, In proceedings of *Colloque AIM*, Paris, France.

Naumann, J. D. and G. B. Davis. (1978). A contingency theory to select an information requirements determination methodology, In proceedings of *Second Software Life Cycle Management Workshop, IEEE*, New-York.

Punter, T. and K. Lemmen. (1996). The MEMA-model: towards a new approach for Method Engineering. *Information and Software Technology*, **38** (4), 295-305.

Ralyté, J. (2001). Vue stratégique sur l'ingénierie des méthodes, In proceedings of *INFORSID 2001*, Geneva, Swiss, pp. 43-66.

van Slooten, K. and B. Schoonhoven. (1996). Contingent Information Systems Development. *Journal of Systems and Software*, **33** (2), 153-161.

Zhu, Z. (2001). Towards an integrating program for information systems design: an Oriental case. *International Journal of Information Management*, **21** (1), 69-90.

Zhu, Z. (2002). Evaluating contingency approaches to information systems design. *International Journal of Information Management*, **22** (5), 343-356.

# PARAMETRIC COST MODEL TO SUPPORT END-OF-LIFE MANAGEMENT OF VEHICLES

**Yasmine Ladjouze and Shahin Rahimifard**

*Advanced Manufacturing Systems and Technology Centre*
*Loughborough University*
*United Kingdom*

Abstract: The End-of-Life Vehicles Directive has been introduced by the European Commission to cope with the environmental effects of nine million vehicles that reach the end of their useful lives each year in Europe. One of the main features of this directive is the principle of 'Producer Responsibility' whereby the vehicle manufacturers are responsible for the take-back and recycling of all their vehicles from 2007. In response to this Directive, there has been significant research targeted at developing technologies aiming to improve vehicles recyclability, but very little focused on the costing issues related to the processing and treatment of vehicles at the end of their lives which remains a major source of concern for regional, national and European organisations and Automotive Manufacturers.

Keywords: Automobile industry, Waste treatment, Vehicles Recovery, Parametrization.

## 1. INTRODUCTION

Since the introduction of the notion of Sustainable Development by the World Commission on Environment and Development in 1987, waste prevention has become a high priority for societies, governments and industries. In this context, a European Directive related to End-of-Life management of vehicles has been proposed, that establishes the producer responsibility, through which Vehicles Manufacturers are required to reduce/eliminate the waste by taking back and processing their vehicles at the end of their useful lives. In addition from 2006, vehicles have to meet very ambitious recycling and reuse targets set by the Directive. It comes without saying that handling all those waste vehicles will engender significant expenses from the producers. However at present, limited research has been directed at modelling the costs of the activities involved in the treatment of End-of-Life Vehicles (ELV). Such models are of crucial importance to Vehicles Manufacturers. It will allow them to assess the economic impact of each activity involved in the processing of their End-of-Life products and will support the decision making tasks involved in investment and development of infrastructures required to implement the ELV Directive.

In this paper, the authors discuss the design and generation of cost models based on parametric equations to outline and highlight the economics of the operations involved in End-of-Life Management of Vehicles. In addition, a computational viewpoint of such ELV cost model has also been presented.

## 2. BACKGROUND

End-of-Life Vehicles are vehicles that are *"waste"* according to the Directive 75/442/EEC (European Council 1975), in which *waste* has been defined as "any substance or object which the holder disposes of or is required to dispose of pursuant to the provisions of national law in force". The environmental burdens associated to ELVs are numerous (Staudinger & Keoleian 2001). To begin with, soil and air contaminations can occur through leakages of fluids during the disassembly process. Furthermore, hazardous substances such as mercury and lead that are very harmful to the environment may be released during the shredding and disposal operations. In addition at present, significant proportions of Automobile Shredder Residue (ASR) and tyres have been sent to landfill or incinerated, which represents further source of pollution.

The ELV Directive consists of thirteen articles. Each one of them deals with an appropriate issue of the treatment of End-of-Life vehicles. Vehicles affected by the ELV Directive are passenger vehicles with up to eight seats plus the driver, light goods vehicles up to 3.75 tonnes unloaded weight and three-wheel motor vehicles. The Directive obligates the manufacturers to limit the use of hazardous substances in the components of their new vehicles and to facilitate disassembly, reuse and recycling processes. Vehicles must also be produced with a higher quantity of recycled materials to develop the market of recycled materials. The Directive imposes for all ELVs to be taken to licensed treatment facilities and a certificate of destruction to be issued to the last owner. The producers will have to pay for all costs related to the collection and take back of the ELV to the authorized treatment facility. The Directive also sets recycle and recovery rates of 85% by January 1st 2006 and 95% by January 1st 2015.

The weight of a car on average consists of 70.2% of ferrous metals, 21.1% of non metals (33% of which are plastics) and 8.7% of non-ferrous metals (Bellmann & Khare 1999). Today it is estimated that about 75% in Europe and 85 % in Japan of the weight of a car is recycled or recovered. When compared to other manufactured products, this recycling rate is very high as the disassembly and recovery of vehicles are well developed activities (Kanari et al. 2003). However this is not enough to meet the recovery targets set by the Directive as until

recently in Europe, there was an obvious lack of knowledge about 'Design for Disassembly' and 'Design for Recycling' from the Automotive Industry (Kazmierczak et al. 2004). These two concepts will become facilitators to the implementation of the ELV Directive as soon as they are understood and integrated in the development process (Das 2001). In the 1990's, the waste from ELVs amounted to only 1% of the tonnage of municipal solid waste (Bellmann & Khare 2000). This has however significantly increased as currently an average of 180kg to 250 kg per vehicle is sent to landfills (Selinger et al. 2003). This is due to the fact that in many cases, the recycling of materials such as ASR and composite materials at present are far less economical than the mere disposal (Jonhson & Wang 2002).

## 3. A PARAMETRIC COST MODEL FOR ELV MANAGEMENT

Nowadays, cost reduction is one of the major drivers in any successful business. Indeed, in modern global market, a product and its lifecycle costs have the same importance as the product quality and functionality (Roy et al. 2001, Asiedu & Gu 1998). But in order to reduce costs, the various company activities have to be measured in detail and when this is not possible, they should be estimated and assessed (Seo et al. 2002). This research has utilised a novel approach to the modelling of the ELV treatment based on six activities of collection, assessment, dismantling, shredding, recycling and disposal. A parametric cost model has been generated to be used as the basis of the development of a decision support costing tool, as described in section 4. The activities considered are collection, assessment, disassembly, shredding, recycling and disposal as shown in figure 1.

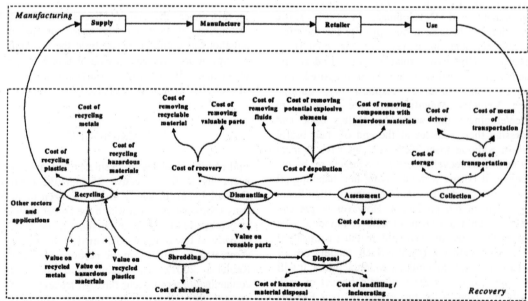

Fig. 1. Costs involved in the management of an ELV

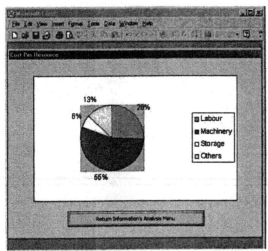

Fig 4. Percentage of types of costs in the treatment of an ELV

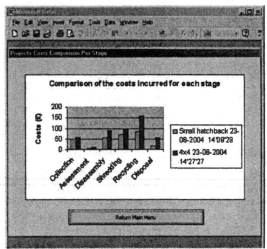

Fig. 5. Comparison of costs for two models of ELV

## 5.1. Analysis Of Experimental Results

A number of observations have been made based on the analysis of the experimental results as outlined below.

Firstly, the ELV experiments showed that the most expensive activities in the treatment of an ELV are currently the disassembly, shredding and recycling activities. With regards to disassembly, this could be due to the cost of the labour intensive processes in current capacities. In the case of shredding and recycling, the high costs could mainly be due to the lack of development of appropriate technologies or simply because at present there is limited capacity available to handle the large volume of recycling processes required for ELV treatment.

Secondly, cost of labour and cost of machinery represent over 80% of the total cost of ELV treatment. This highlights a need for further investigation in improved design of such machinery and where possible, the automation of various ELV processes.

Thirdly, at present the only source of revenue from ELVs is based on resale of parts/components and recycled materials. It should be noted that due to reduction of the costs of new parts, the revenue from reuse of parts is forecasted to decrease. In addition, currently identifying appropriate market for some of the recycled material that cannot be reused within Automotive Industry, itself represents a further source of costs. These are clear indications that the revenues based on current practices of ELV treatment are set to decrease and there is a requirement to consider new methods of generating revenue from ELV treatments. Possible sources of potential revenue have been identified such as the innovative use of ELV materials (e.g. use of tyres for road aggregation) and energy reclamation from ELV wastes (e.g. ASR).

Fourthly, the experimentation has highlighted a significant difference in the costs of treatments of various vehicles models, depending on total weight, range of material used, level of customisation and etc. Therefore, the authors argue that any environmental vehicle tax currently under investigation and trials has to be in proportion with the ELV cost rather than the simple model used at present i.e. a percentage of original price of the vehicle.

### Table 3. Data set established for four types of cars

| Type | Small | Family | Luxury | 4*4 |
|---|---|---|---|---|
| Weight (kg) | 1109 | 1360 | 1683 | 2440 |
| % Metal per weight | 78% | 73% | 70% | 65% |
| % Plastic per weight | 8% | 9% | 10% | 11% |
| | | | | |
| **Collection** | | | | |
| Duration of the journey (h) | 1.50 | 1.50 | 1.50 | 1.50 |
| Driver Hours rate (£/h) | 7.00 | 7.00 | 7.00 | 7.00 |
| Distance covered (miles) | 20 | 20 | 20 | 20 |
| Charge per mile (£/mile) | 6.00 | 6.50 | 7.00 | 7.50 |
| Number of day of storage (d) | 3 | 3 | 3 | 4 |
| Rate of storage per day (£/d) | 6.00 | 6.00 | 7.00 | 7.00 |
| | | | | |
| **Assessment** | | | | |
| Time of assessment (h) | 0.50 | 0.50 | 0.75 | 0.75 |
| Assessor rate per hour (£/h) | 15.00 | 15.00 | 15.00 | 15.00 |

| Type | Small | Family | Luxury | 4*4 |
|---|---|---|---|---|
| **Disassembly** | | | | |
| Cost of Dismantling (£) | 45.00 | 60.00 | 67.50 | 75.00 |
| Cost of Depollution (£) | 10.00 | 10.00 | 12.00 | 15.00 |
| Revenue on Disassembly (£) | 89.00 | 108.44 | 120.85 | 100.13 |
| | | | | |
| **Shredding** | | | | |
| Cost of Shredding (£) | 70.00 | 80.00 | 90.00 | 100.00 |
| | | | | |
| **Recycling** | | | | |
| Metal recycling cost (£/kg) | 0.084 | 0.084 | 0.084 | 0.084 |
| Plastic recycling cost (£/kg) | 0.106 | 0.106 | 0.106 | 0.106 |
| Recycled metal price (£/kg) | 0.124 | 0.124 | 0.124 | 0.124 |
| Recycled plastic price (£/kg) | 0.034 | 0.034 | 0.034 | 0.034 |
| | | | | |
| **Disposal** | | | | |
| Landfilling rate per kg (£/kg) | 0.055 | 0.055 | 0.055 | 0.055 |

Table 4. Repartition of costs per model of ELV

|  | Small | Family | Luxury | 4x4 |
|---|---|---|---|---|
| Collection (£) | 48.50 | 48.50 | 51.50 | 58.50 |
| Assessment (£) | 7.50 | 11.25 | 11.25 | 11.25 |
| Disassembly (£) | 55.00 | 70.00 | 79.50 | 90.00 |
| Shredding (£) | 70.00 | 80,00 | 90,00 | 100,00 |
| Recycling | 86.86 | 96.37 | 116.80 | 160.87 |
| Disposal (£) | 17,04 | 23,66 | 31,12 | 58,66 |
| Total (£) | 284.90 | 329.78 | 380.17 | 479.28 |

Table 6. Repartition of revenues per model of ELV

|  | Small | Family | Luxury | 4x4 |
|---|---|---|---|---|
| Disassembly (£) | 89.00 | 108.44 | 120.85 | 100.13 |
| Recycling (£) | 94.31 | 108.69 | 129.43 | 175.09 |
| Total (£) | 183.31 | 217.13 | 250.28 | 275.22 |

Table 5. Percentage of costs in ELV treatment per model

|  | Small | Family | Luxury | 4x4 |
|---|---|---|---|---|
| Labour | 26% | 27% | 27% | 23% |
| Machines | 55% | 55% | 54% | 55% |
| Storage | 6% | 5% | 6% | 6% |
| Others | 13% | 13% | 13% | 16% |

Table 7. Final cost and revenue per model of ELV

|  | Small | Family | Luxury | 4x4 |
|---|---|---|---|---|
| Cost (£) | 284.90 | 329.78 | 380.17 | 479.28 |
| Revenue (£) | 183.31 | 217.13 | 250.28 | 275.22 |
| Profit/Loss (£) | -101.59 | -112.65 | -129.89 | -204.60 |

## 6. CONCLUSION

The targets of recycling and reuse that are set by the ELV Directive will soon provide major operational and design challenges for Vehicle Manufacturers. The review of contemporary practices by this research has highlighted that European Vehicle Manufacturers are not well prepared for these challenges and require further investigation of best practices and appropriate investment to set up suitable infrastructure for compliance to ELV Directive.

Another issue highlighted by this research is that the cost of recycling some of the materials e.g. plastics and Automotive Shredder Residue is more than the revenue obtained from the recycled material. This presents an interesting dilemma for Vehicle Manufacturers which have to recycle the materials at loss to meet recycling targets.

The ELV costing tool developed by this research is shown to provide a suitable decision support tool on the range of decision tasks concerning the management of ELVs. Based on the experimental studies carried and using this ELV Costing Tool, the authors argue that the Vehicle Manufacturers inevitably have to incur losses in order to comply with ELV Directive for the first few years of its introduction. However, appropriate investment and innovative approaches to the recovery, reuse and recycling of parts and materials within ELVs present a significant potential hidden economic value which can lead, with appropriate management, to a profitable activity.

## REFERENCES

Asiedu, Y. and P. Gu (1998). Product life cycle cost analysis. *International Journal of Production Research*, **36**(4), 883-908.

Bellman, K. and A. Khare (1999). European response to issues in recycling car plastics. *Technovation*, **19**(12), 721-734.

Bellman, K. and A. Khare (2000). Economic issues in recycling end-of-life vehicles. *Technovation*, **20**(12), 677-690.

Das, S.K. (2001). Product Disassembly and Recycling in the Automotive Industry. In: *Mechanical Life Cycle Handbook: Good Environmental Design And Manufacturing*. (M. S. Hundal). Mercel Dekker, Inc, New York USA

European Council (2000). Directive 2000/53/EC of the European Parliament and the Council of 18 September 2000 on End-of-Life Vehicles. *Official Journal of European Communities*, 21 October 2000, **269**, 34-42.

European Council (1975). Council Directive of 15 July 1975 on waste (75/442/EEC). *Official Journal*, 25 July 1975, **L 194**, 39-41.

Johnson, M.R. and M.H. Wang (2002). Evaluation policies and automotive recovery options according to the European Union Directive on end-of-life vehicles (ELV). *Proceedings of the I MECH E Part D Journal of Automobile Engineering*, **216**(9), 723-739.

Kanari N., J.-L. Pineau and S. Shallari (2003). End-of-life Vehicle recycling in the European Union. *JOM*, **55**(8), 15-19.

Kazmierczak K., J. Winkel and R. H. Westgaard (2004). Car disassembly and ergonomics in Sweden: Current situation and future perspectives in light of new environmental legislation. *International Journal of Production Research*, **42**(7), 1305-1324.

Roy, R., S. Kelsevjo, S. Forsberg and C. Rush (2001). Qualitative and quantitative cost estimating for engineering design. *Journal of Engineering Design*, **12**(2), 147-162.

Seo, K.-K., J.-H. Park, D.-S. Jang and D. Wallace (2002). Approximate estimation of the product life cycle cost using artificial neural networks in conceptual design. *International Journal of Advanced Manufacturing Technology*, **19**, 461-471.

Staudinger, J. and G.A. Keoleian (2001). Management of End-of-Life Vehicles (ELVs) in the US. Center for Sustainable Systems, University of Michigan, USA.

ELSEVIER
IFAC
PUBLICATIONS
www.elsevier.com/locate/ifac

# AN INTELLIGENT ARCHITECTURE FOR REAL-TIME MONITORING AND CONTROL IN INDUSTRIAL ENVIRONMENTS

S. I. Kapellaki[+], G. N. Glossiotis[#], N. D. Tselikas[+], G. N. Prezerakos[+,1], I. S. Venieris[+]

[+] *National Technical University of Athens,*

*School of Electrical & Computer Engineering,*
*9 Heroon Polytechniou Str., 157 73 Athens, Greece*
*e-mail: {sofiak, ntsel ,prezerak}@telecom.ntua.gr,  ivenieri@cc.ece.ntua.gr*

[#] *National Technical University of Athens,*

*School of Mechanical Engineering*
*9 Heroon Polytechniou Str., 157 73 Athens, Greece*
*e-mail: gglos@central.ntua.gr*

Abstract: This paper proposes an advanced, flexible and intelligent communication system able to provide optimum management and control in modern industrial environments. The solution focuses on the backbone communication system, which can be used for data acquisition, monitoring and storage. The proposed architecture is based on Distributed Object and Mobile Agents Technologies, i.e. CORBA and Grasshopper and thus results in the successful replacement and decentralization of the old-fashioned traditional systems that are currently used. *Copyright © 2004 IFAC*

Keywords: Mobile Agents Technologies, Mobile Agents, Distributed Object Technologies, industrial environment, data-storage, data-monitoring.

## 1. INTRODUCTION

The fundamental idea of this paper is inspired by the demand of modern industries for regular and automated monitoring of operational conditions of industrial equipment. This is of high interest especially in the field of maintenance engineering, where the development of reliable systems that predict premature or uncontrollable failures in the production line, can compensate for the high repair cost that these incidents induce.

Major feature of these systems is that the vast majority of the physical quantities that contain the diagnostic information for the damaged equipment (e.g. vibrations, noise, mechanical/electrical parameters etc.) are rapidly fluctuating values, often inside the KHz bandwidth. Specially designed stations accomplish the data acquisition of all these crucial characteristics; while storage and management of the acquired information is carried out by dedicated for this reason PCs. The high sampling frequency that the nature of the data acquisition problem imposes, results in a huge amount of data that the engineer has to evaluate in the minimum period of time.

The target of this work is to present a flexible and intelligent communication system able to provide the optimum management and control in an industrial

---

[1] Dr. Prezerakos is also with the Technological Education Institute (TEI) of Piraeus, Dpt. of Electronic Computer Systems, 250 Thivon Av. & Petrou Ralli, 122 44 Athens, Greece

environment. The backbone communication system is based on Distributed Objects and Mobile Agents technologies (DOT, MAT), in order to absorb all their advanced characteristics, such as integration of heterogeneous computing environments, independency of different hardware/software platforms, flexibility and distribution in the network topology, communication transparency, easy extension and expansion of the whole system etc.

Thus, in the proposed architecture, the passive role of the industrial PCs, which are standing at the edge of the system and are responsible to collect all the appropriate mechanical parameters, is ceased. The inflexible nature of the centralized monitoring and management system is also suspended. Contrariwise, all the nodes are equally neuralgic, since they are able to act if it is required and necessitated by the distributed intelligence of the system.

The advantages of the proposed approach are highlighted and it is shown how these can be consistently inherited to a larger-scale system concept. More analytically in Section 2 the State-of-the-Art in real-time monitoring and control in industrial environments as well as on distributed technologies and platforms is provided. Section 3 includes the conceptual model and the detailed description of the proposed architecture. A concrete scenario for further understanding is also presented in this section. An initial evaluation of the system describing its pros and cons is cited in Section 4. Conclusions of the work are summarized in Section 5.

## 2. STATE-OF-THE-ART

### 2.1 Automated real-time monitoring and control in industrial environment

Maintenance costs account for an extremely large proportion of the operating costs of machinery. In addition, machine breakdowns, and consequence downtime, can severely affect the productivity of factories and the safety of production process. It is therefore becoming increasingly important for industries to monitor their equipment systematically, in order to reduce the number of breakdowns and to avoid unnecessary costs and delays by repair.

Modern industries tend to develop specially organised maintenance programs based on the integration of monitoring systems. These systems cannot be directly set as a genuine part of the plant automation system because of the high sampling rate that the nature of the data acquisition problem imposes. Moreover, since the installation of measuring systems for the total machinery would prove to be ineffective, the effort is given to the proper design of monitoring systems that deal with the most critical equipments inside a plant.

A new "middleware layer networking architecture" is formed inside the already complicated automation network of an industry. This network consists of several measuring stations randomly situated inside the plant area, and has to be resilient, reliable and interactive with the industrial network. The optimal design of this network is the key for its success.

### 2.2 Computational Environment and Infrastructure

The design of a system that, at the adequate sampling frequency, combines the monitoring of only those parameters that every moment is critical, along with an efficient storing method, can prove to be of great importance for the work of engineer.

The afore-mentioned concept can be effectively realized under the framework of utilizing a state of the art communication process, inside the complex environment of an industrial plant. Based on the advantages that open architecture data acquisition software provides, one can interfere on the data acquisition parameters (sampling frequency, amount of data etc.) and the way the information is stored.

The configuration of a typical monitoring system consists of four basic areas:
a) The surrounding area from which the signals are transmitted by the sensors.
b) The measuring stations, responsible for the data acquisition, A/D conversion, and logging of the acquired data.
c) The central PC station for the co-ordination of all the individual measuring stations, and
d) Remote areas from which end users can have access to the whole system through the central PC.

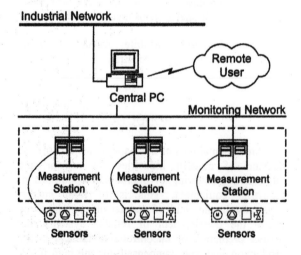

Fig 1. Computational Environment

For every monitored equipment of the factory, which has been selected as part of the maintenance policy, a number of sensors are installed and connected to a basic measuring station. In order to ensure the reliability of the system, it is essential that each measuring station acts independently as a separate node inside a common net, being able however to constantly interact with the net, through a central PC.

Every measuring station delivers three basic features regarding the monitoring process: (a) Temporary data acquisition, (b) Temporary data logging, and (c) Preliminary fault diagnosis by specially designed algorithms. The whole process is programmed in Real-time software, in such way that the reliability and adaptability of the system is ensured.

The dual part of the central PC comprises both the interaction with the main industrial network, and the role of the core node inside the network of the several measuring stations. As a core node it provides, among others, the ability for the control, monitor and access to the set of parameters for each individual station.

The permanent data logging, graphical representation and further process of the acquired data, at the measuring stations, is taking place also at the central PC. The PC collects data by each substation in periodic intervals, without interrupting the measuring procedure as the collection is executed in lower priority. Furthermore, it provides access to the sampled data through the local industrial network for operators inside the plant, and optionally interaction with remote users.

### 2.3 Distributed Object and Mobile Agent Technologies

Software development in telecommunications is targeted to heterogeneous computing environments that integrate new systems with legacy components. The classical client-server model is being gradually replaced by multi-tiered systems. Distributed Object Technology (DOT) is based on Object Oriented technologies, which allow extensive code reusability, by basing the process of creating new objects on already available and distributed ones. In addition, interpreted code allows abstracting from the hardware/software platform used as runtime environment. This way the code is rendered independent of the underlying hardware and software platforms.

One of the dominant Distributed Object Technologies and probably the most well known one is the Common Object Request Broker Architecture (CORBA). CORBA provides a common programming environment across operating systems and allows the development of object-oriented distributed applications (Venieris, 2000). The interoperability protocol that CORBA uses (the Internet Inter-Orb Protocol, IIOP) allows clients running in one ORB to transparently invoke methods on objects within another one.

A major benefit of this technology is that applications written in different ORBs can work together, preserving current software investments (OMG, 2002). Separation of interface and implementation is also of major significance in CORBA. Interfaces between objects are written in the Interface Definition Language (IDL), which is a neutral programming language. The interfaces are translated to the appropriate programming language in the form of client side 'stubs' and server side 'skeletons' implementations. Transparent object detection is also supported in CORBA. There are several different CORBA services supporting object detection. The unique Interoperable Object Reference (IOR), by which each object is characterized, as well as the unique name, with which each object can be registered in the Naming Service of CORBA are the most common mechanisms for object discovery and interconnection (Venieris, 2000). This way all objects can invoke methods and can be connected with remote ones by just calling the desired IOR or CORBA-object name.

On the other hand, mobile agents are intelligent and autonomous software entities, able to migrate and execute their logic in several computational nodes. In order to support the agents' hospitality and execution to distributed nodes, a Mobile Agent Platform (MAP) is required. MAPs provide a set of useful services (e.g. event, communication, indexing etc) and consist a fully operational environment for agents. Mobile Agents can be considered as a middleware-oriented technology enhancing distributed computing. Actually, a MAP is a supplementary middleware layer, exploiting capabilities and features of alternate Distributed Object Technologies, such as CORBA or RMI (Remote Method Invocation) (Bellavista, 2000), (Delgado, 2002).

Some important benefits offered by mobile agents are:
- Communication and Execution State transparency
- Autonomous and intelligent execution (Intelligence / Service on-demand)
- Programming and Communication Flexibility
- Adaptability to specific needs and conditions
- Life cycle management
- Interoperability
- Robustness and Fault-tolerance

A MAP consists of a set of Application Programming Interfaces (APIs) supporting the programming of mobile agents. These APIs exploit underlying middleware (or even plain protocol) capabilities and mechanisms. In other words, a MAP conceptualizes underlying middleware capabilities and offers to programmer a more abstracted view via MAP related APIs. For example, the communication of two agents residing in different nodes may use transparently established (by the MAP) CORBA naming services or the corresponding RMI registers.

The decision of using mobile agents in a prototype depends on several parameters. It is true that using mobile agents technology, advanced programming and communication patterns can be achieved. Apart from the traditional client – server mode, it is often required to apply a more sophisticated pattern (e.g. multicast). Services offered by the MAP enforce the implementation and deployment of advanced communication patterns for distributed software entities. Considering that MAPs deploy also underlying middleware functionalities, if a client – server relationship is the target, the utilization of mobile agents may be unreasonable.

Another strong point of using mobile agents is their embedded intelligence and environment-awareness. In such architectures, it is important to have ambient software entities, able to adapt themselves to specific computational environment. Cases that a software entity is required to detect and identify peculiarities of its environment and to report them to remote units, can be served by mobile agents. However, this scenario can be also served by distributed objects with remote method invocations exploiting a DOT, such as CORBA. So, what for to use a supplementary middleware layer (the MAP) overloading the local environment? The mobile agent can invoke as many server operations as needed to complete its tasks, without transferring intermediate data and responses across the network. The network latency is reduced. A more complex scenario fits perfectly to mobile agents; A single intelligent agent (or type of agent) - that migrates across several nodes, adapts itself to specific conditions and performs specific tasks according to local situations – yields several architectural benefits (e.g. programming cost, communication overhead etc.).

Asynchronous patterns are also supported by mobile agents. This characteristic is important for systems that are not always connected. Combining asynchronous inter-agent communication with the ability of mobile agents to transfer code and files, value added architectures are emerging.

## 3. THE PROPOSED ARCHITECTURE

### 3.1 Conceptual Model

The system's structure has to be practical, user-friendly and flexible to adopt future improvements. It is necessary to cooperate with other conventional control and monitoring platforms (e.g. PLC, SCADA etc.), or other industrial/computerization systems (e.g. ERP) that make use of the predictive maintenance advantages. Aiming to the coverage of the plethora of sensors that already exist and the future anticipated, emphasis has to be given in the expandability of the methods in use both from software and hardware aspect too. For this reason, the usage of MAT in an industrial environment

seems to be ideal (Parunak, 2000). The basic principles of agent-based data acquisition technology are (Hu, 2003):

1. Agent senses environment information.
2. After analysis and reasoning, the agent deduces a global assessment and decides on a response strategy to the outside environment.
3. In a multi-agent system (MAS), collaborating, negotiation, and competition among agents must take place before an agent system can decide on matters of availability and reliability and choose how to respond to the environment.
4. Repeat step 1 to 3 until the agent achieves appointed tasks.

The incorporation of MAT calls for a new conceptual model regarding the IT infrastructure of the industrial environment. Such a model helps to define the jurisdiction of the various architecture-enabling components as well as the boundaries required for information exchange between said components. It should be stressed that although MAT is a revolutionary concept when it comes to industrial use it is no panacea either; as such the conceptual model helps to identify areas where using MAT really makes sense. The proposed model is depicted in Figure 2.

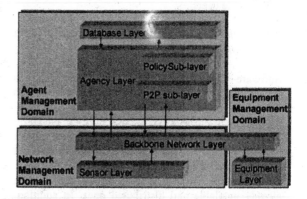

Fig. 2. Conceptual Model

*Sensor Layer:* The various sensors that perform measurements along with their controllers and the corresponding A/D converters belong to this layer. Measurements are transferred outside this layer via the layer immediately above that is the Backbone Network Layer. From a networking point of view a sensor controller is a network node constantly transmitting information towards one or more IP addresses and a specific TCP port. Taking into account the size, processing capacity and power consumption of sensor devices, the Sensor Layer is not agent-enabled.

*Backbone Network Layer:* This layer is comprised of the respective hardware as well as routing / forwarding policies that control information flow between all other layers. As such it is the virtual glue that holds all the layers together. In a way similar to the Sensor Layer, the Backbone Network Layer is not agent-enabled. This is a practical approach since current network equipment (routers, switches) is agent-unaware, moreover network maintenance and expansion should not be concerned with agent-related issues.

With respect to the management, the two layers above are part of the Network Management Domain. When it comes to industrial control applications, the network management domain is responsible for maintaining paths where information can freely flow from one node to another. Adding agent-awareness to this layer would render it backwards incompatible with the majority of existing industrial installations.

*Agency Layer:* This layer is composed of the various agents that undertake the main tasks of the system such as controlling the sensors, gathering and processing measurements and detecting critical situations. This layer includes two sublayers. The p2p sublayer is responsible for the periodic discovery of neighboring network nodes and the corresponding agents hosted there. The need for having such a sublayer arises from the requirements of direct agent-to-agent communication as well as the requirement for agent migration upon certain conditions. The details of the decisions regarding these requirements are dictated by the policy sublayer. This sublayer hosts special agents that provide the rules for decision-making within a certain node. Peer to peer communication helps to create a more sophisticated form of a network: a cluster of awareness (Prayurachatuporn, 2001).

*Database Layer:* Information coming from the sensors is stored at several nodes within the architecture forming a distributed database schema. These nodes can be either dedicated or co-located with agents. There should be at least one agent responsible for each database; on the other hand a database may store information coming from more than one agent. In any case a database should not have full access to the backbone network; it should be configured to communicate only with the responsible agents via an open API (e.g. SQL). The main motivation is again reduced complexity and backwards compatibility since agent-aware databases are uncommon in industrial environments. On the other hand the responsible agent may query the database for certain data and carry them along during migration to a different node.

The Agency and the Database layers belong to the Agent Management Domain. This can be seen as an overlay network above the regular telecommunications infrastructure. The Agent Management Domain is mainly responsible for the initial deployment of agents according to the initial architecture design. The database layer is not agent-aware, however it is included in this domain since a respective agent can be configured to automatically undertake database management functions.

*Equipment Layer:* The equipment that is controlled and/or measured composes this layer. Again, due to backward compatibility reasons, the equipment layer is not agent enabled and as such it belongs to a separate management domain. The Equipment Layer interfaces with other entities through standard industrial protocols.

### 3.2 The Proposed Architecture

The proposed architecture builds on several concepts proposed by Wijata et al (2000) but emphasizes on the actual equipment measuring process instead of the measurement of the network performance. The proposed architecture is depicted in Figure 3.

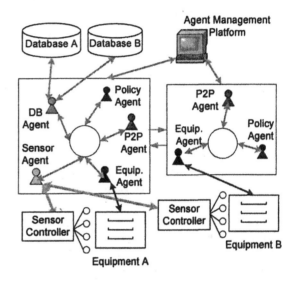

Fig. 3. Overall System Architecture

With respect to the sensor layer, it is envisaged that sensors are controlled in groups by simple controlling devices. These controlling devices are composed of the electrical interfaces to the sensors, the A/D converters plus a simple processor and network card with a TCP/IP/Industrial Ethernet protocol stack. The controlling device communicates to the network via two streams, one input stream used for receipt of control information via a predefined port and one output stream used for data transmission (Umezawa, 2002). The agency layer is composed of several agent-aware nodes, which are hosted inside a protected software space (or agency) provided by the Grasshopper Platform, which uses Distributed Object Technology – and especially CORBA – as the underlying communication bus (Grasshopper Agent Platform). The proposed architecture recognizes the following categories of agents.

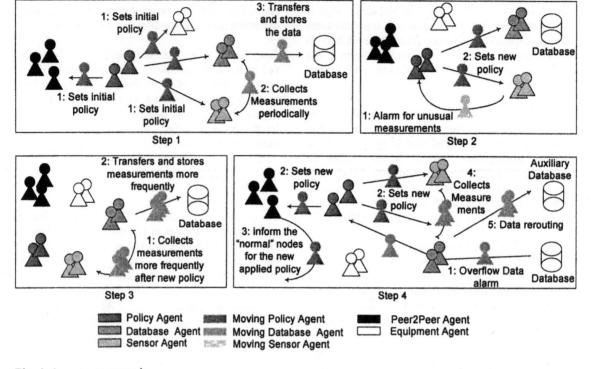

Fig. 4. A concrete scenario

*Sensor Agent.* It is responsible for configuring the sensor controllers as well as receiving the measurements. Moreover, based on the processing of measurements, it can react to potentially critical situations e.g. by increasing the sampling frequency or by raising an alarm towards another agent.

*Policy Agent.* It defines the exact method of reaction of the other agents upon identification of a critical situation. The notion of policies has been identified by Bunch (2004) where the use of ontologies for defining notifications is proposed. This concept can be extended via a dedicated agent (the policy agent) incorporating an ontology for policy definition in conjunction with a reasoning engine thus being able to decide on a course of action when a critical situation occurs. In principle, the policies dictated by the policy agent impact the operation of all other agents in the node. Policies are usually static and pre-defined by some external administrative entity i.e. the responsible engineer. However, one could envisage the incorporation of self-learning algorithms into the policy agents

*Equipment Agent.* It is responsible for communicating with the actual industrial equipment that is measured. Its basic mission is to adjust the various operational parameters upon identification of a critical situation. For this purpose, it exchanges messages with sensor agents in the same or different nodes. Upon detection of a critical situation and the subsequent decision for the enforcement of a relevant policy, the equipment agent controls the respective equipment over standard industrial protocols.

*Database Agent.* It is responsible for storing and retrieving data and events to/from the database. For this purpose, it receives data and notifications from one or more sensor agents as well as messages from one or more equipment agents. The storage/retrieval policy is dictated by the Policy Agent.

*P2P Agent.* This agent is responsible for the periodic discovery of other agents in the various nodes of the architecture. Any agent wishing to communicate with another agent or to migrate to a separate node may refer to a p2p agent in order to get address of the respective address information.

It should be noted that since the proposed architecture is fully distributed an agent-aware node should be equipped with at least one agent. This agent can make use of other agents residing in different hosts if it has been configured with the address of at least one p2p agent. All agents can migrate to a different node if needed e.g. if the system needs to be load-balanced or in cases of malfunction of the host.

### 3.3 A concrete scenario

In this section a concrete scenario is described, in order to explicitly explain the proposed architecture and make clearer the interaction between the different entities.

We assume that the observed industrial environment consists of multiple dedicated agent-aware nodes. In Figure 4 such a node is shown. The nodes are launched and the communication between them is established via the P2P Agents. Thus the Policy Agents are free to apply the desired policies at the edge of each node in both Sensor and Equipment Agents. The Sensor Agents from all three nodes are retrieving different kind of measurements (e.g. temperature), while the Equipment Agents are stand by and ready to interfere with the corresponding equipment.

Let's also assume that the Policy Agents have applied a constant measurement sampling to each Sensor Agent (e.g. one measurement per minute). This results to a periodical interaction between Database and Sensor Agents (Fig. 4: Step 1, normal situation). The latter have retrieved the appropriate measurement, while the former are responsible to transfer and store these measurements to the system's Database acting as freighter agents. In case that the measurements of a value present apparent variation between each other (e.g. a critical augmentation of the temperature is observed), the Policy Agents are notified by the corresponding Sensor Agents, since they also periodically interact with them. Thus, an alarm is generated and the Policy Agents can apply a new policy to the Sensor Agents, in order to verify the unusual phenomenon (Fig. 4: Step 2, application of a new policy). For example, a new policy can be a more frequent sampling e.g. ten measurements per minute instead of one per minute. About this new policy in the node, both Sensor and Database Agents are also informed. Hence, they are "instructed" to interact more frequent with each other, as well as with the Database (Fig. 4: Step 3, more frequent monitoring and storage). As a result, a large amount of data collected by the problematic node is stored in the Database. A new potential problem arising by the previous treatment can be an imminent data overflow in the Database, because of the bursting measurement storage. Thus, a new policy can be applied by the Policy Agents, which can instruct either the corresponding Sensor ones (via the Peer2Peer Agents), to cut off all the interactions and data transfers by the "normal" nodes or the corresponding Database Agents to reroute-forward the data from the "problematic" node to an auxiliary Database of the system (Fig. 4: Step 4). A more extreme – but possible – reaction could also be the instruction of the Equipment Agents for stopping the corresponding equipment of the "normal" nodes at all and focus exclusively on the problematic area.

## 4. INITIAL EVALUATION

The incorporation of MAT in industrial environments is ideal for tackling a series of traditional problems such as:

*Distributed Management:* Traditional industrial architectures rely on a central computing unit for information storage and processing and for managing the measurement / control network. This situation requires that the central unit maintains at all times an accurate picture of the entire system which is very costly due to the high number of associated parameters. MAT can introduce a peer-to-peer direct communication between agent containers that reduces the role of a central computing unit, which now serves basically as an agent repository.

*Resilience:* Traditional industrial architectures cannot handle very well the problem of faults. First of all, malfunctioning of the central unit renders the whole system unusable. Moreover, similar problems in computing units connected to the sensor network results in data loss. The problem can be ameliorated by using redundant or clustering architectures. However this has a significant impact on network complexity as well as cost. Mobile Agents offer two significant advantages when it comes to fault tolerance. First of all, they can migrate to another node if the current one show signs of malfunction. Moreover, an agent can automatically take over (some or all of) the tasks of another agent upon request or if the node where the latter agent resides stops functioning. Both cases are handled automatically, without the need for central intervention and reduce the need for redundant resources.

*Scalability:* The resulting architecture is highly scalable. System expansion takes place by simply adding more sensors and denoting an agent-enabled node as being responsible for the new sensors. If the denoted node cannot handle the task it will automatically try to involve other nodes in the task until the system reaches a balanced state. This way, and up until the point where the overall system capacity becomes critical, the system is self-configured.

*Backwards compatibility:* While this is not a generally-applicable characteristic of MAT, the proposed architecture makes sure that well-established protocols for communicating with entities like sensors, equipment and databases are respected. This way the introduction of agents is transparent to the rest of the infrastructure.

However the resulting reduction in complexity within the equipment and sensor layers is moved to the agency layer. Agents must implement a double means of communication in order to communicate between them (using message-based communication) as well as with the other layers (using well-established protocols). Moreover, the available policies and the initial agent distribution must be carefully preconfigured in a system-wide scale. Taking these facts into account, it seems that for rather small-scale environments the benefits of introducing MAT do not outweigh the complexity of the implementation. For large-scale implementations however, where management and maintenance complexity is in any case high, the advantages offered by MAT surely justify the effort necessary for its introduction.

## 5. CONCLUSIONS

To conclude, the paper proposes a flexible and intelligent communication architecture based on State-of-the-Art technologies. The advantages and the characteristics of these technologies are visibly inheritable and the proposed architecture exploits all of them. Thus, the final system, which is brought in the prominence, presents distributed management, better resilience, backward compatibility and extensibility in comparison with the inflexible traditional ones. Further implementation steps will be very helpful, in order to come into more precise and quantitative results for the proposed platform.

## REFERENCES

Venieris I., Zizza F., Magedanz T (2000). *Object Oriented Software Technologies in Telecommunications*. John Wiley & Sons, ISBN: 0-471-62379-2.

OMG, CORBA/IIOP 3.0 Specification (2002), www.omg.org

Bellavista P. et al. (2000), CORBA Solutions for Interoperability in Mobile Agents Environments. *International Symposium on Distributed Objects and Applications*, Belgium.

Delgado J. et al (2002). An Architecture for Negotiation with Mobile Agents. *MATA 2002*. Barcelona, Spain.

Umezawa T., Satoh I. And Anzai Y. A. (2002). Mobile Agent-Based Framework for Configurable Sensor Networks. *MATA 2002*. Barcelona, Spain, pp. 128–139.

Grasshopper Agent Platform, http://www.grasshopper.de

Hu, Y. M., Du, R. S., Yang, S. Z (2003). Intelligent Data Acquisition Technology Based on Agents. *Int J Adv Manuf Technol*. Vol. 21. pp. 866–873.

Prayurachatuporn S., Benedicenti L. (2001). Increasing the reliability of control systems with agent technology. *ACM SIGAPP Applied Computing Review*. Vol. 9 , Issue 2, pp. 6-12.

Wijata Y.I, Niehaus D., Frost V.S. (2000). A Scalable Agent Based Network Measurement Infrastructure. *IEEE Communications Magazine*, pp. 174-183.

Parunak, H. (2000). A Practitioners' Review of Industrial Agent Applications. *Autonomous Agents and Multi-Agent Systems*, 3, pp. 389-407.

Bunch L. et al. (2004). Software Agents for Process Monitoring and Notification. *ACM Symposium on Applied Computing*, pp. 94-99, Nicosia, Cyprus.

ELSEVIER

IFAC

PUBLICATIONS
www.elsevier.com/locate/ifac

# PROCESS MODEL FOR AL-ALLOYS
# WELDING OPTIMIZATION

**A.D.Zervaki and G.N. Haidemenopoulos**

Department of Mechanical and Industrial Engineering, University of Thessaly,
383 34 Volos, Greece

Abstract: A process model that correlates the softening of the Heat Affected Zone (HAZ) of aluminum laser welds, with the process parameters used, has been developed. This was achieved through the modeling of the microstructural evolution (dissolution and coarsening) of the strengthening precipitates during the weld thermal cycle, by employing the methodology of computational thermodynamics and kinetics. In this way the volume fraction and average precipitate size were calculated under extremely non-isothermal conditions. These microstructural characteristics were incorporated in relevant models in order to evaluate the hardness variation. Calculated hardness profiles in the HAZ are in good agreement with the experimental values. *Copyright © 2004 IFAC*

Key words: modeling, process models, metals, temperature calculations, Finite element method.

## 1. INTRODUCTION

Laser Beam Welding has been recently applied successfully (Tempus, 2001), in heat treatable aluminum alloys for the welding of airframe components. However despite the limited HAZ dimensions, characteristic of the laser welding method, a drop in HAZ hardness is apparent (Zervaki, 2004). Softening in the HAZ is a common and more pronounced effect when welding with conventional welding processes. Kou (2003) observed a hardness minimum in the HAZ of gas-tungsten-arc welded 6061 alloy welded in the artificially aged (T6) or naturally aged (T4) conditions and attributed the softening to coarsening of $\beta''$ precipitate, the basic strengthening precipitate, and formation of the coarser $\beta'$ precipitate. Similar results have been reported by Malin (1995) on gas-metal-arc welded 6061-T6 aluminum alloy. The aim of the present work is to simulate the softening reactions (dissolution and coarsening) in the HAZ of laser-beam-welded 6061-T6 aluminum alloy. A finite-element based heat flow analysis of weld thermal cycles combined with a computational diffusional kinetics analysis of dissolution and coarsening was performed. The DICTRA (DIffusion

Controlled TRAnsformations) methodology was involved, as proposed by Agren (1992) as well as by Engstrom et al. (1994), a software for handling diffusion in multicomponent, multiphase systems based on the numerical solution of the diffusion equations with local thermodynamic equilibrium at the phase interfaces. Several models have been developed in the recent years to describe diffusional phase transformations in aluminum alloys. The majority of the models deal with isothermal transformations (Vermolen, and Vuik, 2000; Bratland, et al.,1997). Relatively few research efforts have been directed towards modeling of non-isothermal transformations as those encountered in welding (Myhr and Grong 2000; Ion, Easterling, Ashby 1984; Bjorneklett et al., 1998; Nicolas, and Deschamps, 2003). The DICTRA methodology, mentioned above, has been applied by Agren (1990) for the modeling of carbide dissolution in steels under isothermal conditions. Also DICTRA has been applied for the solution of coarsening problems under isothermal conditions (Gustafson, 2000; Gustafson, et. al,1998). In the present paper DICTRA is used for the simulation of dissolution, reprecipitation and coarsening during the weld thermal cycle in the HAZ of 6061-T6 laser welds. The thermal cycles in the

HAZ were calculated by the finite element method using the general purpose finite element program ABAQUS.

The calculated thermal cycles were used as input for the DICTRA simulations. The major assumptions made are the following:

(a) the HAZ is divided in two parts (Fig.1): $HAZ_1$ where the maximum temperature of the welding cycle exceeds 400°C and where only dissolution during heating and reprecipitation during cooling can occur. In $HAZ_2$ the maximum temperature does not exceed 400°C and only precipitate coarsening can occur.

(b) Only the equilibrium precipitate $\beta$-$Mg_2Si$ is considered in the simulations, since kinetic data for the metastable phases (GP-zones, $\beta''$, $\beta'$) are not included in the relevant databases.

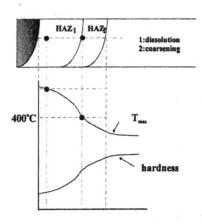

Fig. 1. Temperature and corresponding hardness profile in the HAZ due to dissolution and coarsening.

## 2. EXPERIMENTAL WORK

Bead-on-plate experiments, with a 5 kW $CO_2$ laser was performed. A focusing lens of 127mm focal length was used, while Nitrogen shielding gas was supplied coaxially to the laser beam through a nozzle of 4mm diameter. Sheets with dimensions 100x100x1.6 mm were positioned under the laser beam. The specimens were sandblasted, in order to obtain a rough well absorbing surface. Focal point position was –1mm from the surface of the workpiece.

Details on the experimental conditions used, as well as, on the results of metallurgical evaluation followed, are given elsewhere (Zervaki 2004).

Typical macrostructure appearance of the cross section, of the laser weld bead is presented in Fig 2 and the relevant microhardness profile obtained across a line 0.5 mm below the surface is given in Fig. 3. The hardness drop within HAZ, due to the dissolution and coarsening of the strengthening precipitates, was modeled by the methodology described below.

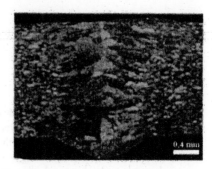

Fig.2. Typical laser weld macrostructure Experimental conditions:4.5kW power, 4.8 m/min weld speed.

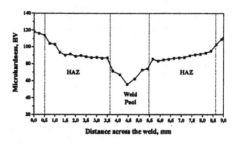

Fig.3. Microhardness profile on the cross section of the weld.

## 3. CALCULATION OF THERMAL CYCLES.

The thermal cycles was the input for the simulation of dissolution and coarsening in the sections to follow. The general purpose finite element program ABAQUS was employed for the solution of the problem.

The governing differential equation describing heat flow by conduction is given by:

$$\rho c_p \frac{\partial T}{\partial t} = \dot{Q}_G + \frac{\partial}{\partial x}\left(k\frac{\partial T}{\partial x}\right) + \frac{\partial}{\partial y}\left(k\frac{\partial T}{\partial y}\right) + \frac{\partial}{\partial z}\left(k\frac{\partial T}{\partial z}\right) \quad (1)$$

where $T$ is the temperature(K), $k$ the thermal conductivity (W/mK), $\rho$ the density (kg/m³), $c_p$ the specific heat (J/kgK) and $\dot{Q}_G$ the internal energy production rate per unit volume (W/m³).

Heat transfer due to convection in the weld pool is ignored.

The boundary conditions take into account heat losses due to convection and radiation.

Heat loss due to convection is:

$$-Q_c^s = h_c \left( T - T_0 \right) \qquad (2)$$

where $h_c$ is the convection coefficient and $T_o$ the ambient temperature.

Heat loss due to radiation is:

$$-Q_r^s = A\sigma \left( T^4 - T_0^4 \right) \qquad (3)$$

where $A$ is the absorption coefficient (%) and $\sigma$ the Stefan-Boltzman constant.

The latent heat of fusion $L$ has been taken into account in the calculations through the definition of the effective specific heat $c_{eff}$:

$$c_{eff}(T) = c_s(T) + \frac{L}{T_L - T_s} \qquad (4)$$

where $c_s$ is the specific heat of the solid and $T_L$ $T_S$ are the liquidus and solidus temperatures respectively.

The temperature dependence of the thermophysical properties was also taken into account. The values were taken from the thermophysical property database THERSYST(1997). The laser heat source was modeled according to method proposed by Voss et al. (1998), as a moving gaussian energy distribution (in the x-y plane) attenuated in the z-direction by the Beer's-Lambert coefficient in a effort to model the keyhole effect. Therefore each node i in the mesh with coordinates $x_i$, $y_i$, $z_i$ under the beam, receives a power density:

$$q_i(x,y,z,t) = q_0 \exp\left\{-c\left[\left(r_b - ut + x_i\right)^2 + y_i^2\right]\right\} \exp\left(-\beta z_i\right) \qquad (5)$$

where:

$$q_0 = \frac{PA}{\pi r_b^2} \qquad (6)$$

is the maximum power density at the center of the gaussian, $P$ is the laser power, $u$ is the laser travel speed, $r_b$ is the radius of the laser spot, $A$ the absorption coefficient defined above and $\beta$ is the Beer's-Lambert coefficient. Calculated weld thermal cycles are shown in Fig.4 for various positions (z=0 corresponds to the surface, while z=1 mm corresponds to −1 mm below the surface of the workpiece), in the HAZ of 6061-T6.

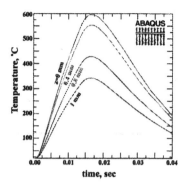

Fig. 4: Thermal cycles for various z positions in the HAZ.

It is obvious that each point in the HAZ undergoes a different thermal cycle as a function of its position in relation to the laser beam. The major parameters characterizing the thermal cycle at each point in the HAZ are the maximum temperature ($T_{max}$), the cycle duration ($\tau$), and the heating and cooling rates ($H_R$, $H_C$) respectively. These parameters were calculated for certain points in the HAZ and were used as an input in the simulation of dissolution and coarsening discussed in the next sections.

## 4. SIMULATION OF DISSOLUTION DURING LASER WELDING OF 6061-T6.

The dissolution problem is treated using the DICTRA methodology. The geometrical model is shown in Fig. 5b. The rod morphology of the $\beta$-Mg$_2$Si precipitate requires the use of cylindrical geometry. Due to symmetry reasons only the prescribed calculation area in Fig.5b is considered. In the geometrical model $r_\alpha$ and $r_\beta$ are the radius of the $\alpha$ and $\beta$ phase regions respectively. The geometry of Fig 5b follows the cell model proposed by Grong (1997), shown in Fig.5a, where each $\beta$-particle is surrounded by its own hexagonal cell and the dissolution region for the $\alpha$-phase is represented by an inscribed cylinder with volume equal to that of the hexagonal cell.

Major assumptions are the following:

- The problem was considered to be one-dimensional where dissolution takes place only in the radial direction.
- Because Mg diffuses much slower than Si in the $\alpha$-phase, it was considered that only Mg diffusion controls the dissolution rate
- The $\beta$-precipitate was considered stoichiometric and therefore no diffusion was considered within the $\beta$-phase

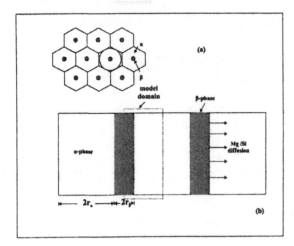

Fig. 5 Geometrical model used in dissolution

The initial compositions of the α and β phases are calculated by the Thermo-Calc software and obey the mass balance equations:

$$r_\alpha^0 C_{Mg}^{\alpha,0} + r_\beta^0 C_{Mg}^{\beta,0} = C_{Mg}^0$$
$$r_\alpha^0 C_{Si}^{\alpha,0} + r_\beta^0 C_{Si}^{\beta,0} = C_{Si}^0 \qquad (7)$$

where

$r_\alpha^0$ and $r_\beta^0$ are the initial sizes of the α and β phases respectively

$c_{Mg}^{\alpha,0}$ and $c_{Mg}^{\beta,0}$ are the initial Mg contents of the α and β phases respectively

$c_{Si}^{\alpha,0}$ and $c_{Si}^{\beta,0}$ are the initial Si contents of the α and β phases respectively

$c_{Mg}^0$ and $c_{Si}^0$ are the Mg and Si alloy contents respectively

The Mg diffusion in the α-phase $(0<r<r_\alpha)$ is described by the following equation

$$\frac{\partial c_{Mg}^\alpha}{\partial t} = \frac{1}{r}\frac{\partial}{\partial r}\left(rD_{Mg}^\alpha \frac{\partial C_{Mg}^\alpha}{\partial r}\right) \qquad (8)$$

where $c_{Mg}^\alpha$ and $D_{Mg}^\alpha$ are the Mg content and the diffusion coefficient of Mg in α-phase respectively. The flux balance at the α/β interface is described by the equation

$$u_{\alpha/\beta}\left(C_\beta^{\alpha/\beta} - C_\alpha^{\alpha/\beta}\right) = D_{Mg}^\alpha \left(\frac{\partial C_{Mg}^\alpha}{\partial r}\right)_{\alpha/\beta} \qquad (9)$$

where

$u_{\alpha/\beta}$ is the velocity of the α/β interface and $C_\alpha^{\alpha/\beta}, C_\beta^{\alpha/\beta}$ are the Mg concentrations of the α and β phases at the α/β interface

For this closed system the boundary conditions are

$$\left.\frac{\partial C_{Mg}}{\partial r}\right|_{r=0} = 0 \quad \text{and}$$

$$\left.\frac{\partial C_{Mg}}{\partial r}\right|_{r=r_\alpha+r_\beta} = 0 \qquad (10)$$

The initial condition is given by:

$$C_{Mg}^\alpha(r,0) = 0.98 \text{ for } 0 \le r \le r_\alpha \qquad (11)$$

where 0.98wt% is the alloy Mg composition. The initial equilibrium volume fraction of the β-phase was calculated by Thermo-Calc and is $f_o$=1.63% .

The problem, described by equations (7-11) was solved by the DICTRA methodology. During the weld thermal cycle the volume fraction of the β-Mg₂Si phase changes. The variation of $f$ with time from $t=0$ up to $t=\tau$ ,where $\tau$ is the duration of the thermal cycle, for $T_{max}$=595°C, is given in Fig.6. The thermal cycle starts with dissolution during heating and ends with reprecipitation during cooling. The extent of dissolution increases with the thermal cycle duration while full dissolution ($f=0$) commences for cycles longer than $10^{-2}$sec. The amount of reprecipitation depends on the extent of dissolution, since dissolution increases supersaturation in Mg and Si and, therefore, increases the driving force for precipitation during the cooling part of the thermal cycle. For a specific $T_{max}$, the general shape of these curves indicates that the final volume fraction of β-phase (at the end of the thermal cycle) first decreases for short cycles and then increases for longer cycles.

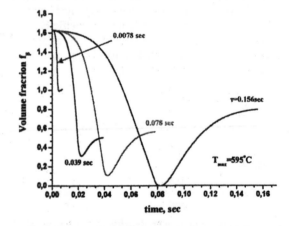

Fig. 6. Variation of volume fraction $f_\beta$ as a function of weld cycle duration and $T_{max}$.

## 5. SIMULATION OF COARSENING DURING LASER WELDING OF 6061-T6.

For these simulations the coarsening module in DICTRA was employed. According to this method coarsening of the dispersion can be described by considering one spherical particle, which has the maximum size of the dispersion prior to the application of the welding cycle. According to the LSW theory of coarsening (1959, 1961), the maximum size, $r_p$, is 1.5 times the mean dispersion size, $\bar{r}$. The geometrical model is shown in Fig. 7. The spherical particle of β-phase is embedded in a sphere of matrix α-phase. At the interface between α and β local thermodynamic equilibrium between α-phase and β-phase with radius $r_p$ is assumed. In this case a Gibbs-Thomson contribution is added to the Gibbs free energy of the particle, which is:

$$\frac{2\gamma V_m}{r_p} \qquad (12)$$

where $\gamma$ is the interfacial energy of the β particle in the α-phase and $V_m$ the molar volume. At the spherical cell boundary the α-phase is in local equilibrium with β-phase particle of the mean size $\bar{r}$, so the contribution to the Gibbs energy in this case is:

$$\frac{2\gamma V_m}{\bar{r}} \qquad (13)$$

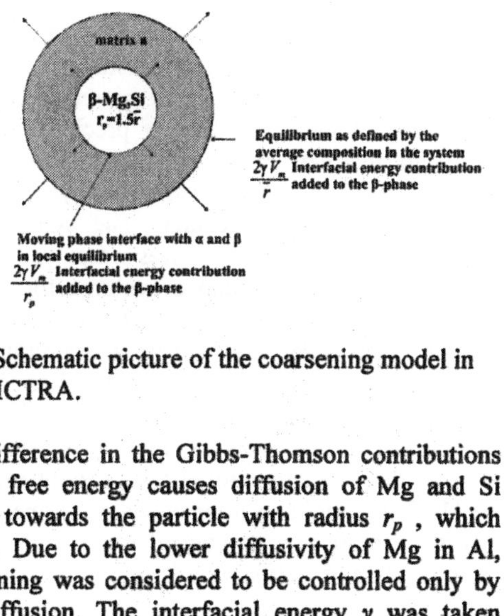

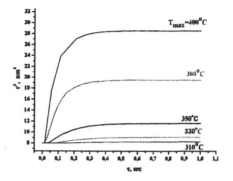

Fig.7 Schematic picture of the coarsening model in DICTRA.

The difference in the Gibbs-Thomson contributions to the free energy causes diffusion of Mg and Si atoms towards the particle with radius $r_p$, which grows. Due to the lower diffusivity of Mg in Al, coarsening was considered to be controlled only by Mg diffusion. The interfacial energy $\gamma$ was taken 0.5Jm⁻². In order to maintain constant volume fraction of β-phase and the initial overall alloy composition, the α-phase cell grows accordingly. Characteristic results of the coarsening simulations are shown in Fig. 8 for the case where the initial mean dispersion size is 2nm. The figure depicts the variation of cube mean size ($\bar{r}^3$) as a function of

cycle duration τ. The mean particle size increases with cycle duration, the change being more rapid for short cycles. For cycles longer than about 0.3sec, particle growth is very slow. As expected, coarsening kinetics is faster at higher $T_{max}$.

Fig. 8. Variation of $r^3$ with time

## 6. COMPARISON WITH EXPERIMENTAL HARDNESS PROFILES IN THE HAZ.

In this section a comparison is being attempted between calculated hardness profiles and experimental ones. The calculation of hardness is based on the hardness of the base metal reduced by an amount depending on the extent of dissolution or coarsening in the HAZ. These in turn depend on the final values of $f$ and $r$ in the HAZ.

### 6.1 Hardness profiles in HAZ₁

Fig. 9 depicts the calculated and experimental hardness profiles in HAZ₁. The experimental profile was measured after laser welding with the following conditions: laser power 4.5kW, laser travel speed 4.5m/min and focal distance –1mm.

The hardness of the base metal is $H_B$=118HV, while for 6061-T6 the precipitate has $f_o$=1.63% and $r_o$=2nm. The experimental hardness profile shows that welding is accompanied by a reduction of hardness in HAZ₁. More specifically the hardness is 88HV at the boundary with HAZ₂ (point F), drops to 83HV in point B and increases to 85HV at the fusion zone boundary (point A). The maximum temperature of the thermal cycle $T_{max}$ increases from 451°C at point F to 595°C at point A. The respective heating and cooling rates for these points were calculated with ABAQUS. The values of $T_{max}$, $H_R$, $H_C$ and τ are the input parameters of the thermal cycle for the simulation of dissolution. The simulation provides the values of $f$ and $r$ at the end of the thermal cycle for points A to F. Precipitation hardening in HAZ1 comes from two contributions:

- Coherency hardening, which is proportional to $f^{1/2}r^{1/2}$, and

- Orowan hardening (obstacle bypassing), which is proportional to $f^{1/2}r^{-1}$

The change in hardness ($\Delta H$) due to dissolution of β-phase relative to the hardness of the base metal ($H_{BM}$) is given by:

$$\frac{\Delta H}{H_{BM}} = \frac{f^{1/2}r^{1/2} - f_{BM}^{1/2}r_{BM}^{1/2}}{f_{BM}^{1/2}r_{BM}^{1/2}} + \frac{f^{1/2}r^{-1} - f_{BM}^{1/2}r_{BM}^{-1}}{f_{BM}^{1/2}r_{BM}^{-1}} \quad (14)$$

where $f_{BM}$=1.63% and $r_{BM}$=2nm for the condition T6. The hardness for each point of HAZ1 is calculated by the expression:

$$H = H_{BM} + \left(\frac{\Delta H}{H_{BM}}\right)H_{BM} \quad (15)$$

The calculated hardness values are plotted with the experimental values in Fig.9. The simulation overestimates the softening of the HAZ by 8-10%. This is attributed to the fact that only dissolution was accounted for the observed softening. Taking into account all the assumptions made for the current simulation, the comparison with the experimental results is satisfactory.

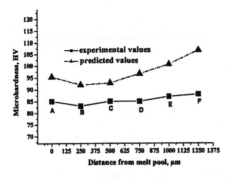

Fig. 9 Comparison of experimental and calculated hardness values in HAZ$_1$.

### 6.2 Hardness profiles in HAZ$_2$

In HAZ$_2$ the maximum temperature T does not exceed 400°C and only coarsening was considered to take place. Fig 10 shows the variation of the hardness profiles (i.e. hardness at points A, B, C, D, E and F of HAZ$_2$).

The experimental hardness profile shows that the hardness drops from 118HV in the base metal to 88HV at the boundary with HAZ$_1$ (point A). The coarsening simulation shows that the size of the β-Mg$_2$Si phase increases from 2nm in the base metal to 3.22nm at point A. In order to calculate the hardness at each point in HAZ$_2$, it is considered that only the Orowan mechanism is active (overaging conditions).

For this case hardening is proportional to $f^{1/2}r^{-1}$.

During coarsening the volume fraction f remains constant for all points of HAZ$_2$. Therefore the hardness of appoints A to F in HAZ$_2$ is

$$H = \frac{H_{BM}r_{BM}}{r} \quad (16)$$

The calculated values are shown in Fig.10. Similarly, taking into account the assumptions made, the comparison between the calculated and experimental values is satisfactory.

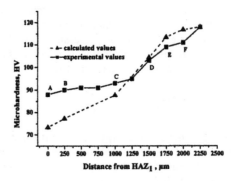

Fig. 10. Comparison of experimental and calculated hardness values in HAZ$_2$.

## 7. CONCLUSIONS

The softening of the HAZ following laser welding of 6061-T6 Al-alloy has been successfully predicted by the simulation of dissolution, reprecipitation and coarsening of the strengthening precipitates during the weld thermal cycle. A finite element based analysis of heat flow was employed for the calculation of thermal cycles. The computational kinetics software DICTRA was employed for the calculation of the variation of volume fraction and average size of the precipitates during the welding thermal cycle under non-isothermal conditions. The calculated hardness profiles in the HAZ are in good agreement with the experimental values. The main conclusion of this work is that it is possible to simulate the microstructure evolution and hardness in the HAZ of aluminum laser welds, thus opening the way for a more precise control and design of laser beam welding of aluminum alloys.

### Acknowledgements

Authors would like to acknowledge General Secretariat for Research and Technology for its financial support through YPER project.

# REFERENCES

ABAQUS, (1997). Hibbitt, Karlsson & Sorensen Inc., Pawtucket, RI.

Agren, J., (1990). Kinetics of Carbide dissolution. In: *Scandinavian Journal of Metallurgy*, Vol. 9, pp.2-8.

Agren, J., (1992). Computer Simulations of Diffusional Reactions in Complex Steels. In: *I.S.I.J International*, Vo32, pp.291-296.

Bratland, D.H., Grong, O., Shercliff, H., Myhr, O.R., Tjotta, S., (1997). Modelling of precipitation reactions in industrial processing. In: *Acta Materialia*, Vol. 45, pp. 1-22.

Bjorneklett, B.I., Grong, O., Myhr O.R., Kluken A.O., (1998). Additivity and isokinetic behavior in relation to particle dissolution. In: *Acta Materialia*, Vol. 46, pp. 6257-6266.

Engstrom, A., Hoglud, l., Agren, J., (1994). Computer Simulation of Diffusion in Multiphase Systems. In: Metallurgical Materials Transactions. Vol. 25A, pp.1127-1134.

Grong, O., Myhr, H., (2000). Modelling of non isothermal transformations in alloys containing a particle distribution. In: *Acta Materialia*, Vol. 45, pp. 1-22.

Grong, O., (1997). Precipitate Stability in Welds. In: *Metallurgical Modelling of Welding* pp.325-334, IOM Communications Ltd, Cambridge.

Gustafson, A., (2000). Aspects of microstructural evolution in chromium steels in high temperature applications. *Ph. D. thesis* KTH, Stockholm.

Gustafson, A., Hoglud, L., Agren, J., (1998). Simulation of Carbo-nitride coarsening in multicomponent Cr-steels for high temperature applications. In: *Advanced Heat Resistant Steels for Power Generation.*(Ed: R. Viswananthnan, J. Nutting), pp. 270-276,IOM Communications Ltd, London.

Ion, J.C., Easterling, K.E., Ashby M.F., (1984). A second report on diagrams of microstructure and hardness for HAZ in welds. In: *Acta Materialia*, Vol. 32, pp. 1949-1962.

Kou, S., (2003). *Welding Metallurgy, pp.359-362*, Wiley,USA.

Liftshitz, I.M. and V.V.Slyozov, (1959) *Soviet Physics JETP*, Vol. 35(8), ,pp. 331-339.

Malin,V., (1995). Study of Metallurgical Phenomena in the HAZ of 6061-T6 Aluminum Welded Joints. In: *Welding Journal* ,Vol. 9, pp. 305-318.

Nicolas, M., Deschamps,A., (2003). Characterisation and modelling of precipitate evolution in an Al-Zn-Mg alloy during non-isothermal heat treatments. In: *Acta Materialia*, Vol. 51, pp. 6077-6094.

Sudman, B., Jonsson,B., Andersson, J.O.,(1985). *CALPHAD, Vol. 9, p. 153.*

Tempus, G., (2001). Werkstoffe fur transport und verkehr, Materials Day, ETH Zurich, Switzerland.

THERSYST, (1997). A thermophysical database for light alloys. Ed. G. Neuer, G. Weiland, IKE, Stuttgard.

Voss, O., I. Decker and H. Wohlfahrt (1998). Considerationsof microstructural transformations in the calculation of residual stresses and distortion of larger weldments. In: *Mathematical Modelling of Weld Phenomena* (Ed: H. Cerjak) pp. 584-596. IOM Communications Ltd, Cambridge.

Vermolen, F.J., Vuik,C., (2000). A mathematical model for the dissolution of particles in multi-component alloys. In: *Journal of Computational and Applied Mathematics*, Vol. 26, pp. 233-254

Wagner, C.,(1961),*Zeitshrift fur Electrochemie*,Vol. 65 (7/8), pp. 581-591.

Zervaki, A.D., (2004). Laser Welding of Aluminum Alloys: Experimental Study and Simulation od Microstructure Evolution in the HAZ. *Ph.d thesis, University of Thessaly, Volos.*

# AUTHOR INDEX

| Title/Year of publication | Editor(s) | ISBN |
|---|---|---|
| **2002 continued** | | |
| Periodic Control Systems (W) | Bittanti & Colaneri | 0 08 043682 X |
| Modeling and Control in Environmental Issues (W) | Sano, Nishioka & Tamura | 0 08 043909 8 |
| Computer Applications in Biotechnology (C) | Dochain & Perrier | 0 08 043681 1 |
| Time Delay Systems (W) | Gu, Abdallah & Niculescu | 0 08 044004 5 |
| Control Applications in Post-Harvest and Processing Technology (W) | Seo & Oshita | 0 08 043557 2 |
| Intelligent Assembly and Disassembly (W) | Kopacek, Pereira & Noe | 0 08 043908 X |
| Adaptation and Learning in Control and Signal Processing (W) | Bittanti | 0 08 043683 8 |
| New Technologies for Computer Control (C) | Verbruggen, Chan & Vingerhoeds | 0 08 043700 1 |
| Internet Based Control Education (W) | Dormido & Morilla | 0 08 043984 5 |
| Intelligent Autonomous Vehicles (S) | Asama & Inoue | 0 08 043899 7 |
| **2003** | | |
| Proceedings of the 15th IFAC World Congress 2002 (CD + 21 vols) | Camacho, Basanez & de la Puente | 008 044184 X |
| Modeling and Control of Economic Systems (S) | Neck | 0 08 043858 X |
| Mechatronic Systems (C) | Tomizuka | 0 08 044197 1 |
| Programmable Devices and Systems (W) | Srovnal & Vlcek | 0 08 044130 0 |
| Real Time Programming (W) | Colnaric, Adamski & Wegrzyn | 0 08 044203 X |
| Lagrangian and Hamiltonian Methods in Nonlinear Control (W) | Astolfi, Gordillo & van der Schaft | 0 08 044278 1 |
| Intelligent Control Systems and Signal Processing (C) | Ruano, Ruano & Fleming | 0 08 044088 6 |
| Guidance and Control of Underwater Vehicles (W) | Roberts, Sutton & Allen | 0 08 044202 1 |
| Analysis and Design of Hybrid Systems (C) | Engell, Gueguen & Zaytoon | 0 08 044094 0 |
| Intelligent Manufacturing Systems (W) | Kadar, Monostori & Morel | 0 08 044289 7 |
| Control Applications of Optimization (W) | Gyurkovics & Bars | 0 08 044074 6 |
| Fieldbus Systems and Their Applications (C) | Dietrich, Neumann & Thomesse | 0 08 044247 1 |
| Intelligent Components and Instruments for Control Applications (S) | Almeida | 0 08 044010 X |
| Modelling and Control in Biomedical Systems (S) | Feng & Carson | 0 08 044159 9 |
| **2004** | | |
| Advances in Control Education (S) | Lindfors | 0 08 043559 9 |
| Robust Control Design (S) | Bittanti & Colaneri | 0 08 044012 6 |
| Fault Detection, Supervision and Safety of Technical Processes (S) | Staroswiecki & Wu | 0 08 044011 8 |
| Technology and International Stability (W) | Kopacek & Stapleton | 0 08 044290 0 |
| System Identification (SYSID 2003) (S) | Van den Hof, Wahlberg & Weiland | 0 08 043709 5 |
| Control Systems Design (C) | Kozak & Huba | 0 08 044175 0 |
| Robot Control (S) | Duleba & Sasiadek | 0 08 044009 6 |
| Time Delay Systems (W) | Garcia | 0 08 044238 2 |
| Control in Transportation Systems (S) | Tsugawa & Aoki | 0 08 0440592 |
| Manoeuvring and Control of Marine Craft (C) | Batlle & Blanke | 0 08 044033 9 |
| Power Plants and Power Systems Control (S) | Lee & Shin | 0 08 044210 2 |
| Automated Systems Based on Human Skill and Knowledge (S) | Stahre & Martensson | 0 08 044291 9 |
| Automatic Systems for Building the Infrastructure in Developing Countries (Knowledge and Technology Transfer) (W) | Dimirovski & Istefanopulos | 0 08 044204 8 |
| Intelligent Assembly and Disassembly (W) | Borangiu & Kopacek | 0 08 044065 7 |
| New Technologies for Automation of the Metallurgical Industry (W) | Wei Wang | 0 08 044170 X |
| Advanced Control of Chemical Processes (S) | Allgöwer & Gao | 008 044144 0 |